Topics in Applied Physics   Volume 72

Topics in Applied Physics   Founded by Helmut K.V. Lotsch

Volumes 1–56 are listed on the back inside cover

Glassy Metals III

Amorphization Techniques, Catalysis, Electronic and Ionic Structure

Edited by H. Beck and H.-J. Güntherodt

With Contributions by
A. Baiker   A.S. Bakai   H. Beck   J. Eckert
H.J. Fecht   H.-J. Güntherodt   P. Häussler
W.L. Johnson   K. Samwer   L. Schultz

With 145 Figures

Springer-Verlag Berlin Heidelberg GmbH

Professor Dr. *Hans Beck*

Institut de Physique, Université de Neuchâtel, 1, Rue A.-L. Breguet,
CH-2000 Neuchâtel, Switzerland

Professor Dr. *Hans-Joachim Güntherodt*

Institut für Physik, Universität Basel, Klingelbergstraße 82,
CH-4056 Basel, Switzerland

ISBN 978-3-662-31194-3

Library of Congress Cataloging-in-Publication Data. (Revised for vol. 3) Glassy metals. (Topics in applied physics;
v. 46,) Beck's name appears first on vol. 3. Includes bibliographical references and indexes. Contents: 1. Ionic
structure, electronic transport, and crystallization -- 2. Atomic structure and dynamics, electronic structure,
magnetic properties -- 3. [without special title] 1. Metallic glasses. I. Güntherodt, H.-J. (Hans-Joachim),
1939-   . II. Beck, H. (Hans), 1939-   . III. Series. TN693.M4G57 1981 620.1'4 80-28575
ISBN 978-3-662-31194-3   ISBN 978-3-540-48154-6 (eBook)
DOI 10.1007/978-3-540-48154-6

Typesetting: Macmillan India Ltd., Bangalore 25
SPIN:10068872   54/3140/SPS - 5 4 3 2 1 0 - Printed on acid-free paper

Preface

The first two volumes of this series, Glassy Metals I and II, appeared in 1981 and 1983. Although several key questions on metallic glasses or amorphous metals are still open, such as structural details and modelling, electronic transport and the formation of glassy refractory alloys, the hard core of researchers has focused on obtaining a more general view of amorphization also in relation to melting, on the strong interrelation between the electronic and ionic structures in the amorphous and liquid states, on new routes to obtain amorphous metals such as mechanical alloying and, very recently, the ultrasonic preparation technique, on potential applications such as catalysis and new glassy Al- and Mg-alloys, and on the relevance of the cluster concept to describe the structure of amorphous metals. The very promising real-space study of the amorphous structure by STM (Scanning Tunneling Microscopy) has not yet reached convincing atomic resolution; however, it points towards a cluster structure. The most exciting highlight is the field of novel bulk metallic glasses with a low critical cooling rate of 10 K/s or less. These easy-glass-forming alloys are highly processable metallic glasses, where heterogeneous nucleation seems to be an important factor in crystallization.

The editors thank the contributors for their collaboration and for the effort to review with such excellence their topic of research.

Neuchâtel, Basel
February 1994

H. Beck
H.-J. Güntherodt

Contents

Contributors

Baiker, Alfons
  Department of Chemical Engineering and Industrial Chemistry,
  ETHZ-Zentrum, Universität Str. 6, CH-8092 Zürich, Switzerland

Bakai, Alexandre S.
  Kharkov Institute of Physics and Technology, Ukrainian Academy
  of Sciences, 310108 Kharkov, Ukraine

Beck, Hans
  Institut de Physique Université de Neuchâtel 1, Rue A.-L. Breguet,
  CH-2000 Neuchâtel, Switzerland

Eckert, J.
  Zentralabteilung Forschung und Entwicklung, Siemens AG,
  27E ME TPH 12, Postfach 3220, D-91000 Erlangen, Germany

Fecht, H.J.
  Institut für Physik, Universität Augsburg, Memminger Strasse 6,
  D-86159 Augsburg, Germany

Güntherodt, Hans-Joachim
  Institut für Physik, Universität Basel, Klingelbergstrasse 82,
  CH-4056 Basel, Switzerland

Häussler, Peter
  TU Chemnitz-Zwickau, Fachbereich Physik, Physik Dünner Schichten,
  D-09009 Chemnitz, Germany

Johnson, William L.
  Keck Laboratory, CALTEC, Pasadena, CA 91125, USA

Samwer, Konrad
  Institut für Physik, Universität Augsburg, Memminger Strasse 6,
  D-86159 Augsburg, Germany

Schultz, Ludwig
  Zentralabteilung Forschung und Entwicklung, Siemens AG,
  27E ME TPH 12, Postfach 3220, D-91000 Erlangen, Germany

1. Introduction

H. Beck and H.-J. Güntherodt

Metallic glasses (MGs) are the subject of an increasing research effort, spurred
by both science and technology. The research in this field helps our understanding of noncrystalline matter in general. The interpretation of the properties of
MGs imposes a particular challenge since the understanding of crystalline solids
has, in the past, generally been based upon their crystal periodicity. A fascinat-
ing theoretical concept based on translational invariance has been developed to
deal with the lattice dynamics and electronic structure of crystalline matter. No
such general theory has yet been developed for the disordered state. Amorphous
metals can be prepared by a variety of methods: 1) Evaporation of metals in
vacuum and condensation of their vapor on a cooled substrate. 2) Sputtering, by
which the atoms are removed from the source under bombardment with
energetic inert gas atoms. 3) Chemical or electroless deposition, a method in
which ions in aqueous solution are deposited onto substrates by chemical
reactions. 4) Electrodeposition, where the chemical reaction requires the presence of an external potential. 5) Rapid quenching from the liquid state.
Amorphous alloys prepared by the latter method are the so-called metallic
glasses or glassy metals.

The general framework of the physics of glassy metals has been presented in
the introduction of [1.1, 2], where the state-of-the-art has been reviewed including the main preparation methods, and various experimental and theoretical
approaches to the characterization and understanding of their physical properties. The main emphasis of this volume is on specific topics such as general
underlying principles of amorphization including mechanical alloying, catalysis,
the similarities of electronic and ionic structure in the liquid and amorphous
state and their interrelation, and the cluster concept.

1.1 Amorphization in Metallic Systems

It has been well known for a couple of decades that noncrystalline metallic
alloys can be made by vapour- and melt-quenching. Recent results show that an
amorphous phase can also be formed directly when a crystalline metallic alloy is
subjected to various types of disordering processes. Solid-state amorphization
can be induced through a variety of methods including absorption of atomic
hydrogen, thermal interdiffusion reaction along the interface separating

Topics in Applied Physics, Vol. 72
Beck/Güntherodt (Eds.)
© Springer-Verlag Berlin Heidelberg 1994

constituent metal layers, irradiation, mechanical alloying etc. As such, a crystalline phase can be driven into a disordered amorphous state as long as the kinetic constraints inhibiting stable-phase formation can be maintained.

In Chap. 1 the theoretical and experimental aspects of the transformation from the crystal into the glassy state by solid-state reactions are covered. The fundamentals of the thermodynamics of metastable phase formations is described along with kinetic requirements for amorphization. The solid-state amorphization, which is characterized by the loss of long-range order, can be understood as a melting process under non-equilibrium conditions. Glass can be considered, thermodynamically, a frozen, highly undercooled liquid, and, therefore, the transition from crystal to glass is compared with the melting process. Thus, solid-state amorphization experiments offer now the opportunity to study the melting transition in more detail and reach a more global understanding of this complex problem.

1.2 Mechanically Alloyed Glassy Metals

Chapter 2 reviews recent results of the formation of glassy metals by mechanically alloying elemental metallic powders. Mechanical alloying is usually performed in a high-energy ball mill in an inert argon atmosphere. The metal powder particles are trapped by the colliding balls, heavily deformed and cold welded. Characteristically, layered particles are formed. Further milling refines the microstructure more and more. Finally, a true alloying takes place. The distinct X-ray diffraction peaks of the metallic elements disappear and typical diffuse patterns of the amorphous state become apparent. The amorphization process taking place during mechanical alloying is a solid-state reaction process and differs completely from rapid quenching. Therefore, the comparison range within which glass formation is possible differs for the two routes. Amorphous metals can be formed by mechanical alloying mainly in the central part of the phase diagram, i.e. also in the range of high-melting intermetallic phases. Eutectic compositions of the equilibrium phase diagram do not play a role; here it seems to exist some similarity with irradiation-induced amorphization. However, mechanically alloyed and rapidly quenched glassy metals are structurally very similar, both topologically and chemically. Mechanical alloying offers also an attractive technique for forming amorphous metals with possible technological applications, especially with regard to new permanent magnets.

1.3 Catalysis

Initially the motivation for using glassy metals in catalysis originated from some of their unique properties, such as high density of low coordination sites and defects, fine tuning of the electronic properties, chemically homogeneous and structurally isotropic, metastable structure, and good electrical and thermal conductivity, as outlined in Chap. 3.

Metallic glasses have been used in catalysis in two ways, namely, in investigations carried out on as-quenched glassy metals and in those where the glassy metals were subjected to different pretreatments and served merely as precursors to catalytically active materials. The use of glassy metals as catalyst precursors has been shown to open up new possibilities for the preparation of supported metal catalysts with unusual chemical and structural properties. This potential resides mainly on the high reactivity and isotropic nature of these materials compared to their crystalline counterparts. Several efficient supported metal catalysts are compared to conventionally prepared supported metal catalysts in Chap. 3.

1.4 Interrelation Between Electronic and Ionic Structure in Metallic Glasses

Chapter 4 discusses the interrelations between the electronic and ionic structure in metallic glasses. Evidence is given for a strong influence of conduction electrons on the ionic structure and vice versa, for that of ionic structure on the conduction electrons. This affects phase stability, the electronic density of states, dynamic properties, electronic transport and magnetism. A scaling behavior of many properties versus Z, the mean electron number per atom, is the most characteristic feature of these alloys. Remarkable properties occur for concentrations where the peak of the structure factor and the diameter of the Fermi sphere are very close. A high resistivity and a negative temperature coefficient of electrical resistivity coincides with a good glass-forming ability. Crystalline alloys strongly dominated by conduction electrons are the so-called electron phases or Hume–Rothery phases. Similar theoretical concepts as applied to crystalline Hume–Rothery alloys are discussed for the amorphous state. It turns out that the amorphous state has many similarities to the liquid state and can be considered an undercooled liquid.

1.5 Polycluster Concept of Amorphous Solids

Contrary to a periodic lattice, the symmetry of which is entirely characterized by its space group, the description of an amorphous structure faces the problem of no long-range order and of a local coordination that varies from site to site. The polycluster description of amorphous solids is a classification based on a set of Coordination Polyhedra (CP) for each atom determined by its nearest neighbors. Topologically equivalent polyhedra form classes and are referred to as "regular". Atoms having different ("irregular") CP are then assigned to the boundary regions between clusters of regular atoms. This description of the amorphous structure permits the identification and classification of point and line defects. Moreover, the low-lying structural excitations ("two-level systems") discussed in Chap. 8 of [1.1] can be identified with rearrangement motions within the polycluster structure. The polycluster concept, outlined in Chap. 5, also allows for a unified understanding of other phenomena, such as particle diffusion, liquid–glass transition and elastic and plastic behavior.

References

1.1 H.-J. Güntherodt, H. Beck (eds.) Glassy Metals I, Topics Appl. Phys., Vol. **46** (Springer, Berlin, Heidelberg 1981)
1.2 H. Beck, H.-J. Güntherodt (eds.) Glassy Metals II, Topics Appl. Phys., Vol. **53** (Springer, Berlin, Heidelberg 1983)

2. Amorphization in Metallic Systems

K. Samwer, H.J. Fecht, and W.L. Johnson

With 26 Figures

The transformation of the crystalline into the glassy state by solid-state reactions is extensively reviewed in its theoretical and experimental aspects. First, we give some historical background and describe the thermodynamics of metastable phase formations, adding as well the kinetic requirements for the amorphization process. Then we discuss the different experimental routes into the amorphous state: hydriding, thin diffusion couples, and other driven systems. In the discussion and the summary, we close the gap between the melting phenomena and the amorphization and provide a tentative outlook.

2.1 Introductory Remarks

2.1.1 Historical Background

For over 30 years, it has been known from the work of *Buckel* and *Hilsch* [2.1] that a noncrystalline metallic solid can be made by the condensation of metal vapors onto a cooled substrate. This results was found accidentally while investigating the influence of lattice defects on superconductivity. Their work was dramatically extended some years later by the pioneering success of *Duwez* and co-workers through the rapid quenching of a melt [2.2]. Again, the interest here was not to look for amorphous systems. Instead, *Duwez* et al. were investigating alloys with extended solid solutions. In order to make these alloys, equipment for rapid quenching had to be constructed and used. It is the merit of both groups to have developed different methods for the preparation of noncrystalline alloys and have carried out the first investigations of their properties. In the meantime, rapidly quenched metals have been the subject of several international conferences.

Over the thirty years which followed those early studies, the techniques of vapor quenching and melt quenching have been extensively developed and elaborated for the purpose of producing a wide variety of amorphous alloy phases. Plasma sputtering, chemical-vapor deposition, and electron-beam evaporation are but a few of the methods in use for the production of amorphous alloys films from the vapor phase, while melt spinning, liquid atomization, and splat cooling are examples of implementations of the liquid-quench approach. During this period, there has been a growing awareness that, under certain

Topics in Applied Physics, Vol. 72
Beck/Güntherodt (Eds.)
© Springer-Verlag Berlin Heidelberg 1994

circumstances, an amorphous phase can be formed when a crystalline solid is subjected to various types of disordering processes. For example, when certain intermetallic compounds are irradiated with energetic charged particles, damage to the crystalline lattice can lead ultimately to a transformation into the amorphous state. A review of such experiments has been published [2.3]. The amorphous phase produced by this method appears to be structurally very similar to that obtained by sputter deposition or melt quenching of the same alloy. Other experiments have suggested that radiation-induced amorphization and melting of a crystalline compound are manifestations of the same first-order phase transition. Both exhibit features of a first-order phase transformation (e.g., nucleation and growth of the disorder phase) [2.4], they differ in that the disordered product phase is below its T_g in one case and above in the other. A disordered crystal can thus be said to "melt" to a glass.

Nearly ten years ago, a third way into the amorphous state was found accidentally by *Yeh*, *Samwer* and *Johnson* [2.5]. Starting with an intermetallic crystalline compound (Zr_3Rh), it was shown that the absorption of hydrogen induces amorphization. Although there were several hints in the literature [2.6], no systematic studies had been carried out up to that time. In these experiments, the amorphous phase starts from grain boundaries and grows into the grain interior until the material is fully amorphous. The crystal–glass transition is caused by the large diffusivity of only one component of the alloy (here hydrogen) and the simultaneous hindrance of the crystallization into a new crystalline structure. Since the starting materials are polycrystalline, the grain boundaries provide, first, the fastest way for the hydrogen to enter the Zr_3Rh-phase and, second, act as inhomogenous *nucleation* sites in the amorphous phase. The latter is still not fully verified and will be discussed later (Sect. 2.3.2).

In another still more striking example, *Schwarz* and *Johnson* [2.7] found that simple thermal interdiffusion of two elemental polycrystalline metals (La and Au) can result in the formation and growth of an amorphous alloy. This solid-state amorphization by interdiffusion of initially crystalline elemental metals has subsequently been found to be rather ubiquitous [2.4, 8]. Again the early stages of the amorphization reaction are still not fully understood (Sects. 2.2.2–2.3.2) especially for GaAs/metal-diffusion couples. As another example of solid-state amorphization, a high-energy ball mill can produce large amounts of amorphous material when elemental mixtures of metal powders are mechanically alloyed. This technique is very promising for the formation of numerous new metastable phases and will be described in a later chapter by *Schultz* and *Eckert*. The high-energy ball mill can also amorphize initially homogeneous intermetallic compounds as first noted by *Schwarz* et al. [2.9]. In the meantime, questions arise with regard to the large amount of impurities from the container walls and the balls due to the grinding process, and the main contribution for the amorphization process. In Sects. 2.2.4 and 2.3.3 we will try to note the problems. The role of defects in these processes is well known from the ion and electron irradiation experiments as reviewed by *Luzzi* and *Meshi* [2.10].

Finally a new class of crystal–glass transformations are discussed by the homogeneous transition due to an instability of the crystal [2.11]. Again, the whole transformation is an interplay between the thermodynamic driving force and the kinetic constraints of the process. In the latter case, the system has to be a homogeneous phase (no phase separation due to long-range diffusion) although it should not be kinetically frozen-in. In the following discussion, we will analyze some aspects of this problem.

2.1.2 Thermodynamics of Metastable Phase Equilibria and Metastable Phase Diagrams

An equilibrium phase of matter can be characterized as macroscopically homogeneous and stationary. The thermodynamic variables normally used to characterize the state of an equilibrium phase include pressure P, temperature T, volume V, chemical constitution as given by C_i is the atomic concentration of the ith component ($i = 1, 2, 3, \ldots, n$, where n is the number of components), etc. For an equilibrium phase of matter, not all such variables are independent. According to the Gibbs phase rule, for a single-phase substance, any two independent variables chosen from among (P, T, V) together with $n - 1$ independent variables C_i are sufficient to characterize its thermodynamic state. The conditions of equilibrium are sufficient to determine the other dependent variables. For a two-phase system in equilibrium, further constraints must be added. The two phases must share a common temperature, pressure, and chemical potential μ_i for each component. For a two-phase, two-component system, this leaves only two degrees of freedom. By fixing P and T, the equilibrium state is defined. All other parameters are obtained by choosing their values such as to minimize the overall Gibbs free energy, G. It is convenient to express these results through the use of free-energy diagrams. For a one-component system, these diagrams give the variation of G with temperature and pressure for each relevant phase. Figure 2.1 shows a schematic diagram at fixed pressure. In the figure, α and β represent two allotropic modifications of the elemental crystalline solid while L represents the liquid phase. Below its T_g, the liquid becomes a glass (labeled G). An allotropic phase transition occurs at T_a whereby (in equilibrium) α transforms to β. At T_m the α-phase melts to liquid phase. At any temperature other than T_a or T_m, the system is single phase; only at T_a and T_m can two phases co-exist. As pressure varies, T_a and T_m can vary, i.e., $T_a = T_a(P)$, $T_m = T_m(P)$. The Clapeyron equation describes these co-existence curves in the $P–T$ plane.

It is now appropriate to ask what is meant by the free energy of a non-equilibrium phase. For example, what is meant by the liquid curve L in Fig. 2.1 when $T < T_m$? Strictly speaking, thermodynamic functions can only be defined for equilibrium states, and yet it is natural to consider an undercooled liquid. *Turnbull* has discussed the fact that liquid metals can be extensively undercooled for extended observable times [2.12] without crystallization. The explanation for this observation lies in the resistance of the undercooled liquids to formation of crystalline nuclei of critical size. The timescale for nucleation of crystals, τ_N, depends strongly on undercooling as described by *Turnbull*. Near T_m, τ_N can be exceedingly large. When the timescale τ_N is sufficiently long so as to allow the liquid to sample its available phase space, one can still define the entropy and other thermodynamic functions of the liquid. In other words, the liquid continues to behave ergodically when τ_N is sufficiently large. The liquid is in a metastable state for which entropy, free energy, and other thermodynamic state variables can be defined. It is in this sense that one can define the free energy curves of non-equilibrium phases as in Fig. 2.1. These curves represent metastable states of finite lifetime. When this lifetime is sufficient to permit ergodic sampling of available microstates corresponding to the metastable phase, one can define a free energy.

To calculate the thermodynamic functions for pure metals, one needs the thermal heat capacity C_p at ambient pressure. Above the Debye temperature, C_p consists of three parts: a) the Dulong–Petit value of $3k_B$, b) an additional linear increase proportional to temperature, which can also be seen in the thermal expansion coefficient, and c) an additional amount close to the melting temperature T_m, which results from the formation of defects (mainly vacancies). The last part can be approximated for small concentration as

$$C_p^v = [(\Delta H^v)^2 / k_B T^2] \, A \, \exp(-\Delta H^v / k_B T) \, . \tag{2.1}$$

Here ΔH^v, the heat of formation of vacancies, is considered to be constant. The heat capacity of pure liquid metals decreases slightly with decreasing temperature. Using experimental values [2.13], one can calculate the entropy and enthalpy of the crystalline and liquid phase in the stable and metastable state.

The entropy S and the enthalpy H are given by

$$S^L(T) = S_0^x + \Delta S_f + \int_{T_m}^{T} C_p^L / T \, dT \, , \tag{2.2}$$

$$S^x(T) = S_0^x + \int_{T_m}^{T} C_p^x / T \, dT \, , \tag{2.3}$$

$$H^L(T) = H_0^x + \Delta H_f + \int_{T_m}^{T} C_p^L \, dT \, , \tag{2.4}$$

$$\text{and} \quad H^x(T) = H_0^x + \int_{T_m}^{T} C_p^x \, dT \, . \tag{2.5}$$

Here S_0^x and H_0^x are the entropy and enthalpy of the crystal at T_m, and ΔS_f and ΔH_f correspond to the entropy of melting and the enthalpy of melting (latent heat). The entropy and enthalpy are calculated this way for Al and shown in Fig. 2.2 and 2.3. The entropy of the undercooled liquid decreases faster with decreasing temperature than that of the crystal because some entropy of the undercooled melt freezes in with decreasing temperature. This results in a crossover of both curves (extrapolated for the undercooled melt) at an isentropic temperature $T_{\Delta S = 0}$ at approximately $0.25 \, T_m$. Here the difference in the entropy vanishes between the crystal and the undercooled melt. Below this temperature,

Fig. 2.1. Schematic free energy diagram for a one-component system at constant pressure. G refers to the glassy phase, L to the liquid, α and β to allotropic crystalline phases

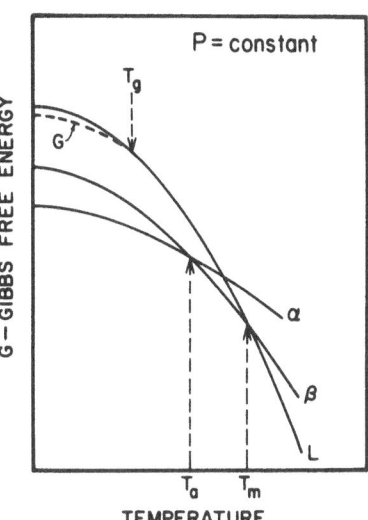

the so-called Kauzmann paradox would occur, that is, the entropy of the undercooled melt would be smaller than the entropy of the ordered system, the crystal. As long as the system is in an ergodic state, the ground state of the system should be ordered at least at very low temperature. This means that the entropy of the crystal should be smaller than that of the undercooled melt. The solution of this paradox is provided in all systems known by either the crystallization of the undercooled melt or the glass transition (except He, which becomes superfluid due to quantum effects). The Kauzmann temperature $T_{\Delta S = 0}$, sometimes called the ideal glass transition temperature, gives the maximum undercooling of the melt. Most glass-forming systems have a higher experimentally observed glass transition temperature for kinetic reasons. Indeed, the situation is very similar to that of the spin glass, as discussed by Anderson [2.14].

At higher temperature one finds the inverse situation. For the hypothetical case that a crystal could be superheated far above the melting temperature, the entropy of such a crystal crosses the curve of the entropy of the stable melt at an "inverse" isentropic temperature, T_i^s. For Al, the entropy of a hypothetical crystal and the entropy of the melt cross at $T_i^s = 1.38\,T_m$. Again one can argue, in analogy to the Kauzmann paradox at low temperatures, that T_i^s is the upper limit for the stability of a superheated crystal against melting because at high temperatures the melt should always have a higher entropy than the crystal. If one uses the classical nucleation theory for such a hypothetical case, it can be argued that for a vanishing difference of the entropy between crystal and melt, the interfacial energy σ_{xe} between crystal and melt would also vanish according to

$$\sigma_{xe} \approx \Delta S \cdot T \, , \tag{2.6}$$

where ΔS is normally the entropy of melting at T_m. If σ_{xe} also vanishes at T_i^s the

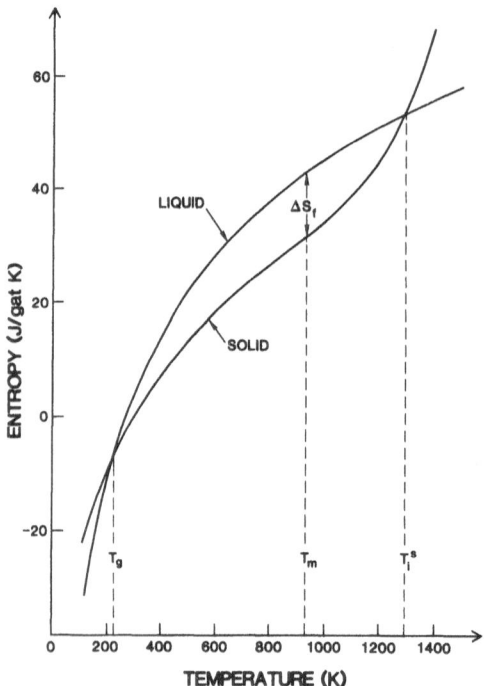

Fig. 2.2. The entropy of liquid and crystalline aluminum as function of temperature in the stable and metastable regime. ΔS_f denotes the entropy of fusion

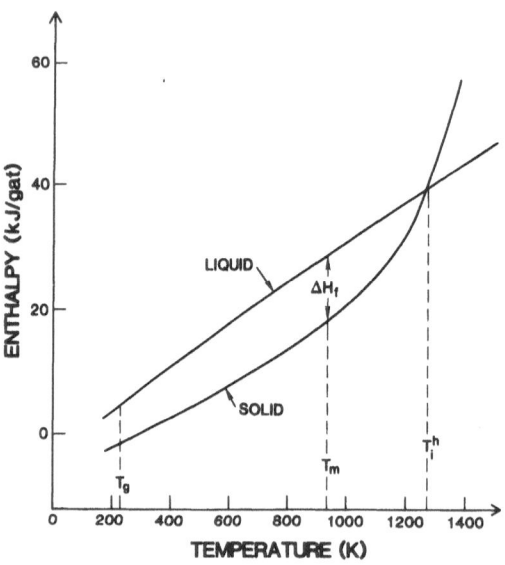

Fig. 2.3. The enthalpy of liquid and crystalline aluminum as function of temperature in the stable and metastable regime. ΔH_f denotes the heat of fusion

superheated crystal no longer exhibits a barrier against melting and the interface itself vanishes. This means that the crystal will melt instantaneously.

In Fig. 2.3 the corresponding curves for the enthalpy are shown. At low temperature one can see that the undercooled melt and the glass (below T_g) have a larger enthalpy compared to the crystal, which results in a still-existing driving force for crystallization as seen in the free-energy diagram in Fig. 2.1. At high temperatures, the curves of the enthalpies for the superheated crystal and the melt cross at an isenthalpic temperature. Above that temperature melting becomes an exothermic process. *Tallon* [2.15] pointed out that these curves might be slightly different, if one considers that certain parts of the entropy (communal entropy) should be subtracted and that the molar volumes of the different states are taken into account. Indeed, the point of identical molar volume might be the correct instability point because there the system becomes unstable against shearing.

For a two-component system, the free-energy diagram has, in addition to T and P, a compositional degree of freedom. A schematic diagram is shown in Fig. 2.4a. Here α and β represent terminal solid solutions, L the liquid phase, and

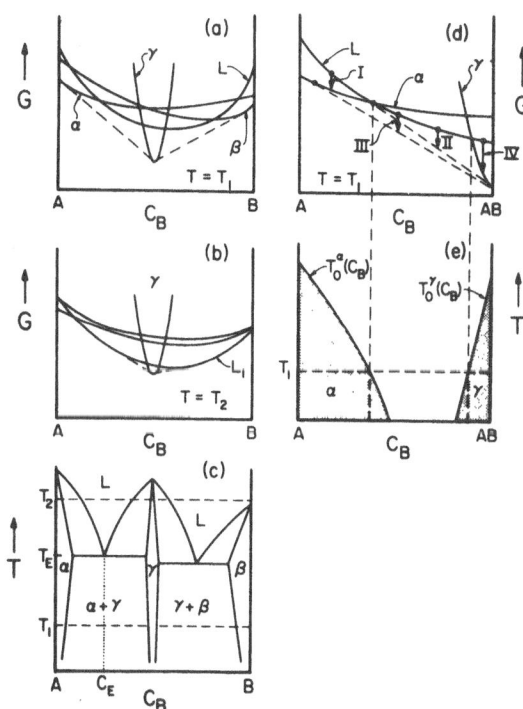

Fig. 2.4. (a) Schematic free-energy diagram of a two-component system at a temperature below solidus; (b) free-energy diagram at a temperature between the solidus and liquids; (c) binary-phase diagram; (d) illustration of polymorphous (I and IV), primary (II), and eutectic (III) crystallization reactions from an undercooled liquid; (e) the T_0 (C) curves for the α and γ phases

γ an intermetallic compound. Figure 2.4a shows free-energy curves at relatively low temperature (below the solidus), while Fig. 2.4b shows curves at relatively high temperature (above the eutectics). Figure 2.4c shows the corresponding phase diagram. The phase diagram contains regions of single-phase equilibrium as well as two-phase regions defined by the common tangents of two free-energy curves, which bound the free energy from below. One can again define the notion of metastable phases. A liquid of eutectic composition C_E can be cooled below the eutectic temperature T_E along the dotted line. Below T_E, this liquid can achieve equilibrium by crystallizing to a two-phase mixture of α and/or γ. *Köster* [2.16] classified systematically the various types of crystallization reactions which may occur in such a situation. The free energy of the undercooled liquid may be lowered by a number of mechanisms, as illustrated in Fig. 2.4d, which one classifies according to the nature of the composition changes involved. Precipitation of any phase without composition change is termed polymorphous. For instance, reaction I and IV in Fig. 2.4d shows polymorphous crystallization of a liquid to the α and γ phases, respectively. Crystallization of a single phase from the liquid accompanied by a composition change is referred to as primary crystallization. Reaction II in Fig. 2.4d illustrates primary crystallization of the γ phase from the liquid state with a composition shift of both the γ phase and the remaining liquid phase. Finally, a crystallization reaction in which two crystalline phases form simultaneously from a homogeneous liquid accompanied by compositional shifts is referred to as eutectic crystallization. Reaction III in Fig, 2.4d is a eutectic reaction, whereby a two-phase mixture of α and γ crystallizes from the liquid state.

Reactions involving a composition change (i.e., primary or eutectic reactions) are inherently more complex than polymorphous reactions in that they involve the development of concentration gradients and therefore atomic transport over substantial distances (distances at least as large as the critical nuclei of the product phases). In contrast, polymorphous reactions involve only topological reconstruction of the atomic arrangement; a given atom needs to move only a distance of the order of the interatomic spacing.

The above observation has led several authors to introduce the concept of $T_0(C)$ curves. Referring to Fig. 2.4d, we note that polymorphous crystallization of any phase is possible only when the free-energy curve of the phase, at the given composition, lies below the liquid at a specific composition. The locus of these crossing points define the $T_0(C)$ curve of the crystal. These relations are illustrated in Fig. 2.4e for the α phase of Fig. 2.4d. One might think of the $T_0(C)$ curve of a crystal as the polymorphous melting curve for the crystal in the $T-C$ plane. In other words, were long-range diffusion was suppressed by kinetic constraints, the single phase crystalline alloy at composition C would melt polymorphously at $T_0(C)$. The $T_0(C)$ curves thus define the thermodynamic limits of homogeneous metastable crystalline phases with respect to melting. These curves will play an important role in our efforts to understand crystal-to-glass transformations. As a preview to this, notice that the $T_0(C)$ curve of the α phase in Fig. 2.4e gives $T_0^{\alpha}(C) \to 0$ as $C_B \to C^*$. The details of this dependence will be discussed later in this section. Beyond C^*, the liquid phase is thermodynamically preferred to the α phase at any temperature. Also recall that as a liquid is undercooled, it ultimately undergoes a glass transition whereby it kinetically freezes. This occurs at a reasonably well defined T_g. Below the $T_g(C_B)$ curve (T_g depends on composition), the $T_0^\alpha(C_B)$ curve then represents the critical line for a polymorphous crystal to glass transformation.

If we want to calculate the real $G(C)$-curve and therefore the $T_0(C)$-lines for a given binary system, several methods are known in the literature. The most direct approach is the semiempirical method of *Miedema* [2.17] and its extension to amorphous systems, which is described in detail in [2.8]. Most binary systems can easily be calculated this way, to get a rough estimate of the composition dependence of the free energy. A more complicated way is described by *Bormann* [2.18], who uses the Calculation of phase diagrams (CALPHAD) method and experimental data (melting point, lattice stabilities of the pure metals, solidus- and liquidus-lines, heat of mixing and entropy of mixing) to calculate a phase diagram of a binary system. Although the pure empirical method of the CALPHAD approach is more detailed, sometimes the lack of input data inhibits the calculation of the phase diagram. The most realistic theoretical approach uses only statistical thermodynamics. Unfortunately this method requires a detailed knowledge of the atomic distributions and their potentials. Only a few attempts have been successfully made in recent years [2.19].

The T_0-concept described above, where the free energy of the liquid and the crystal are equal at a certain temperature for a fixed composition, does not represent in general the equilibrium conditions. As stated above, most systems do not melt congruently. They exhibit a pronounced phase separation into a liquid and a solid phase with very different compositions. This is seen experimentally in most systems (e.g., Zr_3Rh with hydrogen, Zr–Ni multilayer), which amorphize at low reaction rates. The total system remains chemically inhomogeneous during the transformation. It consists of an amorphous phase (liquid-like) and the crystalline phase, which have different compositions. For polymorphous melting (amorphization), the free energy of the two phases *and* the total differentials of the free energies are equal ($dG^x = dG^l$) [2.20]. If follows that with

$$dG^\alpha = V^\alpha\, dP - S^\alpha\, dT + (\mu_A^\alpha - \mu_B^\alpha)\, dc^\alpha \ , \tag{2.7}$$

with V^α: the molar volume of an α-phase and S^α: the molar entropy, μ_A^α and μ_B^α chemical potentials of A in α, and B in an α-phase respectively

$$\Delta V_m(P, T, C_0)\, dP - \Delta S_m(P, T, C_0)\, dT - A_m(P, T, C_0)\, dc = 0 \ , \tag{2.8}$$

with $A^m = (\mu_A^x + \mu_B^L) - (\mu_A^L + \mu_B^x)$ called the *partitionless melting-state affinity* [2.20]. The equations above give the differential form of a *partitionless melting Clapeyron relation*, which defines the polymorphous melting surface $T_0(P, C)$ for a solid solution.

Three intensive variables appear in this non-equilibrium melting equation, yielding the three following partial relationships on the surface $T_0(P, C)$

$$(\partial T/\partial C)_P = -A_m/\Delta S_m \quad \text{(isobaric melting)} , \tag{2.9}$$

$$(\partial P/\partial C)_T = A_m/\Delta V_m \quad \text{(isothermal melting)} , \tag{2.10}$$

$$(\partial P/\partial T)_C = \Delta S_m/\Delta V_m \quad \text{(isoconcentration melting)} . \tag{2.11}$$

The last one corresponds to the well-known Clausius–Clapeyron equation.

If the two elemental crystalline solid solutions in an A–B binary alloy with negative enthalpy of mixing have different crystal structures, A_m cannot be zero and will increase when lowering the temperature. Consequently, the T_0-lines of the stable or metastable solid solutions at constant pressure plunge with the slope $(\partial T/\partial C)_P$. As the affinity can be neither equal to zero nor infinite (only for zero concentration) the slope $(\partial T/\partial C)_P$ can become infinite only if the entropy ΔS_m of melting equals zero. The slope becomes positive for a negative entropy of melting, which would violate the Kauzmann-paradox. Now we can construct a more detailed metastable *polymorphic* phase diagram (Fig. 2.5), which includes the T_0-curve ($\Delta G = 0$), the T_g-curve and the T_i^s-curve ($\Delta S = 0$). The T_0-curve has a negative slope for the system of interest here. It crosses the "ideal" glass transition curve according to the Kauzmann argument at a certain concentration C^* and a temperature T^*. This point represents in the phase diagram a triple point between undercooled melt, glass and supersaturated crystal under the condition that

$$\Delta G(C^*, T^*) = \Delta S(C^*, T^*) = \Delta H(C^*, T^*) = 0 . \tag{2.12}$$

Since the entropy ΔS_m decreases with decreasing temperature and equals zero at (C^*, T^*), the T_0-curve has exactly there an infinite slope. For a binary Zr–Ni system, C^* equals to 11.5 at. % Ni and $T^* = 638$ K [2.20].

Figure 2.5 shows also the concentration dependence of the inverse Kauzmann temperature T_i^s (entropy catastrophe temperature). For the pure metal, T_i^s is much higher than the temperature T_0 as discussed. The T_i^s-line should also decrease with increasing concentration and end in the triple point (C^*, T^*) [2.21] as follows from its definition ($\Delta S = 0$). It is interesting to note that at this point the "real" Kauzmann temperature and the inverse Kauzmann temperature meet. But in real systems, the amorphous phase has an excess entropy (small fraction of the entropy of fusion) when compared to the corresponding crystal, the exact amount determined from the kinetics and timescale of the glass transformation. Therefore, another glass transition temperature line with finite excess entropy must be considered, which will be parallel to the T_g-line (above it) and cross the T_0- and T_i^s-lines not exactly in the triple point.

Under ideal conditions, Fig. 2.5 shows a phase diagram between a supersaturated crystalline solid solution, the glassy state, and the liquid. At the triple point, the entropy of fusion vanishes and melting *becomes* a glass transition. The crystalline phase is strictly unstable beyond this composition (at any temperature). If this argument is valid, then any solid-state process which drives the

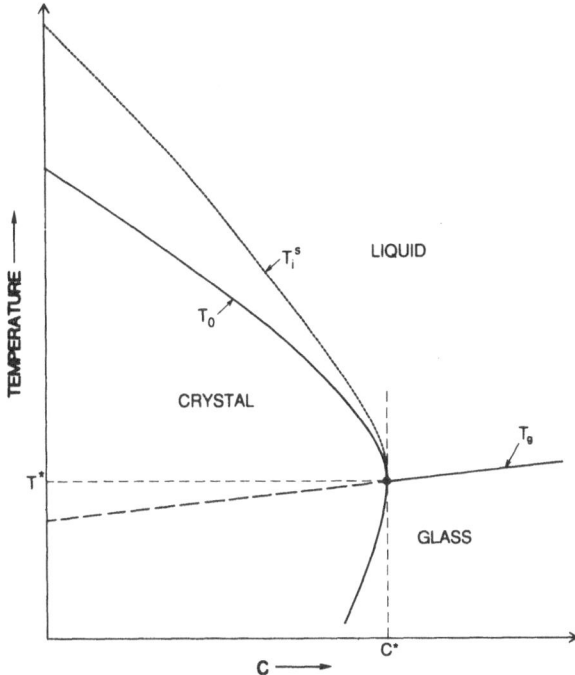

Fig. 2.5. Schematic metastable polymorphous-phase diagram, which includes the instability temperature, T_i^s, the melting temperature, T_0, and the glass temperature T_g, and defines the stability range of the crystal, liquid, and glass phases

composition of the crystalline phase beyond C^* will result in catastrophic amorphization of the unstable crystal. This type of a spinodal argument relates to a discussion by *Egami* and *Waseda* [2.22] in which they show that a solid solution should become topologically unstable beyond a critical concentration limit which depends (in their model) on the size difference of the two atomic constituents comprising the solid solution. Going back to Fig. 2.5, an alloy with fixed composition close to the triple point will vitrify in a second path by heating through the melting or instability line. Below the triple point the lines divide in two branches. The vertical branch is the partitionless melting curve extended below the isentropic temperature curve T_g, with ΔS_m equaling further to zero according to the Kauzmann paradox. If we assume that ΔS_m can be negative below T_g, the T_0-line is curved below T_g. An alloy with a certain constant composition then has two melting points, one of them below T_g corresponding to a re-entrant melting phenomenon [2.23, 24]. But one must note that the crystal-to-glass transition below T_g depends highly on the experimental kinetics of the process. Strictly speaking, the system is far from (metastable) equilibrium and non-ergodic. Therefore, it is ambiguous as to which of the two branches below T_g corresponds to any actual experimental situation. As discussed above, the vertical line gives the maximum allowable concentration in a supersaturated solid solution without undergoing polymorphous amorphization if the system is homogeneous.

Finally, we would like to note that the horizontal axis of Fig. 2.5 need not correspond to an axis of the concentration of B-atoms in an A-matrix. The x-axis can also be interpreted as index for an atomic displacement from its equilibrium position. This would include effects like defects in general, (negative) pressure effects, inverse particle diameter (nanocrystalline materials), mechanical stresses or combinations of them. Experimentally, it will be very difficult to cross directly through the triple points. As known from the other phase transitions of second order without latent heat (like superconductivity without external magnetic field), there will be large fluctuations close to the transition point which follow certain scaling laws [2.25]. In addition, the symmetry of the crystal has to change, reaching the transition line. To use the appropriate Landau theory in the vicinity of the phase transition, we have to define the correct order parameter for the process. Unfortunately, this is an unsolved problem and it will take more experimental results to define the correct order parameter.

2.1.3 Kinetic Constraints and Required Timescales

In the last section, we discussed the thermodynamics of metastable phases. The requirements for the crystal-to-glass transition were given either by the chemical driving force for an inhomogeneous solid-state reaction or by the temperature–composition function for a crystal instability. The latter was based on the constraint that the system has to remain homogeneous. The use of metastable states must be discussed in conjunction with a kinetic constraint. The internal time scale τ_{in} of the system should be long enough to make sure that an ergodic sampling of a set of relevant microstates of the atomic ensemble belonging to the metastable phases takes place, but short enough to enforce the constraint of chemical homogeneity. Since atomic diffusion is most likely the source of a chemical segregation the *external* timescale τ_{ex} is often linked to the atomic motion. The first kinetic constraint following the discussion of thermodynamics is then given by the time window

$$\tau_{in} < \tau_{obs} < \tau_{ex} \, ,$$

where τ_{obs} is the time of observation needed experimentally. In certain cases of metastable phases, τ_{ex} becomes extremely short. As noted before, melting above the thermodynamic melting temperature is very rapid. Only in experiments, where a single crystal of the metal Ag is encapsulated by a coherent Au crystal (to suppress heterogeneous nucleation of the liquid at a free surface) [2.26], can one extend the timescale τ_{ex} to make the superheated metastable crystal accessible to normal observation and study. The timescale increases due to the need for nucleation of the liquid without the solid–vacuum interface.

On the other hand, an undercooling of a liquid is possible for extremely long timescales even for pure metals [2.27]. Figure 2.6 shows the experimental scenario for the amorphization by any method in principal. A system is driven by a certain method (e.g., a crystal in contact with hydrogen gas; two elemental layers; a crystal under the influence or flux of an ion beam) to a thermodynamic state of free energy G_1, which lies higher than the free energy G_a of the liquid or amorphous phase. Under this driving force, it transforms into the amorphous phase within a timescale $\tau_{1 \to a}$. Since the amorphous phase itself is a non-equilibrium phase for $T < T_m$, it will transform into a more stable crystalline phase of free energy G_2 given sufficient time. This time scale $\tau_{a \to 2}$ must be long compared with $\tau_{1 \to a}$ if one is to observe the metastable amorphous phase. It is not clear if the system is always expected to pass through the next lower state as suggested by the so-called Oswald entropy step rule. From an experimental point of view, it is also possible that the initial state of the system transforms directly into the stable state (or some metastable form) of the crystalline phase. The timescale for this process, $\tau_{1 \to 2}$, which might be just a sum of $\tau_{1 \to a} + \tau_{a \to 2}$, must again be long compared with $\tau_{1 \to a}$ if one is to observe for example the crystal-to-glass transformation. So we see that the amorphization needs a thermodynamic driving force and a kinetic boundary condition as well.

The details of each timescale depend very much on the experimental situation. This will be discussed in Sect. 2.2 of this article. Here we give only some general remarks concerning the timescales $\tau_{1 \to a}$ and $\tau_{a \to 2}$ of the solid-state amorphization reaction. For the timescale $\tau_{1 \to a}$, we decide between an in-homogeneous reaction (e.g., multilayers of pure elements will form an amorphous interlayer) and the homogeneous transformation into the amorphous state, where the timescale $\tau_{1 \to a}$ decreases to very low values due to the diffusionless transition. In the first case, the timescale for the amorphization of the entire sample is obviously determined by the diffusion process of one species

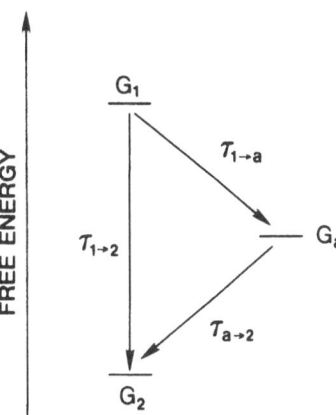

Fig. 2.6. Schematic illustration of the free-energy relationships and timescales involved in crystal-to-glass transformation (see text for further details)

(e.g., Co in Zr/Co-multilayers [2.28]. For most systems investigated up to now at least one component is a fast-diffusing species. Similar results are also found within the amorphous phase [2.28]. Therefore the amorphization process in total is limited by the time required to form the amorphous layer in the diffusion limited regime versus the time required to nucleate and grow the crystalline phase (for details see Sect. 2.2). In the early stages of the process, it appears that also the amorphous phase forms via nucleation in the non-equilibrium crystal. Most often, it nucleates at grain boundaries in much the same way that ordinary melting is found to nucleate at heterogeneities. Very recent results indicate that grain boundaries of one component (e.g., Zr-polycrystals) are filled with the fast diffusing second component (e.g., Co-atoms) up to such a concentration that a crystal-to-glass transition via an instability in this special region of the grain boundary seems possible [2.29].

The timescale $\tau_{a\to2}$ competing with the amorphization transformation has been reviewed several times [2.16]. Again, we should discriminate between polymorphous and non-polymorphous crystallization. As one expects, the timescale for non-polymorphous reactions is much longer, since a chemical segregation process has to take place. Vice versa, this chemical segregation process favors the amorphization by any method known for deep eutectic compositions. Consequently, polymorphous transitions are much faster due to the lack of long-range diffusion processes. Both effects are seen in the activation energy and the enthalpy of crystallization, which usually mirror image each other in a way that E_a is typically a maximum at eutectics and a minimum at compositions where a stable phase occurs in the equilibrium phase diagram, while ΔH_c behaves inversely [2.30]. Microscopically, the details of the crystallization process are not clear because many parameters like free or inner surfaces change the time–temperature–transition diagram very often. In the context of the amorphization reaction, it seems to be important that only one component is mobile, while the other one is more or less inmobile. More experiments are necessary to verify this point.

2.2 Experimental Results

2.2.1 Hydrides

The amorphization of a metallic alloy was first observed and investigated systematically by *Yeh* et al. in 1982 through the absorption of atomic hydrogen into the crystalline Zr_3Rh [2.5]. Hints for the possibility of such a transition had been found earlier by *Österreicher* et al. on $LaNi_3$ [2.6] and *vanDiepgen* et al. on $CeFe_2$ [2.31], but were not analyzed in any detail.

In the case of composition-induced destabilization and vitrification of a bulk crystalline phase, the atomic hydrogen is absorbed at elevated temperatures from a hydrogen gas (H_2) atmosphere or, alternatively, by water electrolysis resulting in solid-state amorphization of the electrode [2.32]. Even though the latter approach allows amorphization of the crystalline structure in much shorter times (10^2 versus 10^5 s) and higher corresponding hydrogen pressures (10^2 versus 10^0 atm) the transformation process itself has been investigated in much less detail than in the case of hydrogen-gas-induced amorphization. Thus, we limit ourselves here to solid-state amorphization by hydrogen via the gas phase.

The first systematic study of such hydrogen-induced amorphization has been reported for the polycrystalline fcc solution of partially ordered Zr_3Rh, a metastable phase with the Cu_3Au-type structure, which was exposed to hydrogen at a pressure of 1 atm. The Zr_3Rh-phase was produced either by cooling of the liquid alloy at a relatively low rate or by crystallization of an amorphous *splat* through thermal annealing at approximately 200°C. Exposure to hydrogen at temperatures between 150°C and 225°C resulted in complete amorphization as demonstrated in Fig. 2.7. The X-ray scans were taken as a function of time (in hydrogen gas) and show that the sample transforms gradually to an amorphous phase. At temperatures above 225°C, the equilibrium phase mixture of the cubic ZrH_2 and fcc Rh is obtained.

Transmission electron microscopy investigations revealed further details on the kinetics of the transition. From Fig. 2.8 it can be seen that the amorphous phase initially forms at the grain boundaries of the original polycrystal. Through wetting of the grain boundaries and subsequent growth into the polycrystalline interior, the grain boundaries become unstable and delocalized, and the amorphous phase grows starting from the grain boundaries into the crystal. The original grains which contain twin boundaries are penetrated effectively along these boundaries by the amorphous phase. Even at high magnification, a very sharp interface is observed between the amorphous phase and the untransformed crystalline material. Thus, this process is similar to melting.

From a thermodynamic point of view, these results can be explained by the common Gibbs free-energy constructions as shown in Fig. 2.9. Since the crystalline Zr_3Rh-phase is metastable with respect to the stable phase mixture, its free energy lies considerably higher, a point also consistent with thermal-analysis results. The diagram shows the free energy of the original cubic phase (ordered Cu_3Au-type) vs. hydrogen concentration (hydrogen-to-metal ratio y) and the free energy of the amorphous phase vs. y. In addition, the free energy of a two-phase mixture of fcc ZrH_2 and rhodium metal – the equilibrium phase mixture obtained at higher temperatures ($> 225°C$) – vs. y is shown. For hydrogen concentrations greater than y_0, no single-phase crystalline material with a free energy lower than that of the amorphous phase is found. In the absence of metal (zirconium–rhodium) interdiffusion, the solid solution of hydrogen in Zr_3Rh can spontaneously pass beyond the relevant T_0-surface during absorption of hydrogen. The sample could *melt* polymorphically provided that interdiffusion of Zirconium and Rhodium be suppressed. In reality, the hydrogen content of the amorphous phase formed at the grain boundaries is larger than the hydrogen content of the solid solution. In other words, the melting is not polymorphous with respect to the relative composition of the two

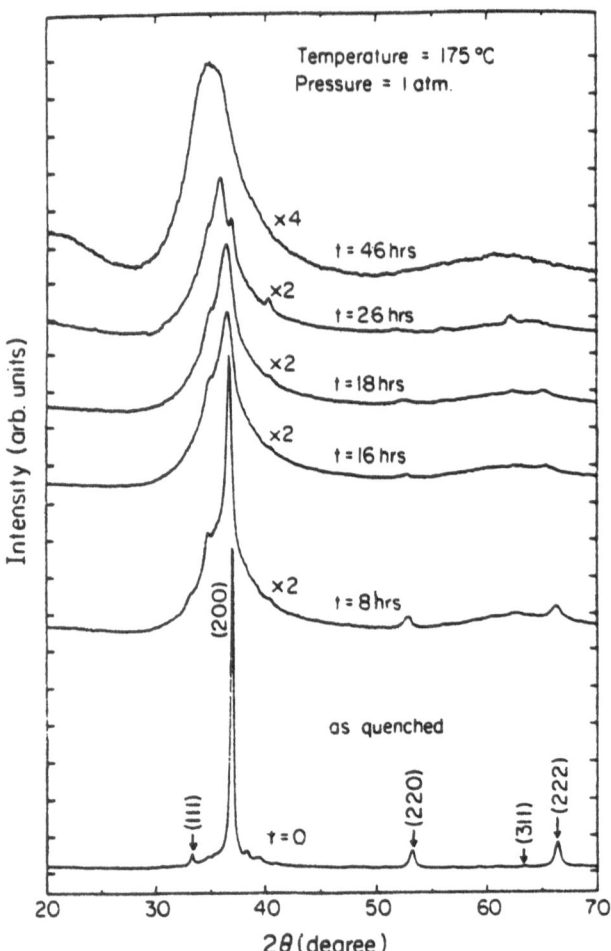

Fig. 2.7. X-ray diffraction patterns of an initially crystalline Zr_3Rh sample (as-quenched), which was hydrided during a 46 h period. The data show the transformation into the amorphous phase

metals. Thus, concentration fluctuations through segregational effects can become sufficiently large within the grain boundaries to form a critical nucleus of the amorphous phase. After nucleation of the amorphous phase the system can attain the local chemical equilibrium condition represented by the common tangent of the fcc and amorphous curves (Fig. 2.9). For sufficiently high overall hydrogen concentration, the entire sample then transforms into the glassy state.

Fig. 2.8. Transmission electron micrographs showing (a) the microstructure and diffraction pattern of the polycrystalline Zr_3Rh, (b) a dark and bright field image of the sample after partial transformation into the amorphous phase and (c) a high resolution image of the partially transformed sample, showing the sharp interface between the amorphous and crystalline regions

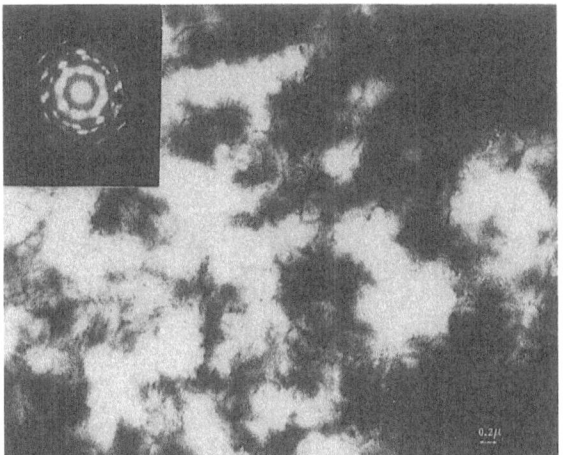

a

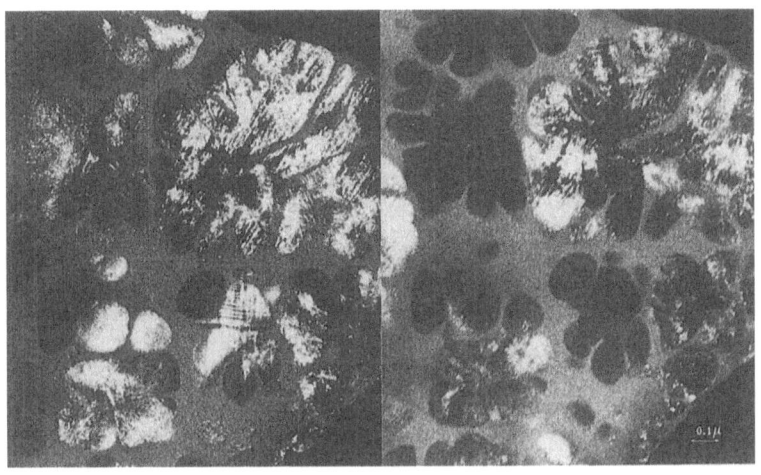

b

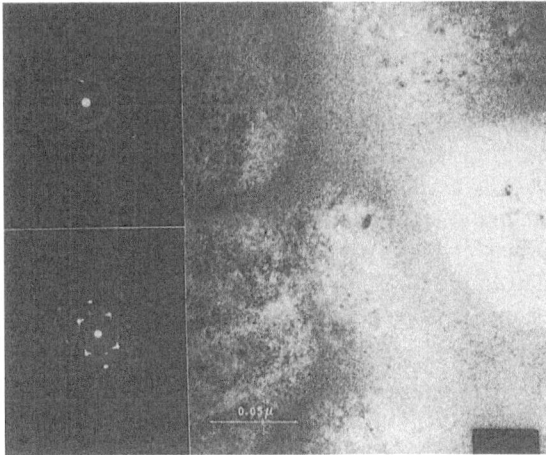

c

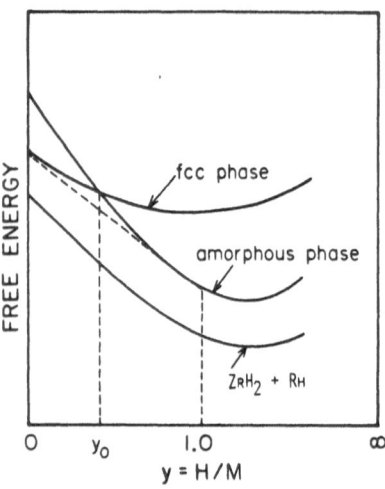

Fig. 2.9. Illustration of the dependence of free energy on hydrogen-to-metal ratio y for the original polycrystalline (fcc) phase, the amorphous phase and the equilibrium two-phase mixture (ZrH_2 + Rh)

Table 2.1. Structures of Laves-phase Fe_2–rare-earth compounds (C15) exposed to hydrogen gas at a pressure of 5 MPa and various temperatures for 86.4 ks, including the crystallization temperatures T_x of the hydrogen-induced amorphous alloys [2.33]

Alloy	Structures after exposure to hydrogen gas			
	323 K	473 K	T_x [K]	773 K
YFe_2	c-YFe_2H_x	a-YFe_2H_x	639	YH_2 + α-Fe
$CeFe_2$	c-$CeFe_2H_x$	a-$CeFe_2H_x$	563	CeH_2 + α-Fe
$SmFe_2$	c-$SmFe_2H_x$	a-$SmFe_2H_x$	620	SmH_2 + α-Fe
$GdFe_2$	c-$GdFe_2H_x$	a-$GdFe_2H_x$	643	GdH_2 + α-Fe
$TbFe_2$	c-$TbFe_2H_x$	a-$TbFe_2H_x$	647	TbH_2 + α-Fe
$DyFe_2$	c-$DyFe_2H_x$	a-$DyFe_2H_x$	650	DyH_2 + α-Fe
$HoFe_2$	c-$HoFe_2H_x$	a-$DHoFe_2H_x$	649	HoH_2 + α-Fe
$ErFe_2$	c-$ErFe_2H_x$	a-$ErFe_2H_x$	651	ErH_2 + α-Fe

The fast diffusion of hydrogen at relatively low temperatures (compared with metal interdiffusion) proves to be what allows observation of a crystal to glass transformation by hydriding. When the heat of formation of an amorphous phase from two-parent crystalline metals is negative, a thermodynamic driving force will exist to form an amorphous phase. If formation of competing crystalline intermetallic phases can be kinetically suppressed, it is possible to observe a crystal-to-glass transition. This transition should thus not be seen as limited to the destabilization of metastable compound phases, but as possible for stable intermetallic compounds as well. Indeed, this has been observed experimentally for a large range of stable intermetallic compounds listed in Table 2.1 [2.33].

The intermetallic A_xB_y compounds (A = a rare-earth metal and Zr, B = Fe, Co, Ni, Rh, Al, Ga, In) were prepared by standard melting and solidification

Table 2.2. Hydrogen-induced amorphization of several types of the AB_2 type intermetallic compounds with C15 structure (Laves phase RM2)

	A = Y	La	Ce	Pr	Nd	Sm	Eu	Gd	Tb	Dy	Ho	Er
B = Fe	●		□			●		●	●	●	●	●
Co			□	●	●	●		●	●	●	●	●
Ni		□	□	□	□	□		□	□	□	□	●

□ SSA occurs both at 323 and 473 K
● SSA occurs around 473 K

Table 2.3. Hydrogen-induced amorphization of the A_3B and A_2B type intermetallic compounds by hydrogen absorption [2.33]

(A) A_3B ($L1_2$ or fcc structure)

	A = Zr	La	Ce	Pr	Nd	Sm
B = Rh	●					
Al	●					
In	●	●	●	●	●	●
Ga					●	●

(B) A_3B (DO_{19}-structure)

	A = La	Ce	Pr	Nd
B = Al	●	●	●	●

(C) A_2B (C23-structure)

	A = Y	Pr	Nd	Sm	Gd	Tb	Dy	Ho
B = Al	●	●	●	●	●	●	●	●

techniques and subsequently ground to a powder of less than 100 mesh. These samples were exposed to hydrogen gas at a pressure below 5 MPa at temperatures between 293 and 773 K for 86.4 ks. The specimen were characterized by X-ray diffraction, TEM, and differential scanning calorimetry (DSC). Samples annealed at 473 K in the hydrogen atmosphere were found to exhibit a broad maximum in the X-ray diffraction patterns. TEM analysis demonstrated a featureless structure in the bright-field image and a diffuse halo in the diffraction patterns characteristic for an amorphous phase. The DSC curves obtained by heating the amorphous phases at a rate of 10 K/min show in general a broad endothermic signal between 450 and 600 K, before crystallization to the stable phases occurs above 650 K as indicated by an exothermic event. The crystallization products in these examples are a mixture of AH_2 and B. However, it should be emphasized here that a rather large number of intermetallic compounds with either C15- (Laves phase, AB_2), C23-(A_2B), DO_{19}-(A_3B) and $L1_2$ structures (A_3B) could be amorphized at temperatures ranging between 323 and 473 K. The results of these investigations are summarized in Tables 2.2–2.3.

Whereas a whole range of these experiments were performed in a hydrogen atmosphere at varying temperature starting with the initial metallic samples, two thermodynamic variables – temperature and hydrogen concentration – are changed simultaneously. Thus, the amorphization reaction cannot be analyzed unambiguously in the thermodynamic sense discussed above. In addition, it has been noted that in most cases the amorphization reaction starts at internal heterogeneous nucleation sites, predominantly grain boundaries due to possible concentration fluctuations at the boundaries [2.34]. However, by detailed TEM analysis, an alternative nucleation mode was observed on Zr_3Al by *Meng* et al. [2.35]. For a hydrided Zr_3Al specimen, homogeneous nucleation of the amorphous phase is observed within crystalline grains indicated by an intimate mix of the crystalline and amorphous phase at the early stage of the reaction.

2.2.2 Diffusion Couples

In this section, the growth of amorphous phases in thin-film diffusion couples will be discussed. Such diffusion couples consist typically of thin layers of two elemental metals deposited in an alternating fashion. The diffusion couples consist of one or more pairs of alternating layers of the two metals. A single pair is referred to as a bilayer, while two or more alternating pairs of metal layers are referred to as a multilayer. The elemental layers are initially crystalline and typically have the equilibrium crystal structure of the parent metal. The diffusion couple is heated to a reaction temperature T_R, at which atomic mobility of the constituent atoms becomes sufficient to permit a diffusional reaction along the interface(s) separating the constituent metal layers. A solid-state amorphizing reaction (SSAR) occurs when an amorphous alloy layer forms and grows beginning along this interface. This phenomenon was first observed in Au/La diffusion couples [2.7]. The amorphous interlayer may be a relatively uniform planar layer which grows in thickness as the reaction proceeds. This was found to be the case for Au/La, Ni/Zr, Ni/Hf, Co/Zr, and numerous other diffusion couple pairs as summarized in earlier review articles [2.4, 8, 37]. In other instances (e.g., Si/Ti, Ni/Ti, Cu/Y), the amorphous phase is found to evolve into a more complicated geometry in which grain boundaries of the original elemental layers act as preferred regions for amorphous-phase formation. This has been referred to in the literature as grain-boundary amorphization.

The discussion of SSAR in thin-film diffusion couples will be divided into two parts. In the first subsection, a review of experimental results will be presented for bilayers and multilayers. The discussion will include cases for which the reactant layers are both metals and cases in which one of the reactants is a non-metal. The second subsection consists of a theoretical discussion of the thermodynamic and kinetic factors which determine the circumstances under which SSAR is observed to occur. Further, the theory of nucleation and growth of the amorphous phase will be outlined.

a) Summary of Experimental Observations of SSAR

The thin film bilayer and multilayer diffusion couples used in the study of SSAR are prepared by sequential deposition of the elements from the vapor phase. Thermal evaporation, electron-beam evaporation, and sputtering have all been used to prepare specimens. Individual layer thicknesses typically range from 5 to 300 nm. The individual metal layers are polycrystalline with grain sizes ranging from 5 to 100 nm. Generally speaking, the crystallite grain size of the pure metals is of the order of but less than the thickness of the individual layers. As such, the individual layers of the diffusion couples generally contain a variety of grain boundaries. Upon heating, the amorphous phase forms along the interface(s) separating the individual metal layers. As mentioned above, the amorphous phase frequently appears as a uniform planar interlayer suggesting that the initial nucleation of the amorphous phase at points along the interface is quickly followed by rapid spreading along the interface. In this, the common case, growth of the amorphous phase along the interface seems much more rapid than in the direction normal to the interface.

Table 2.4 gives a summary of binary metal/metal systems in which SSAR has been reported in thin-film diffusion couples. The entries are divided into several groups, which include binary systems containing a rare earth (RE) and a late transition metal (LTM). This group, the LTM/RE group, includes the Au/La system as its prototype along with other similar systems such as Ni/Er. A second group consists of a late transition metal (LTM) together with an early transition metal ETM. The LTM/ETM group includes binary pairs such as Ni/Zr, Co/Zr, Cu/Y, Ni/Hf, etc. A third group consists of simple metals (e.g., Sn, Pb, etc.) reacted with a late transition metal (e.g., Co, Au, etc.) and is referred to later as the LTM/SM group. All of the above types of binary systems have certain features in common. For example, the phenomenon of *anomalous fast diffusion* (AFD) has been found to occur in experiments where the first member of the pair diffuses into single crystals of the second member. For instance, Au in an anomalous fast diffuser in crystalline La, while Ni and Co are anomalous fast diffusers in crystalline Zr and Hf. As will be seen below, the fast diffusion phenomenon is also common to nonmetal/metal systems which exhibit SSAR (see Table 2.5). For instance, Au, Ni, Co, etc. are anomalous fast diffusers in crystalline Si. ADF is defined as solute diffusion with a diffusion constant exceeding the self diffusion constant of the host solvent by at least several orders of magnitude. The phenomena has been studied extensively by numerous investigators and was well known prior to the discovery of SSAR. Comprehensive reviews have been published by *Le Claire* [2.38] and *Warburton* [2.39]. For anomalous fast diffusion systems, it has been found that the activation energy for diffusion of solute in the matrix is generally significantly smaller than the

Table 2.4. Binary metal/metal systems observed to exhibit solid-state amorphization by interdiffusion or other type of driven mixing (e.g. mechanical codeformation). The table lists the type of experiment, the typical reaction temperature, T_R (for the case of interdiffusion reaction) and references. (B: thin-film bilayer diffusion couples, M: thin-film multilayer diffusion couple, S: interdiffusion of polycrystalline layer of one component with single crystal of another component, MA: mechanical alloying of the metals, MAT: thermal reaction of a mechanically deformed composite

System	Type of experiment	T_R [°C]	References
Late transition metal rare earth systems			
Au–La	B, M	80	2.4, 7
Cu–Er	MA, MAT	150	2.4, 67
Ni–Er	MA, MAT	150	2.67
Late transition metal/early transition metal systems			
Au–Zr	B, M	250	2.91
Co–Zr	B, M	300	2.8, 29, 41, 43, 84
Cu–Ti	MA	—	2.92
Au–Y	B, M	120	2.4, 94
Cu–Y	M, B	200	2.50
Cu–Zr	MA, MAT	—	2.95
Fe–Zr	B, M, MA	300	2.96
Ni–Hf	B, M	350	2.4, 42, 97
Ni–Ti	B, M, MA	350	2.4, 98, 99
Ni–Zr	B, M, MA, MAT, S	320	2.4, 40, 43, 44
			2.47–49, 53, 66
			2.68–70
Simple metal/transition metal systems			
Al–Mo	B	250	2.93
Al–Mn	M	220	2.100
Al–Pt	M	200	2.101
Sn–Co	M	50	2.103
Ni–Mg	M	50	2.102

activation energy for self diffusion of matrix atoms. For most ADF systems, the solute atom has an atomic radius which is at least 10% smaller than the atomic radius of the host atom. It is currently accepted that the fast diffusing solute tends to occupy interstitial sites in the host matrix. The small activation energy is then associated with the existence of easily accessible excited configurations of the interstitial atom. For the present purposes, it is sufficient to note that ADF systems are characterized by a fast moving solute species and a slow moving host matrix, phenomena which has an analog in the SSAR process to be pointed out shortly.

A second feature common to systems which exhibit SSAR is a negative free energy of mixing. SSAR systems tend to form alloys and/or intermetallic compounds in equilibrium. It is this feature of binary systems which provides the thermodynamic driving force for chemical reaction. For many of the SSAR systems, the negative free energy of mixing is quite large and consists of an

Table 2.5. Binary semiconductor/metal systems observed to exhibit solid-state amorphization by interdiffusion reactions. Type of experiment, typical reaction temperature (T_R), maximum thickness of the amorphous layer (X_{max}), and references are listed. (S: interdiffusion of polycrystalline metal with semiconductor single crystal, and A: interdiffusion of amorphous semiconductor layers with polycrystalline metal)

System	Type of expt	T_R [°C]	X_{max} [nm]	References
Si/Nb	S	450	5.5	2.59
Si/Ni	A	280	120	2.55, 57, 58
Si/Rh	A	250	< 10	2.54
Si/Ti	A, S	350	5–15	2.51, 58, 59
Si/V	S	380	2–3	2.59
Si/Zr	S	350	17	2.59
Te/Ag*	A	100	> 100	2.56
GaAs/Ni	S	180	—	2.60
GaAs/Co	S	280	20–30	2.61
GaAs/Pd	S	—	5	2.63
GaAs/Pt	S	60	5	2.60
InP/Pd	S	200	< 10	2.62
InP/Ni	S	200	< 10	2.62

*Diffusion of polycrystalline Ag layer in an amorphous-Te film

enthalpic and an entropic contribution. In those systems with large negative free energies of mixing (e.g., 10–150 kJ/mol), the enthalpic contribution dominates. Systems which exhibit SSAR are thus typically characterized by large negative enthalpies of mixing. (Note that SSAR has been observed in systems with near zero enthalpy of mixing such as Co–Sn.) Since the enthalpy of mixing depends somewhat on the structure of the product phase, it is useful to be more specific. In particular, the liquid (or amorphous phase) is often used as a reference state. This article makes frequent reference to the negative enthalpy of mixing of the parent metals in the liquid/amorphous phase. This is a natural reference quantity since SSAR involves an amorphous product phase. In practice, the free energies and enthalpies of mixing of metal pairs in various phases are evaluated by either the *Miedema* [2.17] or the *Calphad* [2.18] methods discussed in Sect. 2.1.2.

SSAR is observed when the binary diffusion couples listed in Table 2.4 are heated to an appropriate reaction temperature, T_R. Examples of typical values of T_R are given in Table 2.4. It is well known that amorphous metallic alloys tend to crystallize in laboratory timescales upon heating to temperatures close to their glass-transition temperature, T_g [2.16]. For a typical practical timescale (e.g., minutes), one can define crystallization temperature as the temperature at which a significant fraction of an amorphous sample undergoes crystallization in the specified time. The time required for an amorphous phase to crystallize can be identified with τ_{a-2} of Fig. 2.6 (see discussion in Sect 2.1.3). In the low temperature regime (well below T_g), atomic diffusion in amorphous alloys is thermally activated (as is typical of solids) and roughly follows Arrhenius behavior with $D = D_0 \exp(-Q/K_B T)$, where D is the characteristic atomic diffusion constant and D_0 is a characteristic constant premultiplier. The nucleation and growth of more stable crystalline phases over laboratory timescales (seconds to days) requires atomic mobility over distances of the order of the size of a critical crystalline nucleus. For metallic glasses, this requirement is generally fulfilled somewhere in the vicinity of T_g as T_g is approached from below. As such, it is found that formation and growth of amorphous interlayers in diffusion couples generally requires that $T_R \lesssim T_g$. The above factors lead to the concept of a *kinetic window* for SSAR. When the temperature is too low, no atomic diffusion will occur: the diffusion couple is kinetically frozen. The timescale for formation and growth of the amorphous phase (which can be identified as τ_{1-a} in Fig. 2.6 of Sect. 2.1.3) exceeds the experimental timescale. The factors which determine τ_{1-a} for diffusion couples will be discussed at length in this section. Near or above T_g, crystallization of an amorphous product to thermodynamically preferred crystalline phases occurs in relatively short times. SSAR is observed when an intermediate temperature range exists. In this intermediate temperature range, the condition $\tau_{1-a} < \tau_{ex} < \tau_{a-2}$ is satisfied (as discussed in Sect. 2.1.3), where τ_{ex} is the timescale of the experiment. This temperature interval thus presents a *kinetic window* of opportunity with respect to amorphous interlayer growth. The amorphous layer grows in a short timescale compared to available experimental times, but long when compared to timescales for formation of equilibrium crystalline phases.

Figure 2.10a shows a cross-sectional electron micrograph of an amorphous interlayer grown in a bilayer diffusion couple of Ni and Zr [2.40]. The interlayer was grown at a temperature of 300°C for a time period of 6 h. In the micrograph, the amorphous interlayer appears as a featureless gray zone between the polycrystalline layers of Ni and Zr. Note that the amorphous phase forms a rather uniform planar interlayer. At a later time (12 h at 300°C), further evolution of the diffusion couple has occurred as shown in Fig. 2.10b. A second crystalline interlayer has appeared separating the amorphous phase from the polycrystalline Zr. The second phase has the composition $Ni_{50}Zr_{50}$ and is the equiatomic intermetallic compound found in the equilibrium phase diagram of Ni/Zr. Figure 2.11 (taken from *Schroeder* et al. [2.41] show a similar micrograph illustrating the formation of many amorphous interlayers in a multilayer diffusion couple of Co/Zr. Here, the initial layers of Co and Zr are rather thin (15 and 7.5 nm, respectively). One sees that amorphous phase formation and growth occurs at each interface in the multilayer. This latter micrograph was the earliest direct evidence of the uniform planar layer morphology typical of SSAR. In early TEM studies, selected area electron diffraction was employed to prove the amorphous structure of these featureless interlayers. Bright-field/dark-field pairs of electron micrographs were also used to demonstrate the lack of crystalline diffraction in these amorphous interlayers. These studies allayed early criticism that the reaction products of SSAR might be microcrystalline. High-resolution electron micrographs reveal other interesting features of SSAR.

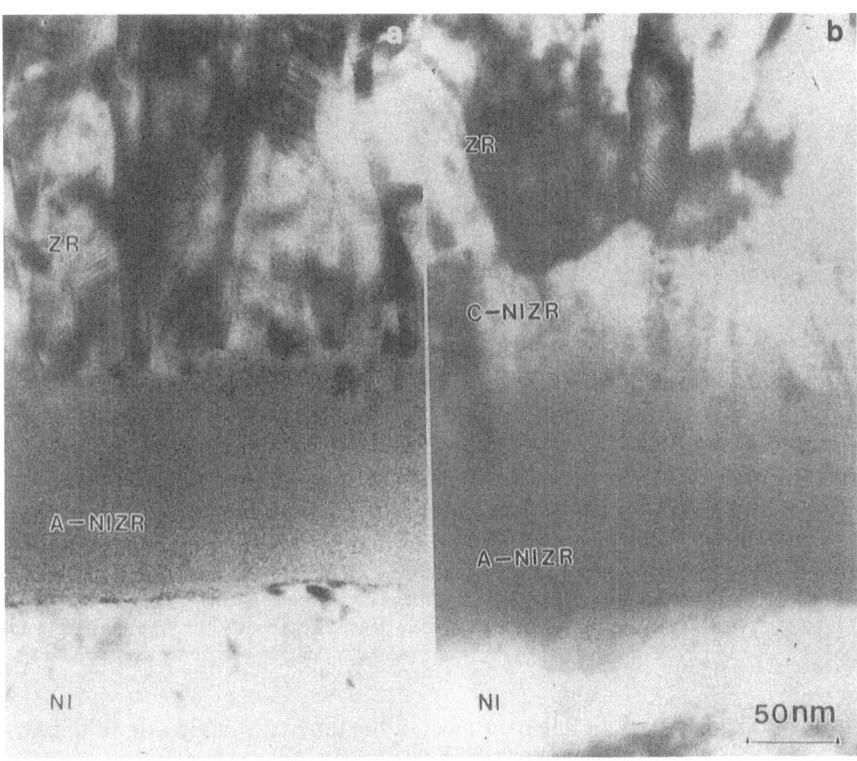

Fig. 2.10. Cross-sectional electron micrograph of a Ni/Zr diffusion couple reacted at (a) 300 °C for 6 h, showing ∼ 80 mm thick amorphous interlayer which has grown by interdiffusion; (b) the same diffusion couple reacted at 300 °C for 18 h, showing both an amorphous interlayer and a layer of the crystalline intermetallic compound Ni/Zr [2.40]

Figure 2.10 illustrates several features of SSAR. There exists, at any given reaction temperature, a maximum critical thickness X_m, to which the amorphous interlayer will grow. For the case of Ni/Zr, this maximum thickness is determined by the onset of growth of the more stable crystalline intermetallic compound NiZr. Generally speaking, growth of the amorphous interlayer is controlled by atomic diffusion across the interlayer. As will be seen in Sect. 2.2.2b, this leads in the simplest case to a growth law of the form $X = (aD(T)t)^{1/2}$, where a is a constant of order unity, X is the thickness at time t, and $D(T)$ is a temperature-dependent diffusion constant, which increases with T in a thermally activated manner. Careful studies using the Rutherford-backscattering technique for composition profiling have been used to study the growth law for uniform planar amorphous interlayers [2.42–44]. Figure 2.12 (taken from [2.79]) shows the thickness, X, of an amorphous interlayer as a function of time for a Ni/Hf diffusion couple reacted at 340°C. Here X was measured by Rutherford backscattering, and has been plotted as a function of

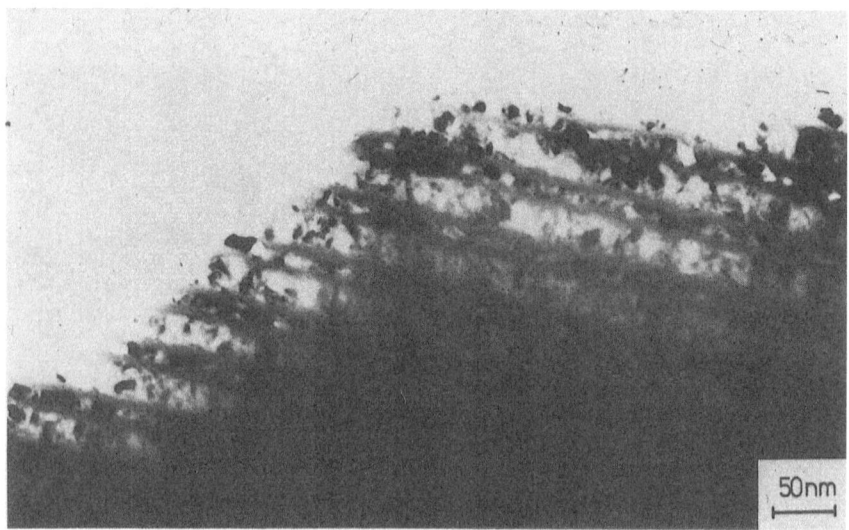

Fig. 2.11. Cross-sectional electron micrograph of Co/Zr multilayer diffusion couple consisting of many alternating layers of Co and Zr. The thin layers showing bright and dark contrast are Ni while the thicker layers are Zr. The period of the multilayer structure is about 50 nm. The diffusion couple has been reacted for 2 h at 210°C. The gray layers separating the Co and Zr are amorphous [2.41]

$t^{1/2}$. This yields a straight line over most of the timescale of growth. In the early stages, the reaction deviates from a $t^{1/2}$ *law* behavior. This will be seen in the following subsection to be related to an interface-limited growth regime. At later stages (typically as X approaches its limiting value), one observes a characteristic slow down of the reaction. A downward deviation from the $t^{1/2}$ law is observed which was originally attributed to stress relaxation in the diffusion couple and/or to stress relaxation effects within the growing amorphous interlayer. Recent detailed studies [2.45] have confirmed this explanation.

The Rutherford backscattering technique can also be used to carry out *Kirkendall marker* experiments, from which one can identify the moving atomic species during interdiffusion. Early experiments of this type were carried out by *Cheng* et al. [2.46] and *Barbour* et al. [2.47]. Later more detailed experiments were reported by *Hahn* et al. [2.48]. These marker experiments clearly established that one of the two atomic species comprising the diffusion couple acted as the dominant moving species during SSAR. In the case of Ni/Zr, it was shown that the atomic mobility of Ni in the growing amorphous interlayer is at least one, and probably several orders of magnitude greater than the atomic mobility of Zr during SSAR. Figure 2.13 (taken from [2.46]) shows the result of a Kirkendall marker experiment in which the position of the amorphous interface is measured with respect to an inert marker. The two dashed lines indicate the theoretical positions of the marker under the assumption that either Zr has no mobility (only Ni diffuses) or Ni has no mobility (only Zr diffuses). To within experimental error, the actual marker shifts agree with the curve predicted by

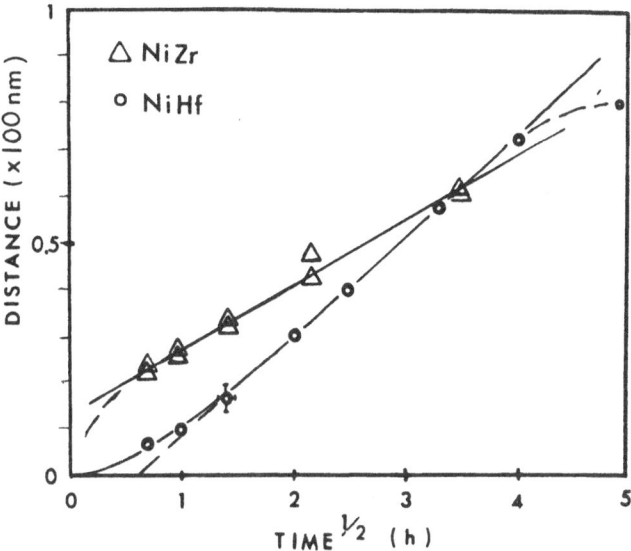

Fig. 2.12. Examples of thickness of an amorphous interlayer growing in bilayer diffusion couples of Ni/Zr and Ni/Hf as a function of $t^{1/2}$. The thickness of the amorphous layer was determined in both cases using Rutherford backscattering analysis. The Ni/Hf bilayer was reacted at a temperature of 340°C [2.79], while the Ni/Zr bilayer at 280°C [2.42]

the assumption that only Ni moves. Several studies of this type have demonstrated that SSAR generally involves a dominant moving species. Further, the dominant diffusing species of a binary pair of metals is invariably that which exhibits ADF behavior as discussed above. The non moving species is that which serves as the non moving host metal in ADF. Naturally, ADF involves diffusion in the crystalline state, whereas the growth of an amorphous interlayer during SSAR involves diffusion within the growing amorphous phase. The relationship between ADF and SSAR will be taken up again later.

Associated with the existence of a dominant moving species in SSAR is the formation of *Kirkendall* voids along the interface of the diffusion couple. In polycrystalline bilayer and multilayer diffusion couples, these voids appear along the interface, separating the amorphous interlayer from the fast diffusing metal (e.g., Ni, Co, etc.). The voids form during the early stages of the reaction and grow as the reaction proceeds. Their appearance in the diffusion couple can be directly related to the existence of a fast moving species (e.g., Ni in Zr/Ni couples) [2.40, 49]. It is argued that the diffusion of Ni atoms in the growing amorphous layer is accompanied by a counter flux of vacancy-like defects. The vacancy-like defects are transported to the amorphous/Ni interface. Here, they can condense to form *Kirkendall* voids or are transported as vacancies into the Ni layer, where they ultimately condense at a free surface or other nucleation site for voids. The formation of voids at the Ni/amorphous interface is found to be suppressed by elimination of grain boundaries along the Ni side of the diffusion couple. No voids are formed when a single crystal of Ni is reacted with polycrystalline Zr. In the absence of Ni grain boundaries along the Ni/amorphous interface, vacancies are apparently transported to the free surface of the Ni layer [2.40, 43, 49].

As mentioned above, the growth of an amorphous phase does not always occur as a uniform planar interlayer. For instance, when Y/Cu multilayers with

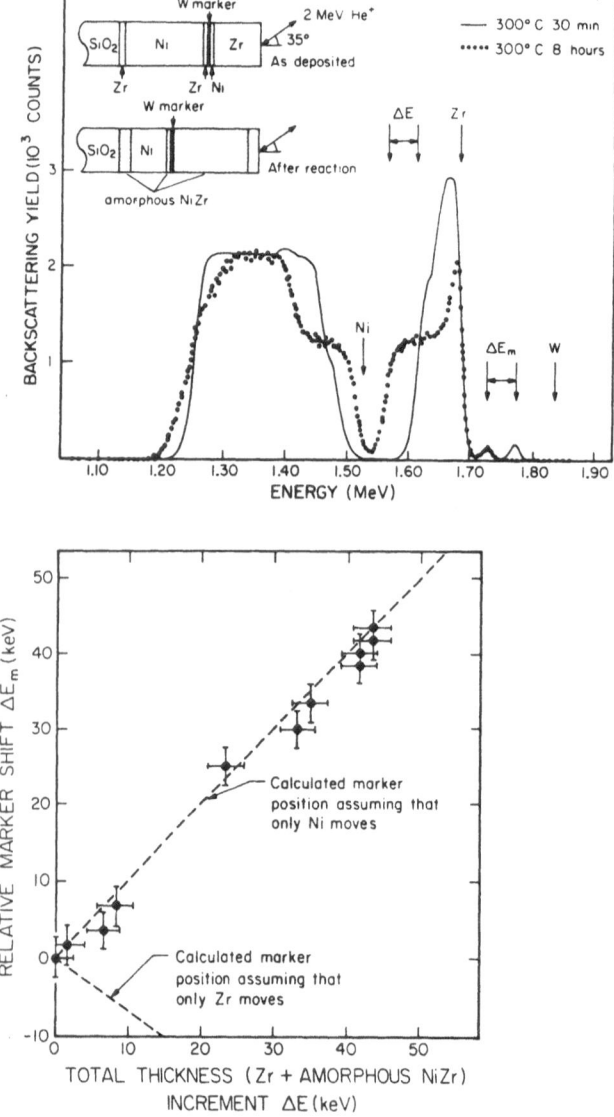

Fig. 2.13. Result of a *Kirkendall* marker experiment carried out by Rutherford backscattering spectroscopy. The upper part shows the marker shifts (ΔE_m and ΔE) for the tungsten marker and the Zr edge from the backscattering spectrum. The lower part compares the experimental data with the expected behavior of the marker shift when only one species moves (dashed lines). Within experimental error, it is found that only Ni moves (see [2.46] for details)

elemental layer thicknesses in the range of 4–30 nm were reacted at temperatures in the range of 60–100°C, the formation of a thin amorphous layer was found to occur simultaneously along grain boundaries of the polycrystalline Y as well as along the interface between Y and Cu grains [2.50]. This leads to a morphology in which individual crystallites of Y are coated with a Cu rich amorphous layer. Similar grain boundary amorphization has been observed in other multilayer systems such as Si/Ti and Ni/Ti [2.51, 52]. In the case of Si/Ti, the initial silicon layers are amorphous, while the Ti layers are polycrystalline and rather thin (∼ 5–15 nm). An amorphous layer is observed to wet rapidly the grain boundaries of Ti during the early stages of the reaction leading to a morphology in which Ti grains are surrounded by amorphous material.

The role of grain boundaries within the two reacting layers in the initial formation of an amorphous phase has been studied in some detail. In the well-studied case of Ni/Zr diffusion couples, experiments were carried out with single crystals of both metals. In one case [2.40, 53], large single crystals of Zr were prepared by recrystallization of a Zr foil. The free surface of these crystals had a typical (0 0 2) orientation (with close-packed planes parallel to the surface). The recrystallized foil consists of a mosaic of such crystals (the crystals are columnar and span the thickness of the foil) with typical sizes (in the plane of the foil) ranging from 20 to 100 μm depending on the details of the recrystallization process. The individual crystals are typically separated by low-angle (low-energy) grain boundaries. The surface of the recrystallized foil was sputter cleaned and thermally treated under UHV conditions to produce a relatively contaminant free surface. Polycrystalline Ni was subsequently deposited onto the clean surface. When this *single crystal Zr* diffusion couple was heated to temperatures between 300°C and 375°C for time periods of hours (Fig. 2.14), no observable reaction occurred along the interface, whereas similar thermal treatments of polycrystalline bilayers resulted (Fig. 2.10) in amorphous interlayer growth to thickness of 100 nm. Upon heating to higher temperatures (> 400°C), sudden formation and growth of crystalline NiZr is observed. Above 400°C, the crystalline NiZr grows to micron-scale thicknesses in timescales of minutes. The absence of grain boundaries on the Zr side of the diffusion couple suppresses the formation of an amorphous interlayer. The low-angle tilt boundaries which remain are apparently also ineffective in initiating amorphous-phase formation. Apparently, the high-angle Zr grain boundaries present in a sputtered or vapor-deposited polycrystalline Zr layer provide a catalytic or heterogeneous nucleation site for initiation of amorphous-phase growth. Without these high-energy grain boundaries, SSAR is not observed. One can also form diffusion couples with Ni single crystals [2.43]. When polycrystalline Zr is deposited onto such single crystals of Ni and the diffusion couple reacted at 300°C, an amorphous interlayer is again readily observed to grow. The absence of grain boundaries on the Ni side of the diffusion couple does not suppress SSAR.

The SSAR phenomenon has also been reported at the interface, separating a crystalline metallic layer and a semiconductor such as Si or GaAs. [2.51, 54–63]. Table 2.5 gives a summary of experimental data on various systems of this type.

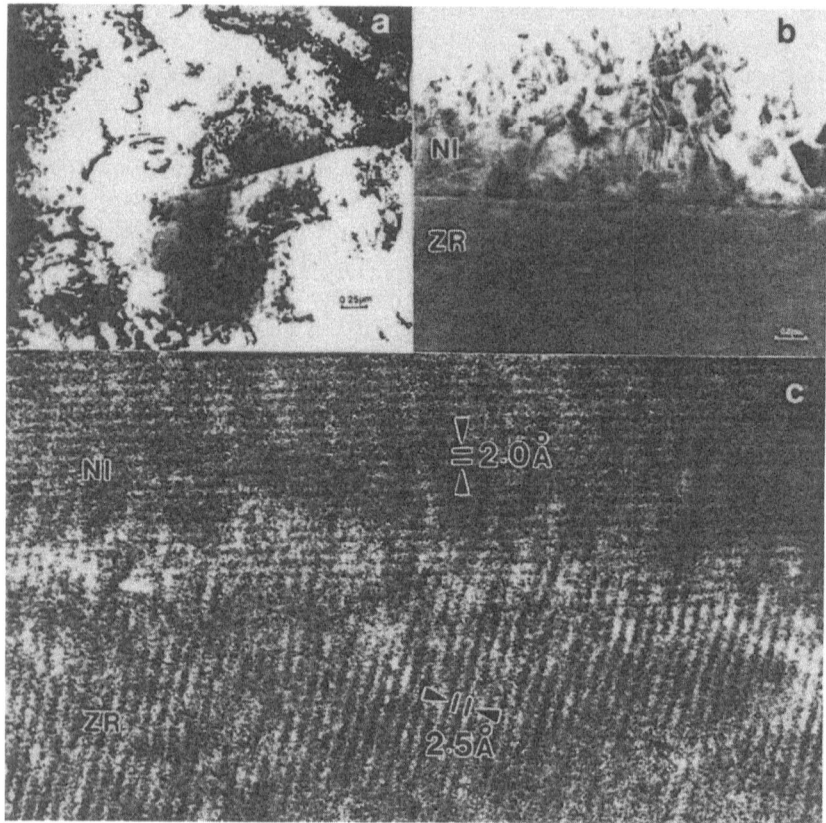

Fig. 2.14. Electron micrographs of a Ni/Zr diffusion couple consisting of a mosaic of large (25–75 μm) single crystals of Zr (a recrystallized foil) and polycrystalline Ni (Sputter deposited). The diffusion couple was reacted at 300°C for 6 h. (a) Plane view showing typical microstructure of the mosaic of Zr crystals, (b) bright field cross-sectional micrograph showing no reaction after 6 h at 300 °C, (c) high resolution image of reacted interface showing Ni (111) and Zr (101) lattice fringes. The interface remains atomically sharp [2.40]

The formation of metal silicides upon thermal reaction of metal layers with silicon crystals is a problem of long standing interest to the semiconductor device industry. Much attention has been given to the problem of understanding the sequence of phases which form and grow during thermal reaction of a metal with a silicon surface. It is interesting to note that despite widespread early studies of silicide formation and growth, it was not until 1983 [2.54] that clear evidence of amorphous-silicide formation was reported. The development of cross-sectional, high-resolution, TEM methods for studying transverse sections of diffusion couples did not occur until the early 1980s. Prior to that time, techniques in use (e.g., X-ray diffraction, depth profiling by backscattering, and plane-view transmission microscopy) lacked sufficient depth resolution to detect ultra-thin (thickness of order of nanometers) interlayers in diffusion couples. In 1976, *Walser* and *Bene* [2.64] suggested that the interface between a silicon crystal and a metal could be viewed as an amorphous membrane. They then suggested that the nucleation of a crystalline silicide occurred within this amorphous membrane. It is ironic not only that an amorphous interlayer phase actually does form in certain metal/silicon diffusion couples, but that it actually grows to significant thicknesses prior to the formation of a crystalline silicide. In such systems, the *Walser–Bene* membrane is more than a theoretical construct.

When silicon is deposited from the vapor phase at ambient temperature, it solidifies as amorphous silicon. Vapor deposited bilayers and multilayers of silicon with metals thus consist of polycrystalline metal and amorphous silicon. The earliest observations of amorphous silicide formation by SSAR were made on such diffusion couples [2.51, 54]. Similar results were also obtained earlier by *Hauser* when Au was diffused into amorphous Te [2.56]. Figure 2.15 shows an example of an amorphous silicide formed by reaction of amorphous silicon with polycrystalline Ni-metal at a temperature of 350°C for reaction times of 2 and 10 s [2.55, 57]. The reaction experiments were carried out by a flash-heating method (see [2.55] for details). In this example, the amorphous phase grows concurrently with a crystalline silicide. The amorphous phase is in contact with amorphous Si and the crystalline silicide in contact with the Ni layer. As in the case of typical metal/metal systems, the amorphous interlayer is planar and uniform. It is also interesting that the interface between amorphous silicon and the amorphous silicide appears to be atomically sharp despite the fact that both phases are amorphous. This suggests that amorphous silicon (a covalently bonded non metallic amorphous phase with fourfold coordinated silicon atoms) is distinctly different from an amorphous silicide (a metallically bonded system with higher atomic coordination number). These two phases are apparently connected by a discontinuous phase transformation.

Following the analogy with metallic bilayers, one would expect that SSAR be possible with silicon single crystals in the case where silicon becomes the moving species (recall the Ni single crystal/Zr experiments described above). On the other hand, when Si is the non-diffusing species (i.e., as Zr in Ni/Zr diffusion couples), use of a single crystal of Si might lead to the suppression of amorphous-phase formation owing to the absence of grain boundaries as nucleation sites for the amorphous phase. In fact, solid-state amorphization has been observed in diffusion couples consisting of silicon single crystals reacted with an early transition metal (e.g., Nb, Zr, V, etc.) [2.58, 59]. In these cases, it is known that Si becomes the dominant diffusing species [2.65]. Note, however, that the maximum thickness of the amorphous layer appears to be very small (2–5 nm). Diffusion couples consisting of silicon single crystals in contact with late transition metals (e.g., Ni, Rh, etc.), where Si is the non moving species, tend not to exhibit solid-state amorphization reactions. The use of an amorphous silicon starting layer (Table 2.5) in contrast, again leads to solid-state amorphization reactions. Summarizing these observations, one sees that for both metal/metal and metal/non metal diffusion couples SSAR is generally suppressed when the

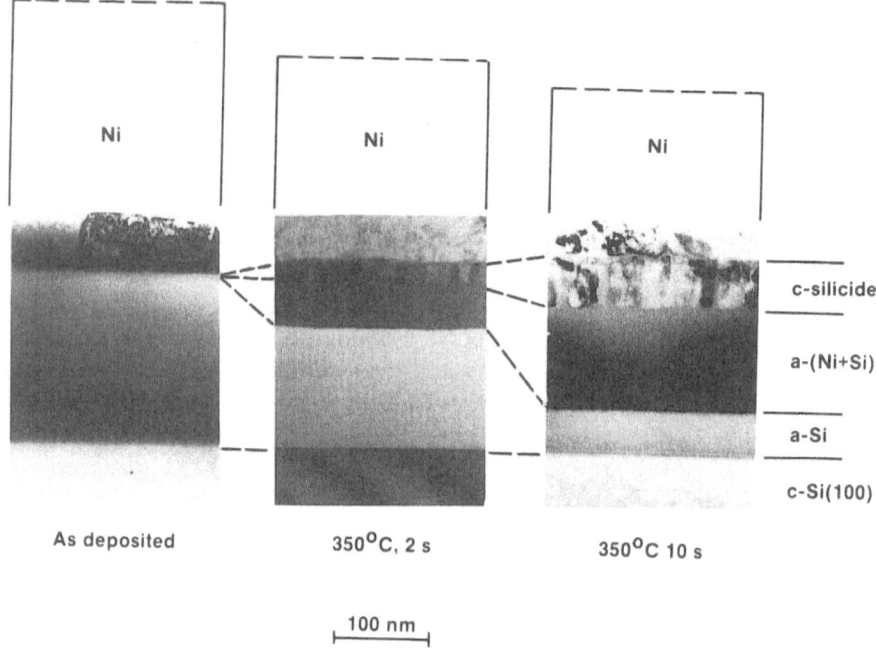

Fig. 2.15. Concurrent growth of amorphous interlayer and crystalline Ni₂Si silicide layer (the layers are adjacent) in an amorphous-Si/polycrystalline-Ni diffusion couple [2.55]

non-diffusing species is initially in the form of a single crystal. This observation is connected to the fact that the amorphous phase requires a suitable nucleation site, a point elaborated upon in the following section.

Table 2.5 is divided into four sections according to the type of diffusion couple. The first includes reactions of amorphous silicon or single crystal Si with a polycrystalline metal film. The second involves a metal reacted with amorphous Te. The third and fourth sections deal with experiments on compound semiconductors (III–V systems) reacted with metals. In this latter section, one notes for example, that late transition metals such as Co, Pt, and Pd form amorphous phases when reacted with single crystals of certain III–V semiconductors such as GaAs and InP [2.61–63]. These reactions are of particular interest in that the reaction appears to be initiated [2.61] at near or slightly above room temperature. When a clean surface of a single crystal of GaAs is covered by deposition of a polycrystalline platinum film at room temperature, an amorphous interlayer is spontaneously formed at ambient temperature. Much higher reaction temperatures are typically required to form crystalline interlayer phases. This low reaction temperature suggests that the interface separating the III–V semiconductor and the metal is rather unstable. Such heterointerfaces may be intrinsically instable against collapse into a glass. This issue will be raised again in the next section.

As a final topic in this section, it should be pointed out that SSAR has been observed in other types of diffusion couples (not made by thin-film methods) fabricated from bulk metals. *Atzmon* et al. [2.66, 67] carried out relatively slow deformation experiments in which thin-but-bulk sheets of two metals (e.g., Ni/Zr and Cu/Er) were codeformed in a conventional rolling mill to form a laminated composite. Successive folding and rolling of the composite leads to an ultrafine laminated microstructure (layer thicknesses in the range of 10–50 nm) consisting of alternating layers of the metals. These laminated composites can then be thermally treated (in the same temperature range as a thin-film diffusion couple) to grow an amorphous interlayer phase along the interfaces of the laminated metal layers *Atzmon* et al., *Wong* et al. [2.68, 69], *Schultz* et al. [2.70, 71] and *Bordeaux* et al. [2.72] have confirmed that such composites can be codeformed to obtain individual layer thickness in the range of 10–50 nm and thermally reacted to grow an amorphous phase. The growth of amorphous interlayers during thermal reaction of such codeformed composites has been found to follow a growth law nearly identical to that found in thin film bilayers and multilayers upon heating [2.68, 71]. In Cu/Er laminated composites, *Atzmon* et al. were able to demonstrate complete amorphization of the bulk composite following extensive and repeated deformation of the sample at ambient temperature [2.67] with no subsequent thermal treatment. To explain these results apparently requires some form of defect enhanced diffusion which together with deformation-induced local-temperature increases could account for the observed amorphous interlayer growth at ambient temperature.

The process of amorphous interlayer formation along a single Zr/Ni interface during codeformation of bulk plates of Zr and Ni in contact was recently studied by *Martelli* et al. [2.73]. They used an impact loading machine to deform a nickel plate pressed between two Zr plates. The plates have thicknesses in the millimeter range. By varying the load, they were able to produce deformation at varying strain rates. Furthermore, by using the junction of the two plates as a thermocouple, they were able to measure the average temperature along the interface during deformation of the plates. The loads are applied during a timescale on the order of 50 ms and produce deformation of both the Zr and Ni plates at typical strain rate of order $10–100$ s^{-1}. Along the interface, a thin and rather uniform amorphous interlayer (thickness $\sim 5–10$ nm) was observed to form. The average interface temperature was observed to rise 60–80 K above room temperature during the deformation. The authors argue, based on the measured temperature increase, that thermal interdiffusion alone cannot explain their results. In a series of experiments at differing loads and strain rates, they show that growth rate of the amorphous interlayer is correlated to the strain rate. This, they argue, demonstrates that interdiffusion is enhanced during plastic deformation by an amount which increases with the deformation rate. Their experiments provide direct and convincing evidence of the role of defect-enhanced diffusion in the observed deformation-induced amorphous interlayer growth.

High-energy ball milling of a physical mixture of metal powders also leads to amorphous-phase formation as first reported by *Yermakov* et al. [2.74] and *Koch* et al. [2.75]. These experiments are discussed at greater length in a separate chapter of this book by *Schultz* et al. As such, only a brief summary of the results is presented here. In these experiments, metal powder mixtures alloyed during collisions of centimeter-size hardened steel or tungsten carbide balls. The balls move at typical velocities of order 1–10 m/s. Powder grains are deformed, cold-welded together, fractured, etc., during the process. This produces particles consisting of laminated domains of the two metals which become progressively refined as the milling proceeds. Deformation of individual grains at strain rates of 10^3 s^{-1} or higher leads to rather localized shear bands (typical widths of order 1 μm) in which temperatures may rise appreciably during a deformation event. The magnitude of the temperature increase within the shear bands is controversial. Estimates range from tens of degrees K to hundreds of degrees [2.89, 90]. If the process is stopped at intermediate stages, amorphous interlayers are found to form at the interfaces of the two metals. Protracted ball milling leads to comminuation of the metal laminae, growth of the amorphous interlayers, and ultimately to a completely amorphous product powder. The relative roles played by thermal diffusion and defect-enhanced diffusion in the growth of the amorphous layers is the subject of ongoing debate. Experiments of the type performed by *Martelli* et al. [2.73] seem to suggest that the inter-diffusion reaction is enhanced by the deformation process itself.

b) Theoretical Discussion – Diffusion Couples

The formation and growth of a new phase in a binary diffusion couple involves both thermodynamics and kinetic principles. One can describe the thermodynamics of the problem in terms of free-energy diagrams. Such diagrams give the free energy of all relevant phases as functions of temperature, composition, and pressure. An analysis of these diagrams permits one to determine the change in bulk Gibbs free energy associated with changes of phase as functions of composition and temperature. All phase changes and processes which lead to a reduction in the free energy of the diffusion couple are thermodynamically allowed. To predict what phases actually form and how they grow, one must also analyze the kinetic aspects of the problem. We discuss the thermodynamic and kinetic aspects of the problem in turn.

Thermodynamics. The construction of free-energy diagrams for solid and liquid phases follows the methods outlined in Sect. 2.1.2 [2.17, 18]. Here, we will use the binary Ni/Zr system as an example for the purposes of illustration. Free-energy diagrams for the various phases have been computed by *Saunders* and *Miodownik* [2.76] using the CALPHAD approach. Figure 2.16 shows the free energies of mixing of the various solid solutions (hcp Zr-base, bcc Zr-base, and fcc Ni-base) as functions of composition at a temperature of 550 K. The free energy of mixing is measured with respect to the free energy of a physical mixture of Ni and Zr. This temperature is typical of the conditions under which an amorphous interlayer grows. The intermetallic compounds $NiZr_2$, $NiZr$, etc., are depicted as line compounds in the free-energy diagram. This amounts to ignoring the homogeneity range of the compounds. The free energy of the liquid phase is also shown as a function of composition at $T = 550$ K. In fact, the liquid is a glass at this temperature since the T_g of liquid Ni/Zr typically lies above 550 K. One readily sees that the free energy of an amorphous phase near the equiatomic composition is lower than that of a physical mixture of Ni and Zr with the same overall composition by an amount $\Delta G = -40$ kJ/mol. An hcp solution would, for example, have $\Delta G \sim -20$ kJ/mol at the same composition. A thermodynamic driving force exist for forming both phases from a physical mixture of Ni and Zr. Both processes are thermodynamically allowed. The formation of the intermetallic compound NiZr is accompanied by a free-energy drop $\Delta G \sim -50$ kJ/mol. This is the lowest free-energy state for an equiatomic mixture of Ni and Zr and thus represents thermodynamic equilibrium.

Now suppose that intermetallic compounds are suppressed in Fig. 2.16. Consider the metastable equilibrium between an amorphous phase and the terminal solutions of α-Zr (bcc Zr) and Ni. Under these conditions, the common tangent construction yields a metastable equilibrium state which consists of a single-phase α-solution for $x_{Zr} > x_4$, a two-phase region of liquid (glass)/α-

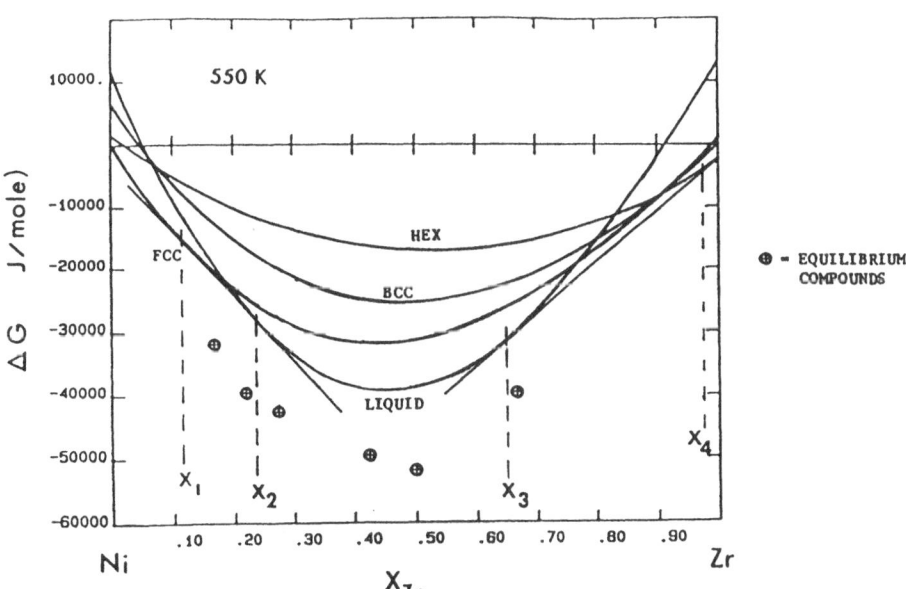

Fig. 2.16. Free-energy diagram for the Ni/Zr system at 300°C showing the free energy of mixing as a function of composition for the hcp, bcc, fcc, and liquid/amorphous phases. Intermetallic compounds are shown by dots and are assumed to be line compounds. The common tangents for the hcp/liquid, and liquid/fcc metastable equilibrium are shown. These define compositions x_1, x_2, x_3, x_4 [2.76]

solution for $x_3 < x_{Zr} < x_4$, a single-phase liquid (glass) for $x_2 < x_{Zr} < x_3$, a two-phase Ni/liquid (glass) region for $x_1 < x_{Zr} < x_2$, and a single-phase fcc Ni-solution for $x_{Zr} < x_1$. Thus, an α-Zr solid solution of composition x_4 can be in chemical equilibrium with an amorphous phase of composition x_3, etc. In fact, when reaction occurs in a diffusion couple, an intermetallic compound forms by a process of nucleation. The nucleation barrier for the compound is related to the interfacial free energy of the interface between the intermetallic compounds and the phase with which it is in contact. Under suitable conditions, the formation of intermetallic compounds is suppressed by this nucleation barrier. When this is true and when chemical interdiffusion is still possible, then the metastable simple eutectic diagram may be appropriate.

As discussed in Sect. 2.1.2, it is often useful to define still another type of nonequilibrium phase diagram. For sufficiently low temperatures, chemical diffusion is suppressed and a system can fall out of chemical equilibrium. Starting with a chemically homogeneous phase, one can consider phase transformations in the absence of chemical diffusion. Such transformations are referred to as polymorphic phase transformations and must occur without change of composition, c. Under this type of metastable equilibrium, the homogeneous phase of lowest free energy becomes the metastable equilibrium phase. If one again ignores intermetallic compounds, one finds (Fig. 2.9) that the single phase of lowest free energy is determined by the crossing of the free-energy curves of the terminal solution phases and that of the liquid (glass) phase. When these crossing points are taken as function of temperature/composition, one obtains T_0-curve for the terminal solution phases. The T_0-curve for a solid solution defines the melting point of the solution at constant composition. In Sect. 2.1.2, the Clayeron equation was extended to cover the case of isobaric melting to a liquid (or glass). The isobaric melting equation, (2.9), derived there describes polymorphic melting. In general, the T_0-curve is a function of pressure as well, $T_0 = T_0(c, P)$. The isobaric, and isothermal melting equations of Sect. 2.1.2 can be used to determine the $T_0(c, P)$ surface.

Kinetics of diffusion-limited growth. Given the free-energy diagram of Fig. 2.16 at an arbitrary temperature, one is in a position to describe the thermodynamic aspects of amorphous interlayer growth. The situation is depicted in Fig. 2.17. Here A and B refer to the pure metals (e.g. Ni and Zr). Several features of the problem are illustrated in the figure. The composition profile shows discontinuities at the interface, separating the glass (amorphous) layer from the crystalline metals. If the interfaces are in chemical equilibrium, then the compositions are of the solutions and amorphous phases along the interface are given by x_1, x_2, x_3, and x_4 of Fig. 2.16. These values depend on the temperature at which growth takes place. The growth of the interlayer can be limited by the mobility of the two interfaces or by diffusive transport over the amorphous interlayer [2.4, 77–80]. The interface mobility can be described by mobility parameters κ_1 and κ_2. These can be defined in terms of the response of the interface to a deviation of the interface from chemical equilibrium (this is the

Fig. 2.17. Schematic illustration of the various quantities used to describe the growth of an amorphous interlayer. The composition profiles, chemical potential profile, etc., are shown. The quantity x_A is the concentration of element A in the diffusion couple (in the text $x_A = 1 - x$), while X_1 and X_2 are the positions of the amorphous/crystal interfaces

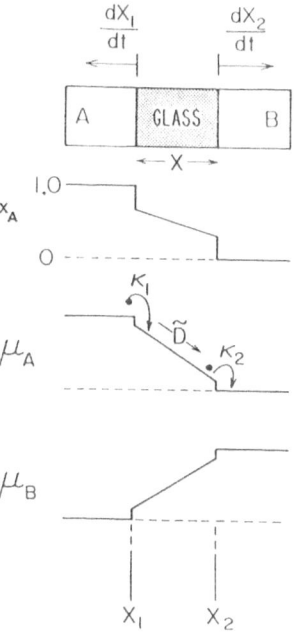

driving force for interface motion). In particular, one can define

$$dX_1/dt = \text{velocity of interface No. } 1 = \kappa_1(x - x_2) ,$$

where x is the composition of the amorphous layer at the interface with A, and x_1 is the composition which the amorphous layer has when in equilibrium with the A solution (Fig. 2.16). One can similarly define κ_2. The chemical potential drop of A and B atoms ($\Delta\mu_A$ and $\Delta\mu_B$) along the interface (Fig. 2.17) is proportional to $(x - x_2)$, and κ_1 is a response coefficient to this chemical-potential drop. The chemical-potential gradient over the amorphous interlayer can also be determined from Fig. 2.22. For interfaces near chemical equilibrium, the drop in chemical potential of an A atom over the amorphous layer

$$\mu_a^A(X_1) - \mu_a^A(X_2) = \Delta\mu_a^A$$

is given by the difference of the intercepts of the two common tangents (from x_1 to x_2 and from x_3 to x_4) with the pure A vertical axis in Fig. 2.16 (where A = Ni or Zr). Here $\mu_a^A(X_1)$ is the chemical potential of A atoms in the amorphous layer along interface No. 1 and similarily for $\mu_a^A(X_2)$. The chemical-potential variation over the growing amorphous layer provide the driving force for atomic diffusion. In the simple case where $\kappa_1/x_1 = \kappa_2/x_4$, the two interfaces can be treated symmetrically. For this case, interdiffusion over the growing interlayer is described by the diffusion equation

$$\partial x/\partial t = \tilde{D}\partial^2 x/\partial X^2 , \tag{2.13}$$

where $\tilde{D}$ is the chemical interdiffusion constant, while interface motion is governed by the continuity equation

$$\tilde{D}\partial x/\partial X = x(X_2)\partial X_2/\partial t \tag{2.14}$$

and the response equation [2.79]

$$\partial X_2/\partial t = (x(X_2) - x_2)\kappa_2 . \tag{2.15}$$

The solutions of these equations have the following asymptotic forms [2.77–80]:

$$X_2 = -\tilde{D}/\kappa_2 + (2a\tilde{D}t)^{1/2} \quad \text{as } t \to \infty \tag{2.16}$$

and

$$X_2 = A\kappa_2 t \quad \text{as } t \to 0 , \tag{2.17}$$

where a and A are numerical constants having values near unity. The interlayer grows linearly at short times with a rate determined by the interface response parameters κ_1 and κ_2. For long times, a $t^{1/2}$ growth law is observed. These regimes are referred to as interface limited and diffusion limited growth. Since κ has the dimensions of velocity, while $\tilde{D}$ has the dimensions (length^2/s), the ratio $\tilde{D}/\kappa_2$ has the dimensions of length. The latter is the characteristic thickness at which the interlayer changes form to interface limited to diffusion-limited growth. Experiments on amorphous interlayer growth have shown that this characteristic length scale is rather small (on the order of several nanometers or less). Thus, diffusion limited growth is generally found over observable ranges of thickness. An example has already been shown in Fig. 2.12 for the NiHf system. Here the thickness of the amorphous layer is plotted as a function of $t^{1/2}$ yielding a straight line over a broad range of growth times. According to (2.16), the slope of the line can be related to $\tilde{D}$, the interdiffusion constant. For the Ni/Zr system, *Kirkendall* marker experiments have shown that Ni is the dominant moving species (Fig. 2.13). As such, $\tilde{D}$ can be related to the intrinsic diffusion constant of Ni in the growing amorphous layer. Assuming $a = 1$, one can determine the diffusion constant of Ni in the amorphous layer. *Hahn* and *Averback* [2.48], *Cantor* et al. [2.81], *Dolgin* et al. [2.79], *Barbour* et al. [2.43], and *Van Rossum* et al. [2.42], have all observed a $t^{1/2}$ growth law and used this method to determine interdiffusion constants in growing interlayers of amorphous Ni/Zr and Ni/Hf. *Hahn* and *Averback* have also measured the intrinsic diffusion constant of Ni, Au, and Cu in an as-deposited amorphous layers of Ni/Zr of similar composition. *Pampus* et al. [2.82], *Schroder* et al. [2.40], and *Dörner* and *Mehrer* [2.28] have made similar studies for the Co/Zr binary system and obtained estimates of interdiffusion constants $\tilde{D}$ as well as intrinsic diffusion constant for Co and other impurities. These diffusion data are summarized in Figure 2.18. The figure is an Arrehnius plot of interdiffusion constants as well as results of tracer diffusion constants for various species. One notes that the interdiffusion constants $\tilde{D}$ obtained for amorphous interlayer growth in NiZr are close in value to the tracer diffusion constant of Ni in an amorphous layer of the same average composition as the growing interlayer (one expects these to differ since interdiffusion involves chemical effects, while tracer diffusion does not). From these diffusion data, one obtains, for instance, an activation energy for interdiffusion (e.g, also for intrinsic diffusion of Ni) of about 1.2 eV/atom for Ni in amorphous $Ni_{\sim 50}Zr_{\sim 50}$. Figure 2.18 also shows data for ADF of Ni and Co in crystalline α-Zr as well as data for the self diffusion of Zr. An upper bound (downward arrows) for the diffusion constant of Zr in amorphous $Ni_{\sim 50}Zr_{\sim 50}$ [2.48] is also shown. This figure provides a useful summary of relative values of these respective diffusion constants and illustrates the points made regarding the similarity between ADF of late transition metals in crystalline α-Zr and the existence of a dominant moving species in the amorphous phase during SSAR.

Such studies as those described above using RBS have convincingly shown that amorphous interlayers grow by diffusion limited growth. Further, such

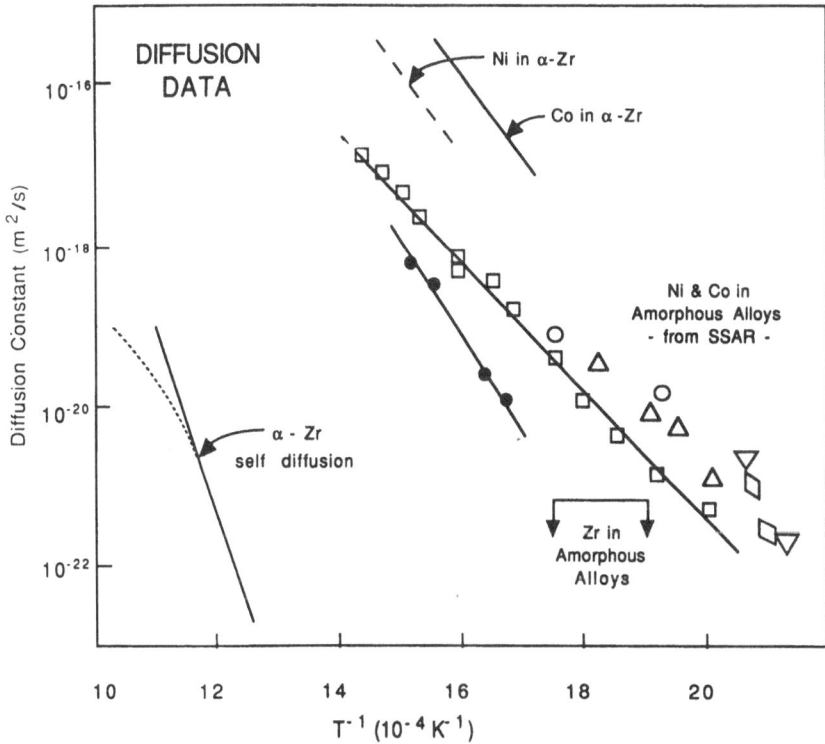

Fig. 2.18. Arrhenius plots of various relevant diffusion constants for Ni/Hf and Co/Zr diffusion couples. The upper dashed and solid line are based on tracer diffusion data for Ni and Co in crystalline α-Zr [2.38, 28]. The lower curve is for self diffusion of α-Zr [2.103]. The open squares are Co tracer diffusion data in amorphous $Co_{89}Zr_{11}$ [2.28]. Solid circles are the interdiffusion constant D obtained by SSAR on Ni/Hf diffusion couples [2.79]. Open circles, upward triangles, downward triangles, and diamonds are interdiffusion constants obtained from studies of SSAR in Co/Zr diffusion couples [2.28]. The latter data were determined from RBS studies and direct cross-sectional TEM observation of the thickness of the amorphous layer. The downward arrows indicate an upper bound for the intrinsic diffusion constant of Zr in the amorphous layer during SSAR

experiments have been useful in determining the magnitude of the chemical interdiffusion constant $\tilde{D}$, as well as the tracer diffusion constant of the fast-moving species (e.g., D_{Ni}^0, D_{Co}^0) in amorphous alloys.

Growth of amorphous interlayers has also been studied by differential scanning calorimetry (DSC) [2.68, 69, 83]. In fact, these studies represent the first use of the DSC technique to study thin-film reactions. In the DSC technique, a diffusion couple is heated at a constant heating rate. One measures the rate at which heat (enthalpy) is evolved by the diffusion couple during the interdiffusion reaction, $\dot{H}$. While an amorphous interlayer is growing, the heat released corresponds to the enthalpy of formation of the growing amorphous layer from the elemental components (e.g., Ni and Zr). In other words, the experiment measures the enthalpy of mixing of the components in the amorphous phase. In practice, a multilayer stack consisting of many diffusion couples is used in order to obtain an adequate signal from the calorimeter. *Highmore* et al. [2.83] studied Ni/Zr diffusion couples in this manner. In such an experiment, the total integrated enthalpy released by the diffusion couple is given by

$$H = \int_{t=0}^{t} \dot{H}(t') \, dt',$$
(2.18)

and is proportional to the total thickness of the grown amorphous layer. On the other hand, $\dot{H}$ is proportional to the rate of growth of the amorphous layer. According to (2.16), if one neglects the small term D/κ (this amounts to neglecting interface limited growth) and assumes diffusion limited growth over the entire range of times, then one should have

$$H \sim (2a\tilde{D})^{1/2} t^{1/2} \sim X \ ,$$
(2.19)

while

$$\dot{H} \sim \tfrac{1}{2}(2a\tilde{D})^{1/2} t^{-1/2} \sim X^{-1} \ ,$$
(2.20)

so that the product $H\dot{H}$ has the form

$$H\dot{H} \sim a\tilde{D} = a\tilde{D}_0 e^{-Q/k_B T} \ .$$
(2.21)

When growth is diffusion limited, if one plots $\ln(H\dot{H})$ vs. T^{-1} from DSC data, one should obtain an Arrhenius plot giving the activation energy for inter-diffusion. This has indeed been found experimentally to yield activation energies in agreement with those obtained from RBS studies. By independently measuring the total thickness of the amorphous layer grown in a DSC experiment, one can actually determine $\tilde{D}$. Figure 2.19, taken from [2.83], shows an example of DSC data used to obtain such information for Ni/Zr multilayered diffusion couples. A linear fit to the data yields an activation energy of 1.05 eV/atom, in good agreement with that obtained by RBS studies of the interdiffusion reaction. This shows that reaction calorimetry can be used to monitor amorphous interlayer growth directly. The straight line obtained in Fig. 2.19 (lower figure) further verifies that growth is diffusion limited.

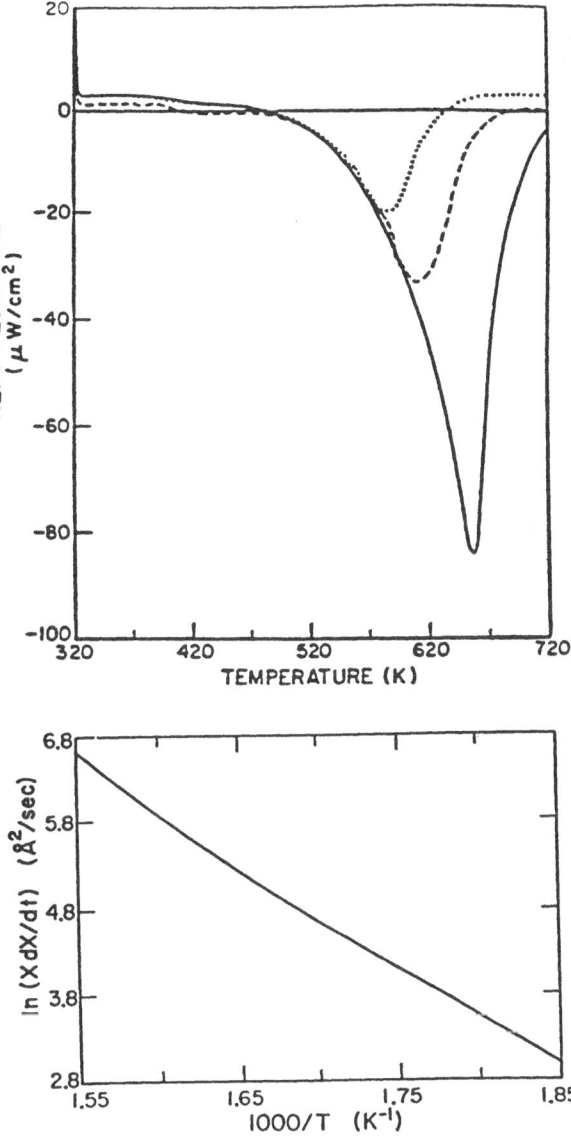

Fig. 2.19. (upper figure) Differential scanning calorimetry scans of various Nr/Zr multilayer diffusion couples heated at a constant rate of 20 C/s. The heat flow rate, $\mathring{H}$, has been normalized by the total Ni/Zr interfacial area in the diffusion couple. The dotted line corresponds to an individual Ni layer thickness of 30 nm and an individual Zr layer thickness of 45 nm, the dashed line to 50 nm/8 nm, and the solid line to 100 nm/100 nm, respectively. (lower figure) a plot of ln $(X\mathring{X})$ (note that H is proportional to X – the proportionality constant is determined by direct measurement of H_{tot} and X_{tot}, the final thickness of the amorphous layers) vs. $(1/T)$ for the third sample in the upper figure. See text for further explanation. The slope of the curve gives the activation energy for interdiffusion of Ni and Zr in the amorphous layer [2.69]

In addition to allowing measurements of $\tilde{D}$, calorimetry studies also allow direct determination of the total enthalpy evolved when Ni and Zr layers are completely converted to amorphous phase. This is the enthalpy of mixing of the amorphous phase (referred to the starting pure metals). *Cotts* et al. [2.68, 69] showed that, for example, the enthalpy of mixing of an amorphous alloy of composition $Ni_{68}Zr_{32}$ is 35 ± 5 kJ/mol. Since the entropy of mixing of this alloy at a temperature of 300°C (the typical growth temperature) is much smaller, one can approximate

$$\Delta G_{mix} = \Delta H_{mix} + T\Delta S_{mix} \approx \Delta H_{mix} .$$ (2.22)

In Fig. 2.20, the results of DSC measurements of the total enthalpy of formation of the amorphous phase for Ni/Zr multilayer diffusion couples of different over total compositions are compared with the values of ΔG_{mix} obtained by the Calphad calculation of *Saunders* and *Miodownik* (see Fig. 2.18). In the experiments, the overall composition is varied by altering the ratio of thicknesses of the individual Ni and Zr layers. To obtain an accurate enthalpy of formation of the amorphous phase, one must show that the SSAR goes to completion. This is true provided that the overall composition lie in the range form x_2 to x_3 (Fig. 2.16). The total enthalpy of mixing data obtained from DSC in Fig. 2.19 are in very good agreement with the calculated enthalpies of *Saunders* and *Miodownik*. It is thus directly established that the larger negative heat of chemical

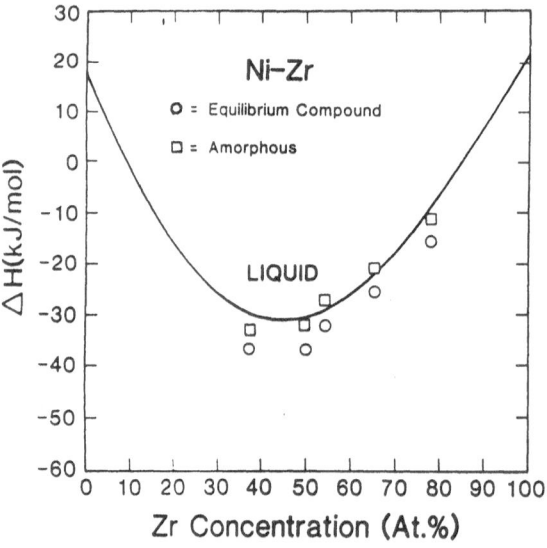

Fig. 2.20. The total enthalpy of formation of the amorphous phase (squares) and equilibrium crystalline phase (circles) as obtained by DSC studies of multilayered Ni/Zr diffusion couples during SSAR. The experimental data are compared with the CALPHAD calculation (solid line) of *Saunders* and *Miodownik* [2.76]. The data points give the experimentally observed enthalpy of formation for diffusion couples of various overall compositions [2.65]

mixing provides the primary thermodynamic driving force for amorphous interlayer growth.

Other types of measurements have been used to monitor amorphous interlayer growth. For instance, one can monitor the in plane resistance of a thin film diffusion couple. It can been shown that the change in resistance due to the growth of an amorphous interlayer is proportional to the interlayer thickness. *Samwer* et al. [2.84], *Dolgin* [2.79], and others have used this method to monitor amorphous interlayer growth, verifying again that growth is diffusion limited. An example is shown in Fig. 2.21 [2.84].

In conclusion, it is now well established that the growth of planar amorphous interlayers is essentially a diffusion-limited process. Here we have focused on the Ni/Zr system and Co/Zr. Such experiments have in fact been performed for a variety of systems and show similar results. Monitoring of diffusion-limited interlayer growth provides a direct measurement of the interdiffusion constant $\tilde{D}$. One can, for instance, obtain the activation energy for interdiffusion from such experiments. One can also directly monitor the enthalpy changes during SSAR. On the other hand, the description of the growth law for an amorphous layer leaves unanswered a more fundamental question. Why does an amorphous layer grow instead of a more stable intermetallic compound? To answer this

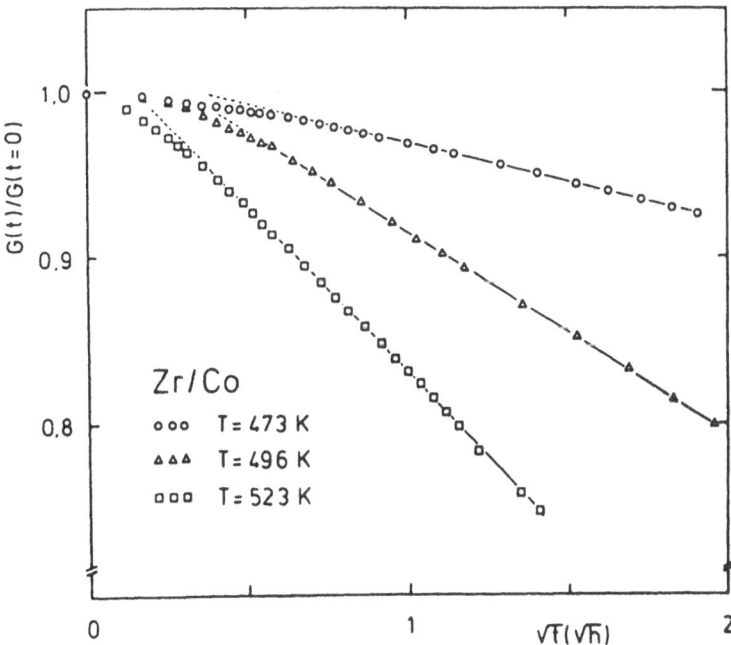

Fig. 2.21. Normalized conductance of Zr/Co bilayer diffusion couples vs. square roots of time for three different temperatures. Note the deviations at short reaction times from a parabolic time law [2.84]

question requires that we examine the nucleation behavior of competing phases in the diffusion couple.

Nucleation and the phase sequence problem. We showed above that many possible reactions can occur in a diffusion couple. Any new phase for which $\Delta G < 0$ is thermodynamically allowed. In fact, the thermodynamic driving force for intermetallic compounds (e.g., crystalline NiZr) is greater than that for the formation of an amorphous phase

$$\Delta G_{c-NiZr} < \Delta G_{a-NiZr}, \tag{2.23}$$

where ΔG is the negative free energy change associated with forming crystalline NiZr (c-NiZr) and amorphous NiZr (a-NiZr). Condition (2.22) is equivalent to saying that the amorphous phase is metastable. Why should a metastable phase form during the evolution of a diffusion couple? To answer this question, one must first examine the nucleation behavior of the competing phases. Other factors such as interfacial mobilities may also play an important role in the phase sequence problem. We shall examine these effects in turn.

A new phase in a diffusion couple can nucleate either homogeneously or heterogeneously. The existence of an initial hetero-interface separating the reactants offers a natural site for heterogeneous nucleation. This interface is typically incoherent since the crystal structures of the reactants are generally not well lattice matched. Typical situations for both homogeneous and heterogeneous nucleation is depicted schematically in Fig. 2.22. Heterogeneous nucleation along an interface is illustrated as well as a situation where a grain boundary exists in one of the reactant crystals along the interface. The classical theory of nucleation for such situations is well studied [2.85]. Here only the salient features of heterogeneous nucleation are emphasized as they relate to the phase-selection problem. The critical contact angles for heterogeneous nucleation along an interface shown in the center diagram of Fig. 2.22 are φ_1 and φ_2. These are determined by the interfacial free energies σ_{AB} = interfacial energy between initial phases A and B, σ_{AC}, and σ_{BC}, the latter being the corresponding interfacial energies between the new phase C and the existing phases A and B. For simplicity and illustration, we can assume $\sigma_{AC} = \sigma_{BC}$ and $\sigma_{AC} < \sigma_{AB}$. Then $\varphi_1 = \varphi_2 = \varphi$, and

$$\cos \varphi = \sigma_{AB}/2\sigma_{AC} . \tag{2.24}$$

If $\sigma_{AC} = \sigma_{AB}/2$, then $\varphi = 0$ and the new phase C wets the interface. If $\sigma_{AC} < \sigma_{AB}/2$, then formation of C will be a driven process. Under either circumstance, the nucleation barrier for the new phase C will vanish. On the other hand, if $\sigma_{AB} \ll \sigma_{AC}$, then $\varphi \to \pi/2$, and we recover the case of homogeneous nucleation. This would be the case if the original interface were nearly coherent, i.e., A/B is an epitaxial structure. Heterogeneous nucleation would be effectively suppressed in this case. In general, the nucleation barrier for C, ΔG_{Nuc}^C depends on φ; it achieves a maximum for $\varphi = \pi/2$ and vanishes when $\varphi \to 0$. These general features are applicable irrespective of whether C is a metastable

Homogeneous Nucleation

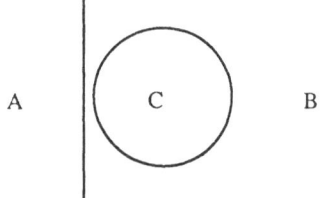

Heterogeneous Nucleation-Interface

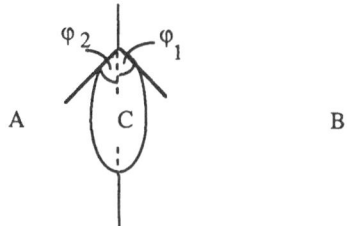

Heterogeneous Nucleation -Grain Boundary

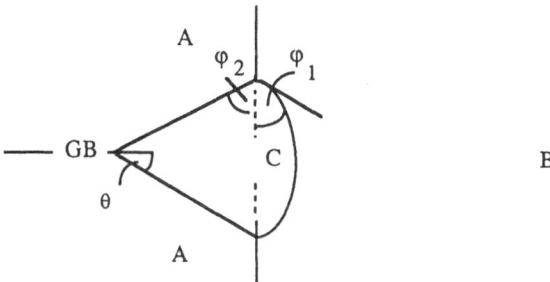

Fig. 2.22. Schematic illustration of the geometries for homogeneous and heterogeneous nucleation of a new phase, C, in a diffusion couple of A and B. See text for further explanation

amorphous phase or a crystalline intermetallic phase. The lower diagram of Fig. 2.22 illustrates a more complex case. The existence of a grain boundary in A, perpendicular to the interface, permits a more complex shaped heteronucleus. With the assumptions above for σ_{AB}, σ_{AC}, and σ_{BC}, and the introduction of a grain boundary energy σ_{GB}, the problem involves an additional contact angle θ. Generally speaking, the nucleation barrier becomes a function of two of these angles. The nucleation barrier for C will vanish when either

$$\sigma_{AC} \leq \sigma_{AB}/2 \tag{2.25a}$$

or

$$\sigma_{AC} \leq \sigma_{GB}/2 \ . \tag{2.25b}$$

In general, for given values of the interfacial energies, the existence of the grain boundary leads to a lower nucleation barrier than for a simple interface. The greater the grain boundary free energy σ_{GB}, the more favored heterogeneous nucleation will be. High-energy grain boundaries are more potent catalytic sites for heterogeneous nucleation that low-energy grain boundaries. Finally, intersections of three grain boundaries (each perpendicular to the original interface), provide still more favorable heterogeneous nucleation sites. In Sect. 2.2.1a it was noted that grain boundary amorphization occurs spontaneously in certain binary diffusion couples containing high-energy grain boundaries in the reactant layers (e.g., recall the case of Cu/Y and Si/Ti referred to in Sects. 2.2.2a and 2.2.2b). Apparently, this occurs because the free energy of a high-angle grain boundary energy in Y (for example) is more than double the interfacial energy of an amorphous Cu/Y phase in contact with an yttrium crystal. There is no nucleation barrier for amorphous-phase formation in this case.

Based on the above discussion, one can understand how the formation of an amorphous interlayer is influenced by the nature of the interface separating the two reactants. The rate at which a new phase nucleates is given in general by

$$\dot{N} = \text{nucleation rate (nuclei/s cm}^3) = KDe^{-\Delta G_N/k_B T} \ , \tag{2.26}$$

where ΔG_N is the nucleation barrier and D is a characteristic atomic diffusion constant which determines the rate at which atoms rearrange diffusively to form the critical nucleus. The constant, K, is determined by the geometry of the problem and other details which we shall neglect here. In the case of an amorphous phase, it is reasonable to assume that D is related to the interdiffusion constant, $\tilde{D}$, of Ni and Zr since formation of a nucleus of the amorphous phase requires a local composition fluctuation but may assume a variety of topological forms. In the case of nucleation of a crystalline intermetallic, this identification may be less clear. To form a nucleus of a crystalline intermetallic compound requires atomic rearrangement of both species (e.g., Ni and Zr) both topologically and chemically. This will likely require atomic mobility of both species, whereas formation of an amorphous nucleus is topologically far less restrictive and may require the movement of only one species (e.g., Ni in the case of Ni/Zr). Both D and ΔG_N are required to determine the nucleation rate. It can be argued that the free energy of the heterointerface between two crystals depends greatly on lattice-matching. If mutual orientations of two crystalline phases exist along which coherent, or near coherent interfaces are formed, then one expects a low interfacial energy. Incoherent interfaces are expected to have much higher energy. For nucleation kinetics to favor formation of an amorphous phase, there must be an absence of low-energy interfaces between the intermetallic compounds and the parent crystalline layers. One might expect energy of the interface between an amorphous phase and parent metal phase to be smaller than that of an incoherent crystalline interface, but larger than that of a coherent interface. This, in turn, would lead to an intermediate value of ΔG_N.

Westendorp et al. [2.53] observed no amorphous interlayer growth in Ni/Zr diffusion couples consisting of a single crystal of Zr in contact with a poly-crystalline Ni layer. *Meng* et al. [2.40] similarly observed that polycrystalline Ni reacted with a mosaic of large ($\sim 50~\mu m$) Zr crystals separated by low-angle tilt boundaries results in no amorphous interlayer growth. By contrast, amorphous interlayer growth is observed when polycrystalline Zr is reacted with either polycrystalline Ni or single crystal Ni. These experiments suggest that Zr grain boundaries characterized by significant misorientation are required to nucleate the amorphous phase. *Kamenetsky* et al. [2.86] and *Koster* et al. [2.87] have reported evidence that triple junctions of Zr grain boundaries along the interface with Ni are the likely nucleation sites for the amorphous phase. They further report that once nucleated at these grain boundary intersections, the amorphous phase spreads rapidly along the Ni/Zr interface. Based on these studies, one concludes that heterogeneous nucleation of the type illustrated in Fig. 2.22 is responsible for the initial formation of the amorphous phase. The fact that high-energy grain boundaries of Zr are required (the low-angle tilt boundaries in the experiments of *Meng* et al. were not effective nucleation sites) suggests that a large value of σ_{GB} (Fig. 2.22b) is required to lower the nucleation barrier sufficiently for the amorphous phase to allow nucleation in the observed temperature ranges ($\sim 300°C$). In the case of Cu/Y [50], the observation of an amorphous phase along the grain boundaries of Y in Cu/Y multilayers sput-tered at ambient temperature but otherwise unreacted suggests either a very low or near zero nucleation barrier along grain boundaries. In summary, one sees that initial nucleation of an amorphous phase (as opposed to the crystalline intermetallic) can either be a consequence of a lower ΔG_N, of a different rate limiting D, or of some combination of both.

Meng et al. [2.40] and *Highmore* et al. [2.83] considered the problem of nucleation of a second intermediate phase along a moving interface separating the first nucleated phase from the parent reactants. In particular, they have examined the formation of the intermetallic compound NiZr along the interface separating a growing amorphous layer from cyrstalline Zr. *Meng* et al. proposed that heterogeneous nucleation of crystalline NiZr along this interface must involve a timescale related to the characteristic dimension of a critical nucleus R_c and to the velocity of the interface $v = dX/dt$ by

$$\tau = R_c/v ~, \tag{2.27}$$

where τ is the time required for the moving interface to advance over the length scale of a critical nucleus. This leads, at a given temperature, to a critical velocity for the moving interface. Below this velocity, nucleation of the intermetallic compound will occur. Diffusion-limited growth at a fixed temperature leads to an interface velocity which falls like $(1/L)$, where L is the total thickness of the amorphous interlayer. Since L increases with time like $t^{1/2}$, one expects the velocity of the Zr/amorphous interface (in the case of Ni/Zr diffusion couples) to decrease like $t^{-1/2}$. At a critical thickness L_c, nucleation of the compound becomes possible. *Highmore* et al. have developed a similar model called the *transient nucleation model* which makes essentially the same prediction. Both models predict a critical thickness for the amorphous-layer growth. Both predictions are based on a minimum growth velocity below which nucleation and growth of the intermetallic compound replaces further growth of the amorphous layer.

Desre et al. [2.88] have proposed a mechanism for the suppression of nucleation of intermetallics in the case that an amorphous layer has already formed. In this model, nucleation of the intermetallic is impeded by the composition gradient in the growing amorphous interlayer. According to Figs. 2.18, 20, this composition gradient is given by

$$G = \text{composition gradient} = (x_3 - x_2)/(X_2 - X_1) = (x_3 - x_2)/L , \qquad (2.28)$$

where L is the total thickness of the amorphous interlayer which has grown. This gradient decreases like $L^{-1} \sim t^{-1/2}$ in the case of diffusion-limited growth. *Desre* et al. showed that this gradient increases the nucleation barrier for intermetallic compounds within the amorphous layer. Their argument applies to both homogeneous nucleation of the intermetallic phase within the amorphous layer as well as heterogeneous necleation of the intermetallic along the interfaces separating the amorphous layer from the reactant phases (interfaces at X_1 and X_2) in Fig. 2.17. They showed that the composition gradient leads to an enhancement of the nucleation barrier of the intermetallic, which increases with increasing G. In turn, the nucleation barrier for intermetallic compounds is thus expected to decrease as the amorphous interlayer thickens. In the case of Ni/Zr diffusion couples, the amorphous interlayer is observed to reach thicknesses of $L \sim 80$–120 nm prior to the formation of the intermetallic NiZr. This critical thickness, defined as the thickness of the amorphous interlayer where nucleation of NiZr occurs, was measured directly from cross-sectional TEM micrographs and found by *Meng* to depend slightly on the growth temperature. *Meng* argued that this dependence was consistent with the transient nucleation model. *Highmore* et al. have argued similarly. *Desre* et al. have claimed that the *composition gradient* model is also consistent with the experimental data.

Another model for the appearance of a second intermediate crystalline phase by *Gosele* and *Tu* [2.80] is based on *kinetic barriers*. In this model, the lack of interfacial mobility (as measured by the parameter κ in Fig. 2.17) could also be a rate-limiting step in the growth of an intermediate phase. In this picture, the initial formation of the amorphous phase is assumed. It is then argued that the interfacial mobility of the amorphous/Zr interface is greater than that of the crystalline NiZr/Zr interface. Crystalline NiZr appears only when the amorphous interlayer growth velocity, $\dot{L}$, slows sufficiently that it falls below the interface-limited velocity of NiZr. In the *Gosele* and *Tu* model, it is assumed that nucleation of intermediate phases is not the limiting step in their formation, but rather their interfacial mobilities. Experimental results discussed earlier have shown that formation of an amorphous interlayer certainly involves a nucleation barrier. Furthermore, studies of the time dependence of amorphous interlayer growth demonstrate clearly the validity of the growth law given by (2.17) and $\tilde{D}/\kappa \sim 1$–3 nm. As such, we must conclude that interface limited growth is confined to very small interlayer thicknesses. If this is true, then the *Gosele* and *Tu* model does not seem applicable. Nucleation barriers, and not *kinetic barriers*, seem more likely to control the phase sequence in Ni/Zr diffusion couples.

2.2.3 Irradiation-Driven Systems

Irradiation can produce metastable crystalline and amorphous phases as is well known from nuclear power plants. In ion implantation a beam of ions within the energy range of 80–200 keV is injected into a thin film or in the surface area of a bulk sample. Ion beam mixing is seen as supplementary to ion implantation. Often bi- or multilayers are irradiated with inert gas ions of high energy ($E > 100$ keV) at low temperatures (77 K) or ambient temperature [2.105]. The beam ions are slowed down by a series of elastic and inelastic collisions. Inelastic collisions transfer the incoming energy to the electrons and nuclei of the target. In the elastic collisions, the energy of the ions is transferred to the atoms, which are kicked out of their lattice position and can carry out further scattering processes. [2.106]. The atomic-mixing process is induced by collision cascades in which atoms are displaced by interatomic collisions. In earlier models of this process, analyses of the mixing process was based on ballistic models. These models assume that the majority of atomic displacements occur as a result of collisions of relatively high-energy particles [2.107]. When an atom is displaced, it leaves a vacant site. The displaced atom either comes to rest at an interstitial site or ultimately recombines with a vacancy. A vacancy/interstitial pair is referred to as a Frenkel defect or Frenkel pair. More recent models are based on the evolution of the collision cascade into a thermal spike [2.108, 109] as originally proposed by *Vineyard* [2.110]. The latter models predict that the majority of atomic displacement occur during the late stages of cascade development when small (of order 3–10 nm) regions become thermalized at temperatures of the order or 10^4 K. These hot regions are subsequently quenched to ambient temperature by heat conduction to the surrounding medium at cooling rates ranging from 10^{10}–10^{12} K/s. In this picture, atoms in local regions along the interface are suddenly heated to very high temperature; atomic configurations evolve at this temperature for short times ($\sim 10^{-11}$ s) and are then quenched to ambient temperature at very high rates. The entire process can be described by an effective diffusion constant which reflects the number of atomic displacements induced per incident ion. The process can be viewed as *athermal diffusion* since the ambient temperature has little to do with the amount of actual diffusive mixing. For the present purposes, we can view the process as an externally driven mixing of atoms along the interface between two materials.

In general, one distinguishes two regimes of ambient temperature in ion mixing. When defects produced by the collision cascade (the Frenkel pairs) are mobile at the ambient temperature of the experiment, one observes thermally activation migration of the defects. This is referred to as radiation-enhanced diffusion. When the ambient temperature is sufficiently low, the Frenkel defects produced in the cascade are configurationally frozen and the chemical configuration is fixed following the cool-down phase of the thermal spike. Under these conditions, diffusional mixing does not occur at the ambient temperature and the process is referred to as *temperature independent mixing*.

Ion mixing of metal layers along an interface frequently leads to the formation of an amorphous interlayer [2.111–114, 97]. As a general rule, metals which do not form extended terminal solid solutions will tend to form an amorphous interlayer when mixing is carried out at low temperature [2.105]. This can be explained in terms of a polymorphic melting diagram in which the melting point of a solid solution with fixed composition is determined as a function of the composition. Ion mixing produces metastable solid solutions with compositions lying beyond the equilibrium solubility limits. For sufficiently extended solutions, the *polymorphic* melting point falls below the ambient temperature and below the T_g of the liquid phase. Under these circumstances, the solution can *melt* to a liquid below its T_g. In other words, the solution under goes a crystal-to-glass transformation.

Recent experiments on the irradiation of intermetallic compounds show that a large elastic softening and delatation strain due to disordering precede the crystal-to-glass transformation [2.115]. The authors measured the degree of long-range chemical order S, the lattice dilatation $\Delta a/a$, and the shear constant using Brillouin-scattering technique. The reduction in the shear constant and the decrease of the long-range order parameter can be described by an elastic instability as described earlier [2.4]. The actual nature of the amorphization transition is not fully understood in these experiments. There is an interesting parallel to earlier experiments by *Linker* [2.116], who implanted B-ions in Nb-thin films and studied the amorphization due to a collection of internal stress until a spontaneous collapse destroys the long-range order. The accumulation of stress can be determined as a function of the concentration of B atoms. More experiments of such kind and computer modelling are important to draw conclusions and similarities to the mechanical-driven systems as described in the next section.

2.2.4 Grinding of Intermetallic Compounds

During the last decade a large number of intermetallic compounds were amorphized by means of mechanical processes [2.117]. Although the process takes place mainly in the same kind of ball mills used for mechanical alloying of elemental powders [2.118], the microscopic details are obviously very different. The first examples of mechanical grinding of intermetallic compounds were presented by *Yermakov* et al. [2.74] in the Y–Co and Gd–Co systems. *Weeber* and *Bakker* [2.119] and more recently *Koch* [2.120] summarized the experimental work and showed clearly that the amorphous phase prepared by mechanical milling has similar properties as the melt quenched metallic glasses of the same nominal composition.

In the following discussion, we try again to separate the thermodynamic from kinetic aspects of the crystal-to-glass transition of compounds by mechanical processes. In terms of the free-energy diagrams as used in Sect. 2.2.2., the mechanical process must provide the difference in the free energy between the intermetallic compound and the amorphous alloy. This increase of the free energy due to the mechanical milling, as pointed out by *R.B. Schwarz* et al. [2.121], can be measured partly by the heat release of the milled material during a DSC-run. Figure [2.23] shows a typical measurement on a ball milled Zr–Co-compound and mechanical alloyed ZrCo-powder of the same composition for comparison. As clearly seen, the mechanically alloyed material has reached the maximum value of the latent heat, which was stored during the process and released during the DSC-scan, in much shorter milling times than the mechanically ground powder [2.122]. This can be easily understood in terms of the driving forces for the amorphization process, which is given for the mechanical alloyed materials mainly by chemical effects (alloying as discussed in Sect. 2.2.2.), while for the ground material the driving force must result from other reasons. As already discussed in the past for ion-irradiated compounds, the defects created by the mechanical process in the ball mill play a major rule for the amorphization process. Deformation is well known to create defects such as vacancies, dislocations, atomic site defects, antiphase boundaries or grain boundaries [2.123]. The maximum energy stored by cold working of metals or alloys is typically less than 2 kJ/mol. *Koike* et al. [2.124] give, for example, such a value for high dislocation densities ($- 10^{14}$ cm^{-2}) in cold worked NiTi-alloys. For the amorphization process a typical value about 40% of the heat of fusion (~ 5–7 kJ/mol) is necessary to reach the free energy of the amorphous phase [2.122]. The creation of chemical effects such as atomic-site defects, antiphase boundaries, and the influence of impurities from the container walls/balls are likely to cause an increase in the free energy. We separate artificially these chemical effects (which are well known from irradiation effects) from others, which we will call *size effects* due to grain size reduction during the milling process. As shown by *Hellstern* et al. [2.125] the increase in the free energy alone is not always sufficient for amorphization, such as for several CSCl-type structures. Although kinetic effects like recovery and recrystallization may play the dominant part, we just state here that the grain-size reduction was not sufficient for the amorphization process. In a very simplified picture for the well-known ZrCo system, we calculate the enthalpy difference for that system due to grain size reduction assuming different values of the interface energy for a grain boundary σ^*. With such a calculation, one obtains a $1/R$-dependence for the enthalpy difference (energy storage) due to the increased surface-to-volume ratio

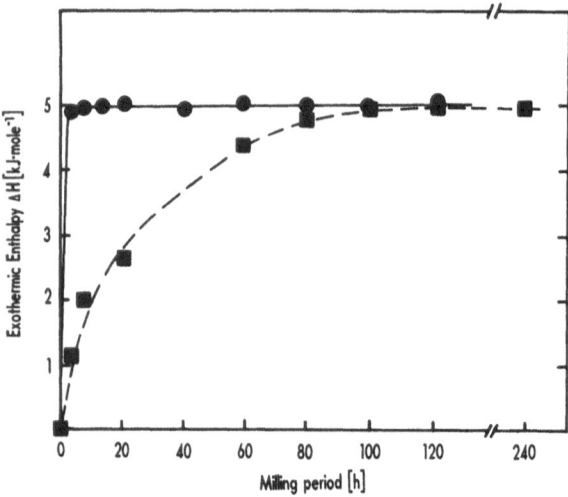

Fig. 2.23. Exothermic enthalpy ΔH as a function of milling period for the intermetallic phase CoZr (quadrats) and the power mixture of 50 at. % Co and 50 at. % Zr (circles)

normalized to the average atomic volume of the specific compound

$$\Delta H \sim \frac{1}{R} \cdot \sigma^* \ . \tag{2.29}$$

A more complete calculation was recently given by *Spaepen* et al. [2.126], who include a geometrical factor. Figure 2.24 shows the calculated enthalpy for the Zr–Co system versus the grain-size radius for two different effective interfacial energies σ^*. For our simplified discussion, we note that a material with $\sigma^* = 0.1$ J/m^2 would not reach the *break-even* line of the enthalpy of the amorphous phase with a grain-size radius as low as 1 nm. For $\sigma^* = 1.0$ J/m^2, this value is reached for $R \approx 6$ mm. In other words, a process of any kind, which provides a grain size of R smaller than 6 nm for this particular material, crosses the limit of the amorphous phase and results thermodynamically in a driving force for the amorphization process. There might be kinetic reasons why the amorphization does not take place; alternatively, a lower limit for the grain-size radius might arise from other effects. Experimentally, a discontinuity between the lowest grain-size radius of the crystalline phase and the "effective" grain size of the amorphous phase (typically $R < 1.2$ nm) was observed [2.127]. Obviously, the splitting of the mechanical grinding process into a part due to chemical effects and a part due to size effects is artificial and both occur in the real process. In a very recent publication by *Bakker* et al. [2.128], the elastic mismatch energy during the order/disorder and the amorphization process is calculated with respect to *Miedema's* values for an extended solid solution. This seems to be an interesting starting point for further experiments.

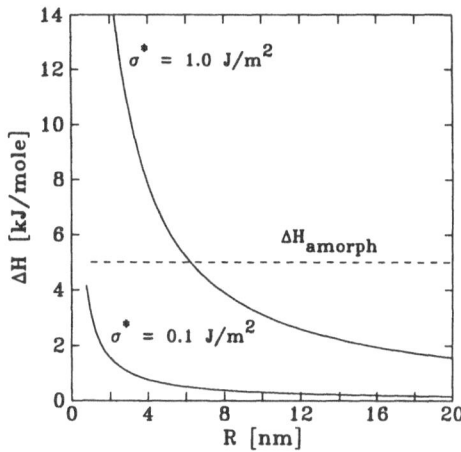

Fig. 2.24. Enthalpy ΔH as a function of grain-size radius R for two different effective interfacial energies δ^* calculated for the ZrCo system. The enthalpy difference between the amorphous and the crystalline phase according to Fig. 2.28 is also given. A driving force towards amorphization is expected for values above ΔH_{amorph}

The kinetic aspect of the mechanical grinding process is still highly controversial. Obviously, the temperature is an important parameter for all diffusion-controlled processes. The overall temperature during ball milling increases up to 100°C, depending on the details of the ball mill and the hardness of the powder used. The local temperature of the powder within shear bands during deformation of a powder grain may rise to much higher values. *Martin* et al. [2.90] recently demonstrated the first attempt to decouple these effects.

2.2.5 Pressure Application

Based on the thermodynamic description given above (Sect. 2.1), the destabilization of a crystalline phase can be induced by excursions in thermodynamic state space, for example by variation of pressure. This results in the generalized Clapeyron equations given by (2.9–11), which describe the dependence of the melting point on pressure. The change of the Gibbs free energy resulting from the pressure application may be found by integrating $d(\Delta G_m) = \Delta V_m \, dP$. Because ΔV_m is not exactly known in an extended pressure and temperature range (especially not for metastable phase), we limit ourselves to discussing the influence of pressure only qualitatively.

It appears from (2.11) that, when ΔV_m is negative, amorphization becomes possible at high pressures as shown experimentally for ice [2.129] and silicon [2.130]. For positive ΔV_m, which describes generally the behavior of metallic systems, the principal features of a polymorphous Clapeyron diagram is presented in Fig. 2.25 in T–P–C space [2.20]. In the T–C plane, an increase in

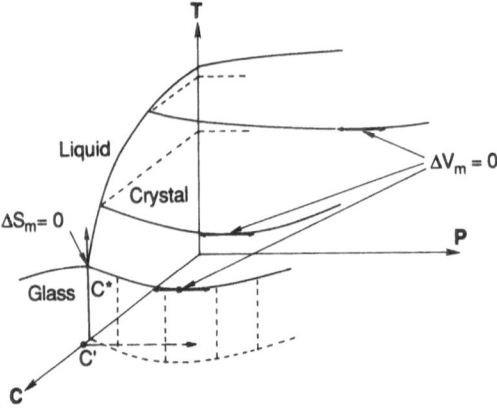

Fig. 2.25. Polymorphous Clapeyron diagram in $T–P–c$ space for a binary solid solution displaying three isothermal sections to the surface $T(P, c)$

pressure would move the T_0 curves to a higher concentration, whereas application of negative pressure would reduce the stability limit of the crystalline phase. Taking into account that ΔV_m is temperature and pressure dependent, it appears that ΔV_m can vanish at a certain temperature and pressure because of the higher compressibility of the liquid or amorphous state in comparison with the crystal. Consequently, a critical point for melting discontinuities where all thermodynamic variables $(\Delta V_m, \Delta S_m)$ vanish becomes possible at high pressures. However, as long as ΔV_m is positive, direct amorphization by pressure application does not seem possible. On the other hand, when ΔV_m becomes negative at high-pressure, solid-state amorphization is possible at sufficiently low temperatures. Such a behavior has been observed for oxides (quartz [2.131]) and polymers [2.132]. To the authors' knowledge, such detailed studies have so far not been carried out for metals and alloys.

However, Fig. 2.25 is suggestive of another possibility for solid-state amorphization. With positive pressure, the solubility range of the crystal increases. Thus, releasing the pressure suddenly at sufficiently low temperatures would result in amorphization of the sample below T_g, if the atoms cannot diffuse on the same timescale as the pressure is reduced. This is indicated in Fig. 2.25 by the projected isothermal section at concentration c'. In fact, a similar type of amorphization reaction by pressure application has been observed for semi-metallic systems when a phase transformation to a high pressure phase is involved. In many binary alloys, high-pressure phases exist which may remain in a metastable state at atmospheric pressure and sufficiently low temperature. As shown by *Ponyatovsky* et al. [2.133], several metallic alloys including Cd–Sb, Zn–Sb, CdSb–ZnSb, Al–Ge and Ga–Sb alloys exhibit solid-state amorphization. The method rests simply on the fact that a high pressure crystalline phase exists, which can be retained in a metastable state at sufficiently low temperature (liquid-nitrogen temperature) for essentially indefinitely extended times. This phase is obtained by thermobaric quenching from the high-temperature regime followed by a release of pressure to 1 atm. By subsequent heating of the metastable crystalline phase from liquid-nitrogen temperature, the crystalline phase is found to become unstable and to undergo a transition to a glass close to room temperature. Further heating results in crystallization of the amorphous phase to the stable phase mixture. Even though this is a good model system to study solid-state amorphization on bulk samples, the process is not well understood. For example, the transition to the amorphous state is connected with a metal-insulator transition together with exothermal heat effects. In addition, the amorphous phases in the Cd–Sb and Zn–Sb alloy systems exhibit a higher density (smaller volume) than the stable crystalline phase mixtures [2.133, 134]. This behavior is different from that of glasses prepared from the liquid state but might be related to the metal-insulator transition.

2.3 Discussion and Summary

2.3.1 Relation Between Solid-State Amorphization and Melting

Solid-state amorphization can be induced through a variety of methods including irradiation, chemical reactions, mechanical-deformation techniques, pressure application, etc., as discussed above. As such, a crystalline phase can be driven into a disordered amorphous state as long as the kinetic constraints inhibiting stable phase formation can be maintained.

Under equilibrium conditions, the transition from a crystalline phase having long and short-range atomic order to a disordered phase exhibiting only short-range atomic order corresponds to the melting transition. As such, solid-state amorphization, which is characterized by the loss of long-range order, can be understood as a melting process under non-equilibrium conditions. Therefore, the radial distribution functions of amorphous alloys produced from either the liquid state or the solid state are identical [2.8, 135]. The difference between glass and liquid becomes pertinent in the dynamics of the system, e.g., in the transport properties, etc. With this in mind the glass can be considered thermodynamically a configurationally frozen highly undercooled liquid and, therefore, the transition from crystal-to-glass has been compared with the melting process.

Despite the strong interest in melting over the past hundred years, a generally accepted theory for the crystal-liquid transition has not been developed [2.136]. Thermodynamically, the melting point T_m is defined as the temperature where the Gibbs free energies or corresponding chemical potentials of liquid and crystal are equal. Some melting theories are based on some sort of instability or catastrophe which is either vibrational [2.137], elastic [2.138], isochoric [2.15, 139], defective [2.140] or entropic [13] in nature. Furthermore, these melting theories rely on a homogeneous melting process throughout the entire crystal, even though it is a well-known experimental fact that melting starts at extended lattice defects, such as surfaces and internal interfaces. Thus, it is observed experimentally that ordinary melting occurs at T_m, (i.e., before an instability develops) and as such, represents a first-order phase transition as indicated by a well-defined crystal/liquid interface and an entropy of fusion on the order of $1k_B$. If the effective heterogeneous nucleation sites for the formation of a liquid nucleus can be eliminated, considerable crystalline superheating can be achieved as shown for a variety of materials including oxides [2.141], pure metals [2.142] and solid helium [2.143].

It is generally agreed that thermally induced vibrations of atoms in solids play a major role in melting [2.144]. The simple vibrational model of Lindemann predicts a lattice instability when the root-mean-square amplitude of the thermal vibrations reaches a certain fraction f of the next neighbor distances. However, the Lindemann constant f varies considerably for different substances because lattice anharmonicity and soft modes are not considered, thus limiting the predictive power of such a law. Furthermore, *Born* proposed the collapse of the crystal lattice to occur when one of the effective elastic shear moduli vanishes [2.138]. Experimentally, it is found instead that the shear modulus as a function of dilatation is not reduced to zero at T_m and would vanish at temperatures far above T_m for a wide range of different substances [2.145]

In addition, the glass transition, e.g., the freezing of an undercooled liquid to a glass is not well understood. For example, it is still a matter of controversy if the glass transition itself is a phase transition at all or just kinetic freezing. Experimental evidence and theoretical models suggest the glass transition to be either first order (free-volume approach [2.146]), second order [2.147], third order [2.148] or no phase transition at all, e.g. kinetic freezing [2.149]. However, there is general agreement that the maximum undercooling of a liquid is limited to the isentropic temperature in order to avoid the paradoxical situation (Kauzmann paradox) where the configurational entropy of the disordered state (liquid) is smaller than the configurational entropy of the ordered state (crystal). In practice, the glass transition sets in the above T_{go} with the deviation from internal equilibrium and related relaxation rates being determined by the applied cooling rates. Therefore, only at infinitely slow cooling would ergodic conditions prevail and *all* configurations in phase space be sampled, resulting in a phase transition at T_{go}. On realistic timescales the system exhibits non-ergodic behavior with the glass having excess entropy in comparison to the crystal.

For a non-ergodic situation, which is a necessary condition for glass formation, it has been shown by molecular-dynamic simulations that the maximum undercooling of a liquid is limited to the iso-volume condition [2.15]. According to this argument, glass formation would set in when the undercooled liquid has the same volume as the crystalline phase. *Tallon* has shown that this condition is identical with the temperature at which only the communal entropy is left in the undercooled liquid. Such calculations have been performed for the idealized case of pure metallic elements where glass formation is extremely difficult. For glass forming alloys, such extrapolations have not been performed but they are under way now [2.150].

Like in any phase transition away from a critical point, the observation of the occurrence of the new phase depends on nucleation and growth. The above concept of the solid-state amorphization being triggered by an instability is difficult to observe experimentally since the nucleation rate of the amorphous phase is a function of the density of lattice defects – predominantly grain boundaries – acting as heterogeneous nucleation sites for the amorphous phase. Thus, the transition will be different depending on whether it occurs homogeneously or heterogeneously such as for most examples where fluctuations lead to the development of a critical glassy nucleus at grain boundaries.

2.3.2 Melting Kinetics

From the experimental evidence so far it becomes clear that heterogeneous nucleation is the rate-determining step for amorphous-phase formation in most experimental conditions. For example, for thin-film diffusion couples this has been discussed in detail in Sect 2.2.2. Other experimental approaches are more difficult to analyze since they often involve several different dynamical effects occurring during the process, such as the development of strong chemical concentration gradients together with high defect densities (ion-beam mixing and mechanical alloying). However, the problem of nucleation has not been addressed in much detail for alternative processes (pressure application, ultraviolet irradiation) and thus is still unsolved.

In terms of nucleation and the evolving structure, solid-state amorphization shows strong similarities with melting. For melting, the role of defects such as the surface and grain boundaries as catalytic heterogeneous nucleation sites for liquid have been known for a long time. In addition, it could be shown that the effective control of such nucleation sites can lead to drastic superheating of crystalline matter above its melting as shown for oxides (SiO_2 [2.141]), metals (Au on Ag [2.26, 142], Pb precipitates in an Al matrix [2.151], and solid He^4 under high pressure [2.143]. However, melting kinetics are very fast: the melting front can move with the velocity of sound. Thus, kinetic superheating in pure metals is extremely difficult to achieve. In contrast, the melting front of solid solutions can move considerably slower and be studied in detail [2.152]. However, the kinetics of glass formation from the solid is many orders of magnitude slower than regular melting due to the slow diffusion rates as discussed above. Thus, solid-state amorphization experiments afford an opportunity to study the melting transition in more detail and reach a more global understanding of this complex problem.

2.3.3 Related Topics

Based on the approach presented here, a first-order phase transition can be reduced to an isentropic transition and possibly to a second-order phase transition by the variation of an additional external parameter. In a theoretical model, such a possibility was investigated by *Imry* describing the effects of microscopic randomly quenched-in impurities [2.153]. There it was shown that pronounced local fluctuations in impurity density develop when the free-energy gain more than offsets the energy cost of the interface produced, thus smearing out the otherwise pure-system transition. Similar arguments apply to our case of melting, if one is close to triple point, where the interfacial energy between crystal and liquid or glass becomes very small or even vanishes.

Furthermore, the metastable-phase diagram proposed for binary alloys distinguishing the crystal, liquid and glass phases is of generic character. The phase diagram indicates the structural stability of non-equilibrium phases and exhibits strong similarities to the magnetic–spin-glass transition including the corresponding Nishimori lines [2.154]. The horizontal axis in Fig. 2.5 however represents in general a tunable parameter, characteristic of the degree of frustration of the system. This parameter can correspond to alloy composition, but also to defect concentration, (negative) pressure, excess free volume, reciprocal particle, or grain size, etc. Above a certain degree of frustration, the ordered crystalline state becomes unstable against liquid-like heterophase fluctuations, and the system collapses to a disordered glassy or liquid-like state as indicated in a universal-phase diagram by the melting instability with intersecting isentropic, isenthalpic, and isenergic lines.

Thus, it is possible experimentally to reduce the regular discontinuous melting transition to continuous-like behavior by the static disorder built into the crystal lattice. With such an approach, the melting point is reduced to the glass transition temperature characterized by the triple point (c^*, T^*) with a vanishing entropy of fusion. This process is further elucidated in Fig. 2.26 by a family of free-energy curves. If we consider the free-energy difference ΔG between crystal and liquid as function of temperature at different degrees of static disorder, we obtain the following situation. For the pure metal $\Delta G = G^l - G^x$ is negative above T_m, positive below T_m and zero at T_m. The entropy of fusion at T_m is given by the slope $\partial \Delta G / \partial T = -\Delta S_f$, which is large and negative at T_m and zero at T_s^i and T_{go} (Fig. 2.26a). Below T_{go} the undercooled liquid becomes unstable with respect to the crystal and above T_s^i the superheated crystal becomes unstable with respect to the liquid based on the Kauzmann and inverse Kauzmann paradox. This regime is indicated by the dashed line of the free-energy curves of Fig. 2.26. With increasing static disorder, the melting point T_o is lowered and ΔS_f at T_o decreases as shown in Fig. 2.26b. With increasing disorder, the melting point T_o is reduced further until it becomes identical with the glass transition temperatures T_{go} as shown in Fig. 2.26c. This point in T–d space is identical with the triple point at a critical concentration c^* and temperature T^*. This condition is characterized by $\partial \Delta G / \partial T = 0$ and $\partial \Delta^2 G / \partial T^2 = 0$. At this point ΔG, ΔS and, consequently, ΔH become zero indicating an instability similar to the inflection point in P–V space for the van der Waals equation of state of a fluid. Such a triple point has also been predicted based on catastrophe theory by investigating the perturbations of catastrophe germs [2.155]. For larger degrees of disorder, the entropy difference ΔS would be

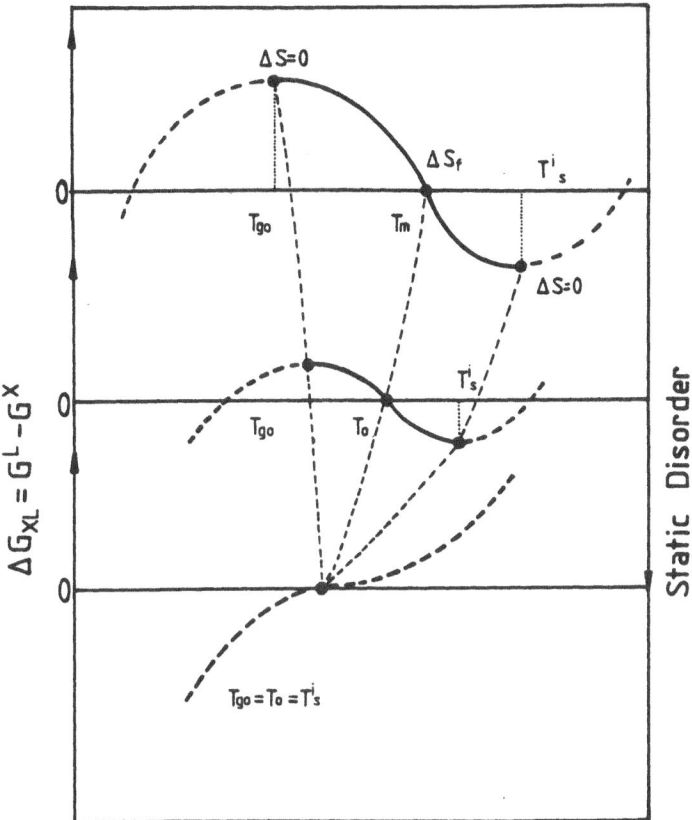

Fig. 2.26. The Gibbs-free-energy difference ΔG between liquid and crystal ($G = 0$) as function of temperature with varying static disorder built into the crystal lattice and indicating the reduction of the equilibrium melting point to an isentropic transformation

negative for all temperatures. Therefore, the crystalline phase becomes intrinsically unstable and ceases to exist.

This generic approach allows to describe non-equilibrium crystalline phases and their stability against melting and glass formation. It applies not only to the case of alloying, but can be extended to characterize the thermal stability of a non-equilibrium crystalline phase as function of defect concentration, negative pressure, excess free volume, reciprocal particle size in the case of small clusters or reciprocal grain size in the case of nanocrystalline materials. Thus, this fertile area of research will not only allow us to develop a deeper understanding of the glassy state but also to improve our knowledge about the melting transition under equilibrium and non-equilibrium conditions.

Acknowledgements. The authors gratefully acknowledge the collaborative Research Grant No. 890239 by NATO Scientific Affairs Division, which made this chapter possible. One of us (W.L.J) acknowledges further the continuous support of his work provided by the U.S. Department of Energy under Contract No. DE-FG03-86ER545242. The Deutsche Forschungsgemeinschaft is gratefully acknowledged for the continuous support in the past of one of the authors (K.S.) via the SFB 126 TP E19. The present support of the Bundesminister für Forschung und Technologie under Contract No. 03 M0040C3 is gratefully acknowledged. Special thanks are given to Professor M. Atzmon, Professor J. Böttiger, Professor R.W. Cahn, Professor P. Desré, Professor H. Gleiter, Professor P. Haasen, Professor G.v. Minnigerode, Professor M.-A. Nicolet, Professor J.H. Pevepezko, Dr. L. Schultz, Dr. R.B. Schwartz, and many of our coworkers, who have all contributed in one or the other way to the development of the field of solid-state amorphization.

References

2.1   W.Buckel, R. Hilsch: *Z. Phys.*, **138** 109 (1954): W. Buckel: *Z. Phys.*, **138** 136 (1954)

2.2   W. Klement Jr., R.H. Willens, P. Duwez: Nature **187**, 869 (1960)

2.3   K.C. Russell: Progr. Mater. Sci. **28**, 229 (1985)

2.4   W.L. Johnson: Progr. Mater. Sci. **30**, 81 (1986)

2.5   X.L. Yeh, K. Samwer, W.L. Johnson: *Appl. Phys. Lett.* **42**, 24 (1983)

2.6   H. Österreicher, H. Clinton, H. Bittner: *Mater. Res. Bull.* **11**, 1241 (1976)

2.7   R.B. Schwarz, W.L. Johnson: *Phys. Rev. Lett.* **51**, 415 (1983)

2.8   K. Samwer: *Phys. Rep.* **161**, 1 (1988) and papers in *J. Less Common Metals* **145**, 1ff (1988)

2.9   R.B. Schwarz, R.R. Petrick, D.K. Saw: *J. Non-Cryst. Sol.* **76** 281 (1985)

2.10   D.E. Luzzi, M. Meshii: *Res. Mechanica* **21**, 207 (1987)

2.11   H.J. Fecht, Z. Fu, W.L. Johnson: *Phys. Rev. Lett.* **64** 1753 (1990)

2.12   D. Turnbull: *Solid State Phys.* **3** 225 (1956)

2.13   H.J. Fecht, W.L. Johnson: *Nature* **334** 51 (1988) and ref. therein

2.14   P.W. Anderson: *Phys. Today*. **41**, Nos. 1, 3, 6, 9 (1988) and **42**, No. 7 (1989)

2.15   J.L. Tallon: *Nature* **342**, 658 (1989)

2.16   U. Köster, U. Herold: in *Glassy Metals I*, ed by H.-J. Güntherodt and H. Beck, Topic Appl. Phys. 46 (Springer, Berlin, Heidelberg, 1980) Chap. 9

2.17   A.R. Miedema, R. Boom, F.R. de Boer; *J. Less Common Metals* **41**, 283: (1975)
       A.K. Niessen, F.R. de Boer, R. Boom, P.F. de Châtel, W.C. Mathens, A.R. Miedema: CHALPHAD **7** 51 (1983)
       G.J. van der Kolk, A.R. Miedema, A.K. Niessen: *J. Less Common Metals* **145**, 1 (1988)

2.18   R. Bormann: Thermodynamik metastabiler Phasen Habilitationsschrift, Universität Göttingen (1988)

2.19   J. Hafner, *From Hamiltonians to Phase Diagrams* Springer Ser. Solid-State Sci:, Vol 70 (Springer, Berlin, Heidelberg 1990)

2.20   H.J. Fecht, P. Desré, W.L. Johnson: *Phil. Mag. B* **59**, 577 (1989)

2.21   H.J. Fecht, W.L. Johnson: *J. Less Common Metals* **145**, 63 (1988)

2.22   T. Egami, Y. Waseda: *J. Non-Cryst. Solids* **64**, 113 (1984)

2.23   A. Blatter, M.V. Allmen: *Phys. Rev. Lett.* **54**, 2103 (1985)

2.24   A.L. Greer: *J. Less Common Metals* **140**, 327 (1988)

2.25   J.-C. Tolédano, P. Tolédano, *The Landau theory of phase transitions, Lecture Notes in Physics*, Vol. 3 (World Scientific Singapore 1987)

2.26   J. Däges, H. Gleiter, J. Perepezko: *MRS Symp. Proc.* **57**, 63 (1987)

2.27   J.H. Perepezko: *J. Mater. Sci. Eng.* **65**, 125 (1984)

2.28   W. Dörner, H. Mehrer: *Phys. Rev. B* **44**, 101 (1991)

2.29   R. Busch, S. Schneider K. Samwer: *Nachrichten d. Akad. d. Wissenschaften zu Göttingen* **1**, 1 (1991)

2.30   Z. Altounian, R.J. Shank, J.O. Strom-Olsen: *J. App. Phys.* **58**, 1192 (1985)

2.31   A.M. van Diepgen, K.H.J. Buschow: *Solid State Commun.* **22**, 113 (1977)

2.32   R.Schulz, A. van Neste, L. Brossard, Y.J. Huot: *Mater. Sci. Eng.* **99**, 469 (1988)

2.33   K. Aoki, T. Masumoto: *The Science Reports of the Research Institute*, Tohoku University, Series A, **42** No.1, 79 (1988)

2.34   W.L. Johnson, H.J. Fecht: *J. Less-Common Metals* **145**, 63 (1988)

2.35   W.J. Meng, P.R. Okamoto, L.J. Thompson, B.J. Kestel, L.E. Rehn: *J. Appl. Phys.* **53**, 1820 (1984)

2.36   A.V. Andreyev, A.V. Deryagin, A.A. Yezov, N.V. Mushnikov: *Phys. Met. Metall.* **58**, 124 (1984)

2.37   W.L. Johnson; In *Rapidly Quenched Metals VI*, ed. by J. Strom Olsen, R.W. Cochrane; *Mater. Sci. & Eng.* **97**, 1 (1988)

2.38   A.D. Le Claire: *J. Nucl. Mater.* **69/70**, 70 (1978)

2.39   W.K. Warburton, D. Turnbull: In *Diffusion in Solids-Recent Developments*, (Academic, New York 1975) Chap. 4

2.40   W.J. Meng, C.W. Nieh, W.L. Johnson; *Appl. Phys. Lett.* **51**, 1693 (1987)
         W.J. Meng, E.J. Cotts, W.L. Johnson: *Mater. Res. Soc. Symp. Proc.*, 77, 223 (1987)
         W.J. Meng, C.W. Nieh, E. Ma, B. Fultz, W.L. Johnson; *Mater. Sci. Eng.* **97**, 87 (1988)
         W.J. Meng: *Ph.D. Thesis*, Dept. of Applied Physics Calif. Inst. of Tech.,

2.41   H. Schroeder, K. Samwer, U. Köster; *Phys. Rev. Lett.* **54**, 197 (1985)
         K. Samwer: In *Amorphous Metals and Non-Equilibrium Processing*, ed by M. von Allmen (Editions de Physique, Couteabeouf, France 1984) p. 123
         H.U. Krebs, K. Samwer: *Europhys. Lett.* **2**, 141 (1986)

2.42   M.van Rossum, M.-A. Nicolet, W.L. Johnson: *Phys. Rev. B* **29**, 5498 (1984)

2.43   K. Pampus, K. Samwer, J. Böttiger: *Europhys. Lett.* **3**, 581 (1987)

2.44   J.C. Barbour, F.W. Saris, M. Nastasi, J.W. Mayer: *Phys. Rev. B* **32**, 1363 (1985)

2.45   K. Karpe, J. Bottiger, A.L. Greer, J. Janting, K. Larsen: *J. Mater. Res.* **7**, 926 (1992)

2.46   Y.T. Cheng, M.-A. Nicolet, W.L. Johnson: *Appl. Phys. Lett.* **47**, 800 (1985)

2.47   J.C. Barbour; *Phys. Rev. Lett.* **55**, 2872 (1985)
         J.C. Barbour: The Diffusion of Nickel in Amorphous Nickel-Zirconium Alloys and the Composition Analysis of Nickel-Silicide Formation in Lateral Diffusion Couples. *Ph.D. Thesis*, Dept. of Mater. Sci. Cornell Univ., (1986)

2.48   H. Hahn, R.S. Averback, S.J. Rothman: *Phys. Rev. B* **33**, 8825 (1986)

2.49   S.B. Newcomb, K.N. Tu: *Appl. Phys. Lett.* **48**, 1436 (1986)

2.50   R.W. Johnson, C.C. Ahn, E.R. Ratner: *Phys. Rev. B* **40**, 8139 (1989): R.W. Johnson, C.C. Ahn, E.R. Ratner. *Appl. Phys. Lett.* **54**, 795 (1989)

2.51   K. Halloway, R. Sinclair; *J. Less Common Metals* **140**, 139 (1988)
         K. Halloway, P. Moine, J. Delage, R. Bormann, L. Capuano, R. Sinclair: *Mater. Res. Soc. Symp. Proc.* **187**, 71 (1990)

2.52   C. Michaelsen, M. Piepenbring, H.U. Krebs: *J. Physique C4*, 151 (1990)
         H.U. Kerbs, D.J. Webb, A.F. Marshall: *J. Less Common Metals* **140**, 17 (1988)

2.53   J.F.M. Westendorp: Ion and Laser Beam Induced Metastable Alloy Formation. Dissertation, Univ. of Utrecht, The Netherlands (1986) A.M. Vredenberg, J.F.M. Westendorp, F.W. Saris, N.M. van der Pers. *J. Mater. Res.* **1**, 774 (1986)

2.54   S.R. Herd, K.N. Tu: *Appl. Phys. Lett.* **42**, 597 (1983)

2.55   E. Ma, W.J. Meng, W.L. Johnson, M.-A. Nicolet: *Appl. Phys. Lett.* **53**, 2033 (1988)

2.56   J.J. Hauser: *J. Physique* **42**, C4-943 (1981)

2.57   D.M. van der Walker: *Appl. Phys. Lett.* **48**, 707 (1986)

2.58   C.V. Thompson, L.A. Clevenger, R. DeAvillez, E. Ma, H. Miura: *Mater. Res. Soc. Symp. Proc.* 187, 61 (1990)

2.59   J.Y. Cheng, M.H. Wang L.J. Chen; *Mater. Res. Soc. Symp. Proc.* 187, 77 (1990)

2.60   V.A. Ushkow, A.B. Fedotov, E.A. Eroteeva, A.I. Rodionov, D.T. Dzhumakulov. Izv. Akad. Nauk SSSR, Neorgan. Mater. **23**, 186 (1987)
         D.H. Ko, R. Sinclair: *Appl. Phys. Lett.* **58**, 1851 (1991)

2.61   F.Y. Shiau, Y.A. Chang: *Appl. Phys. Lett.* **55**, 1510 (1989)
         F.Y. Shiau, Y.A. Chang. *Mater. Res. Soc. Symp. Proc.* **187**, 89 (1990)

2.62   T. Sands, C.C. Chang, A.S. Kaplan, V.G. Keramidas, K.M. Krishman, J. Washburn: *Appl. Phys. Lett.* **50**, 1346 (1987)
R. Caron-Popowich, J. Washburn, T. Sands, A.S. Kaplan: *J. Appl. Phys.* **64**, 4909 (1988)

2.63   F.Y. Shiau, Y.A. Chang: *Appl. Phys. Lett.* **55**, 1510 (1989)

2.64   R.M. Walser, R.W. Bene: *Appl. Phys. Lett.* **28**, 624 (1976)

2.65   M.-A. Nicolet, S.S. Lau: In *VLSI Electronics, Microstructure Science*, **6**, 330 (Academic, New York 1983)

2.66   M. Atzmon, J.R. Veerhoven, E.R. Gibson, W.L. Johnson: *Appl. Phys. Lett.* **45**, 1052 (1984)

2.67   M. Atzmon, K. Unruh, W.L. Johnson: *J. Appl. Phys.* **58**, 3865 (1985)

2.68   G.C. Wong, W.L. Johnson, E.J. Cotts: *J. Mater. Res.* **5**, 488 (1990)

2.69   E.J. Cotts, G.C. Wong, W.L. Johnson: *Phys. Rev. B*, **37**, 9049 (1988)

2.70   L. Schultz: In *Rapidly Quenched Metals V*, ed S. Steeb, H. Warlimont (North Holland, Amsterdam 1985) p. 1585

2.71   L. Schultz: *Mater. Res. Soc. Symp. Proc.* **80**, 97 (1987)

2.72   F. Bordeaux, E. Gaffet, A.R. Yavari: *Europhys. Lett.* **12**, 63 (1990)

2.73   S. Martelli, G. Mazzone, A. Montone, M.V. Antisari: *J. Physique*, C4, 241 (1990)

2.74   A.Y. Yermakov, Y.Y. Yurchikov, V.A. Barinov: *Phys. Met. Mattalogr.* **52**, 50 (1981)

2.75   C.C. Koch, O.B. Cavin, C.G. McKamey, J.O. Scarbrough: *Appl. Phys. Lett.* **43**, 1017 (1983)

2.76   N. Saunders, A.P. Miodownik: *J. Mater. Res.* **1**, 38 (1986)

2.77   B.E. Deal A.S. Grove: *J. Appl. Phys.* **36**, 3770 (1985)

2.78   Y.Y. Geguzin, Y.S. Kagonouskly. L.M. Paritskoya V.I. Solunskiy: *Phys. Met. Metallogr.* **47**, 127 (1980)

2.79   B. Dorgin; Kinetics of the Formation of an Amophous Layer during a Solid State Amorphization. *Ph.D. Thesis*, Dept. of Applied Physics Calif. Inst. of Tech., (1985): see also [8] for a description of interdiffusion reaction kinetics

2.80   U. Gösele, K.N. Tu: *J. Appl. Phys.* **53**, 3252 (1982)

2.81   R.W. Cahn: *Physique* C4, **51**, 3 (1990)
B. Cantor: In Amorphous Metals and Semiconductors, eds. P. Haasen, R.I. Jaffee (Pergamon, Oxford 1986) p. 108

2.82   K. Pampus, J. Bottiger, B. Torp, H. Schröder, K. Samwer: *Phys. Rev. B* **35**, 7010 (1987)

2.83   R.J. Highmore, J.E. Evetts, A.L. Greer, R.E. Somekh: *Appl. Phys. Lett.* **50**, 566 (1987)

2.84   H. Schroder, K. Samwer: *J. Mater. Res.* **3**, 461 (1988)

2.85   J.W. Christian: *The Theory of Transformations in Metals and Alloys* (Pergamon, London 1965) Chap. X

2.86   E. Kamenetsky, W.J. Meng, L. Tanner, W.L. Johnson: In *Analytical Electron Microscopy*, ed. by D.C. Joy (San Francisco Press, San Francisco 1987) p. 83
W.J. Meng: Solid State Amorphization Reactions in Thin Film Diffusion Couples. *Ph.D. Thesis*, Calif. Inst. of Tech. (1988)

2.87   U. Köster, R. Pries, G. Bewernick, B. Schuhmacher, M. Blank-Bewersdorff: *J. Physique*, C4, 121 (1990)

2.88   P.J. Desre', A.R. Yavari: *Phys. Rev. Lett.* **64**, 1533 (1990)

2.89   R.B. Schwarz, C.C. Koch: *Appl. Phys. Lett.* **49**, 146 (1986)

2.90   G. Martin, E. Gaffet: *J. Physique*, **51**, C4-71 (1990)

2.91   F.R. Ding, P.R. Okamoto, L.E. Rehn: In *Beam Solid Interactions and Transient Thermal Processing*, MRS Symp. Proc., 100 69 (1987)

2.92   C. Politis, W.L. Johnson: *J. Appl. Phys.* **60** (1986)
G. Cocco, L Soletta, S. Euzo, M. Magini, N. Cowlam: *J. Physique* C4, 181 (1990)

2.93   E. Ma, C.W. Nieh, M.-A. Nicolet, W.L. Johnson: *J. Mater. Res.* **4**, 1299 (1989)

2.94   R.B. Schwarz, K.L. Wong, W.L. Johnson: *J. Non. Cryst. Sol.* **61/62** 129 (1984)

2.95   M. Atzmon, J.R. Veerhoven, E.R. Gibson, W.L. Johnson: *Appl. Phys. Lett.* **45**, 1052 (1984)
L. Schultz. In *Rapidly Quenched Metals V*, ed. S. Steeb, H. Warlimont (North Holland, Amsterdam 1985) p. 1585

2.96   H.U. Krebs, D.J. Webb, A. Marshall: *Phys. Rev. B* **35**, 5392 (1987) C. Michaelsen, M. Piepenbring, H.U. Krebs: *J. Physique*, C4 151 (1990) H.U. Krebs, D.J. Webb, A.F.

Marshall: *J. Less. Common Metals*, **140** 17 (1988) E. Hellstern, L. Schultz: *J. Appl. Phys.* **63** 1408 (1988)

2.97 M. van Rossum, U. Shreter, W.L. Johnson, M.-A. Nicolet: *Mater. Res. Soc. Symp. Proc.*, 27, 107 (1984)

2.98 G.J. van der Kolk, A.R. Miedema, A.K. Niessen: *J. Less Common Metals*, **145** 1 (1988)
P.I. Loeff, A.W. Weeber, A.R. Miedema: *J. Less Common Metals* **140** 229 (1988)

2.99 B.M. Clemens: *Phys. Rev. B* **33**, 7615 (1986)

2.100 J.M. Firgerio, J. Rivory: *J. Physique* C4, Suppl. to No.14, 51, 163 (1990)

2.101 B. Blainpain, J.M. Legresy, J.W. Mayer: *J. Physique* C4, Suppl. to No. 14 **51**, 131 (1990)
L. Hung, M. Nastasi, J.W. Mayer: *Appl. Phys. Lett.* **42** 672 (1983)

2.102 T. Ben Ameur, A.R. Yavari: *J. Physique* C4, 219 (1990)

2.103 P. Guilmin, P. Guyot, G. Marchal: *Phys. Lett.* **109** A 174 (1985)

2.104 J. Horvath, F. Dyment, J. Mehrer: *J. Nucl. Mater*, **126**, 206 (1984)

2.105 W.L. Johnson, Y.T. Cheng, M. van Rossum, M.-A. Nicolet: *Nucl. Instrum. Methods B* **7/8**, 657 (1985)

2.106 R.S. Averback: *Nucl. Instrum. Methods. B* **15**, 675 (1986)

2.107 P. Sigmund, A. Gras-Marti: *Nucl. Instr. Methods* **182/183**, 25 (1981)

2.108 T.W. Workman, Y.T. Cheng, W.L. Johnson, M.-A. Nicolet: *Appl. Phys. Lett.* **50**, 1485 (1987)

2.109 T. Diaz de la Rubia, R.S. Averback, R. Benedek, W.E. King: *Phys. Rev. Lett.* **59**, 1930 (1987)

2.110 G.H. Vineyard: *Radiat. Eff.* **29**, 245 (1976)

2.111 B.Y. Tsaur, S.S. Lau, S. Hung, J.W. Mayer: *Nucl. Instrum. Methods* **182/183**, 67 (1981)
B.Y. Tsaur: Ion-Beam-Induced Modifications of Thin Film Structures and Formation of Metastable Phases. *Ph.D. Thesis*, Calif. Inst. of Tech. (1980)

2.112 B.X. Liu, W.L. Johnson, M.-A. Nicolet: *Appl. Phys. Lett.* **42** 45 (1983)

2.113 L.S. Hung, M. Nastasi, J. Gyulai, J.W. Mayer: *Appl. Phys. Lett.* **42** 672 (1983)
M. Nastasi: Ion-Irradiation-Induced-Transformations in Intermetallic Alloys: Formation and Stability. *Ph.D. Thesis*, Cornell University (1986)

2.114 Y.T. Cheng, W.L. Johnson, M.-A. Nicolet: *Mater. Res. Soc. Symp. Proc.* **37**, 565 (1985)

2.115 P.R. Okamoto, L.e. Rehn, J. Pearson, R. Bhadra, M. Grimsditch: *J. Less Common Metals.* **140**, 231 (1988)

2.116 G. Linker: *Solid State Commun.* **57**, 773 (1986)

2.117 E. Arzt, L. Schultz (eds): *New Materials by Mechanical Alloying Techniques*, (DGM, Oberursel 1990)

2.118 L. Schultz, J. Eckert: this volume, Chapter 3

2.119 A.W. Weeber, H. Bakker: *Physica B* **153**, 93 (1988)

2.120 C. Koch: *Mater. Sci. Techn.* 15 (VCH, Weinheim 1991) Chap. 5

2.121 R.B. Schwarz, R.R. Petrich, C.K. Saw: *J. Non-Cryst. Solids* **76**, 281 (1985)

2.122 A. Mehrtens, G.V. Minnigerode, K. Samwer; *Z. Physik B* **83**, 55 (1991)

2.123 P. Haasen: *Physical Metallurgy* (Springer, Berlin, Heidelberg 1974)

2.124 Jr. Koike, D.M. Parkin, M. Nastasi: *J. Mater. Res.* **5**, 1414 (1990)

2.125 E. Hellstern, H.J. Fecht, Z. Fu, W.L. Johnson: *J. Mater. Res.* **4**, 1992 (1989)

2.126 L.C. Chen, F. Spaepen: *J. Appl. Phys.* **69**, 679 (1991)

2.127 A. Regenbrecht; Bildungsbereich amopher CoZr- und CuZr- Aufdampfschichten bei Variation der Substrattemperatur. Dissertation Thesis, Universität Göttingen (1989)
A. Regenbrecht, G.V. Minnigerode, K. Samwer: *Z. Phys. B* **79**, 25 (1990)

2.128 D.L. Beke, P.I. Loeff, H. Bakker: *Acta metall. mater.* **39**, 1259 (1991)

2.129 O. Mishima. L.D. Calvert, E. Whalley: *Nature* **310**, 393 (1984)

2.130 D.R. Clarke, M.C. Kroll, P.D. Kirchner, R.F. Cook, B.J. Hockey: *Phys. Rev. Lett.* **60**, 2156 (1988)

2.131 R.J. Hemley, A.P. Jephcoat, H.K. Mao, L.C. Ming, M.H. Manghnani: *Nature* **334**, 52 (1988)

2.132 S. Rastogi, M. Newman, A. Keller: *Nature* **353**, 55 (1991)

2.133 E.G. Ponyatovsky, I.T. Belashn, O.I. Barkalov: *J. Non-Cryst. Solids* **117/118**, 679 (1990)

2.134 E.G. Ponyatovsky, O.I. Barkalov: *J. Mater. Sci. Eng. A* **133**, 726 (1991)

2.135 C.K. Saw, R.B. Schwarz: *J. Less. Common. Metals* **140**, 385 (1988)

2.136   R.W. Cahn: *Nature* **323**, 668 (1986)
2.137   F.A. Lindemann: *Z. Phys.* **11**, 609 (1910)
2.138   M. Born: *J. Chem. Phys.* **7**, 591 (1977)
2.139   D. Wolf, P.R. Okamoto, S. Yip, J.F. Lutsko, M. Kluge: *J. Mater. Res.* **5**, 286 (1990)
2.140   T. Gorecki: *Scr. Metall.* **11**, 1051 (1977)
2.141   R.L. Cormia, J.D. MacKenzie, D. Turnbull: *J. Appl. Phys.* **34**, 2245 (1963)
2.142   J. Däges, J.H. Perepezko, H. Gleiter: *Phys. Lett. A* **119**, 79 (1986)
2.143   J. Jung, J.P. Franck: *Jpn. J. Appl. Phys.* **26-3**, 399 (1987)
2.144   J. Bilgram: *Phys. Rep.* **153**, 1 (1987)
2.145   J.L. Tallon: *Phil. Mag. A* **39**, 151 (1979)
2.146   M.H. Cohen, G.S. Crest: *Phys. Rev. B* **20**, 1077 (1979)
2.147   G. Adam, J.H. Gibbs: *J. Chem. Phys.* **43**, 139 (1965)
2.148   L.V. Woodcock: *J. Chem. Soc. Faraday* **72**, 1667 (1976)
2.149   N.O. Birge, S.R. Nagel: *Phys. Rev. Lett.* **54**, 2674 (1985)
2.150   H.J. Fecht, P. Desre', Unpublished research
2.151   L. Grabaeck, J. Bohr, E. Johnson, A. Johanson, L. Sarholt-Kristensen, H.H. Andersen: *Phys. Rev. Lett.* **64**, 934 (1990)
2.152   W.P. Allen, H.J. Fecht, J.H. Perepezko: *Scr. Met.* **23**, 647 (1989)
2.153   Y. Imry, M. Wotis: *Phys. Rev. B* **19**, 3580 (1979)
2.154   P. LeDoussal, A.B. Harris: *Phys. Rev. Lett.* **61**, 625 (1988)
2.155   R. Gilmore: *Catastrophe Theory* (Wiley, New York 1976)

3. Mechanically Alloyed Glassy Metals

L. Schultz and J. Eckert

With 46 Figures

Mechanical alloying is a nonequilibrium processing tool, which allows to form metastable alloys as glassy metals or quasicrystalline materials. In this chapter, it will be shown that the amorphization process during mechanical alloying is a solid-state amorphization transformation. Besides the details of glass formation and the structural characterization, the glass-forming ranges and the glass-forming systems will be described. Mechanically alloyed amorphized powders can be used advantageously to form intermetallic phases for new hard magnets by crystallization. Similar principles as for glass formation by mechanical alloying also determine the formation of quasicrystalline materials. Of special interest are here the various possible phase transitions between the glassy, the quasicrystalline and the crystalline state during ball milling.

3.1 Background

About twenty years ago, mechanical alloying was developed by *Benjamin* [3.1] as a new technique of combining metals. It circumvents many of the limitations of conventional alloying and creates true alloys of metals or metal–non-metal composites that are difficult or impossible to combine by other means. Examples are high-strength superalloys for jet engines or tungsten carbide–cobalt composites [3.1]. Mechanical alloying is performed in a high-energy ball mill in an inert gas. The metal powder particles are trapped by the colliding balls, heavily deformed and cold welded. Characteristically layered particles are formed. Further milling refines the microstructure even more. Finally, in many cases, a true alloying takes place.

In 1979, *White* [3.2] observed that, by milling elemental Nb and Sn powders, the distinct X-ray diffraction peaks of the elements disappeared and typical diffuse peaks of an amorphous pattern showed up. But these samples did not show the superconducting transition temperature of vapor-quenched amorphous Nb–Sn alloys. In 1983, *Koch* et al. reported on the "Preparation of *amorphous* $Ni_{60}Nb_{40}$ by mechanical alloying" [3.3]. After the detection of amorphization by solid-state reaction in evaporated multilayer films by *Schwarz* and *Johnson* [3.4] (see also Chap. 2), *Schwarz* et al. [3.5] proposed after investigating glass formation in Ni–Ti alloys, that amorphization by mechanical alloying is also based on the solid-state reaction process. Within the last couple

Topics in Applied Physics, Vol. 72
Beck/Güntherodt (Eds.)
© Springer-Verlag Berlin Heidelberg 1994

of years, several authors have reported on the formation of amorphous powders by mechanical alloying in a large number of alloy systems [3.6–10]. The *amorphous* is now no longer written in italics. In this contribution, we give an overview of recent results on the formation of glassy metals by mechanical alloying of elemental powders. We describe in detail the formation process by the microstructural development and by X-ray diffraction results, report on the glass-forming ranges of some TM–TM alloys (TM: transition metal), list some of the alloy systems where this process has been found and where it does not work, and present results on the further characterization of the samples by measuring their physical properties. Finally, we describe mechanical alloying of special systems like boron-containing alloys or RE–TM–X alloys (RE: rare earth metal), which are very promising for the production of novel high-performance permanent magnets as well as the formation of quasicrystals.

3.2 The Mechanical Alloying Process

Mechanical alloying is usually performed under an inert (mostly argon) atmosphere in a high-energy ball mill (Fritsch or Retsch), a vibratory ball mill (Spex), an attritor-type ball mill or, especially for large-scale production up to 1.5 tons of alloyed material, a large conventional ball mill. For the experiments described in this chapter, we used a planetary ball mill, the principle of which is shown in Fig. 3.1 [3.11]. The elemental powders are mixed and poured into a cylindrical milling container together with 10 mm diameter steel balls with a powder-to-ball weight ratio of about 1 : 10. At least one component of the starting powders should be ductile. The second (or third) element can be brittle (like boron). Also prealloyed powders can be used.

The initial basic event of the milling process is a trapping of the elemental powder particles by the colliding balls. They are cold welded and heavily deformed leading to characteristically layered powder particles (Fig. 3.2a). Further milling results in cold welding and deformation of the layered particles and, therefore, in a more and more refined microstructure (Fig. 3.2b). As an example, the rate of refinement of an Fe–50 vol.% Cr mixture processed in a high-speed vibratory mill [3.1] is shown in Fig. 3.3. The lamellar spacing is the mean distance between the Fe/Cr phase boundaries, whereas the smaller weld spacing indicates that also a cold welding between particles of the same element occurs. Due to the initially low hardness of the powders, the lamellar spacing is fast reduced at the early stage of the alloying. Later, the hardness increases caused by deformation and layering and the rate of change of the lamellar spacing becomes nearly logarithmic with time [3.1]. The increased hardness also leads to fracturing of the large particles, so that the initially growing particle size becomes constant or is even reduced. Therefore, in total, the milling process consists of cold welding, deformation and fracturing. It was recently modeled by several groups in detail [3.12–14].

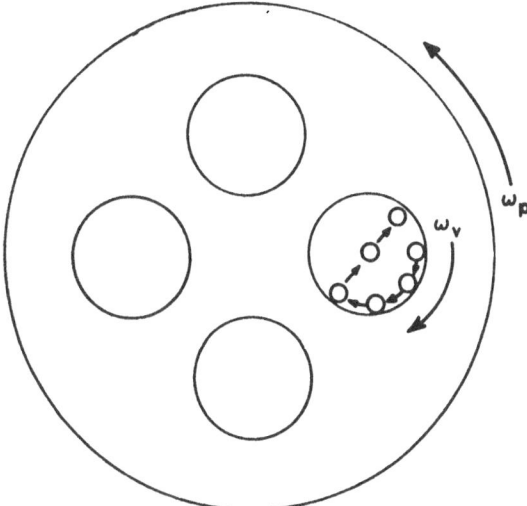

Fig. 3.1. Schematic drawing of a planetary ball mill (Fritsch or Retsch). ω_v and ω_p are the nonindependent angular velocities of the ball-mill plate and of the milling containers, respectively [3.11]

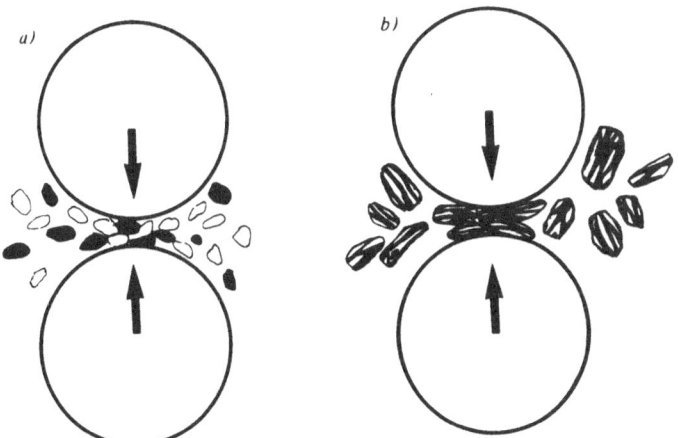

Fig. 3.2. Formation of layered powder particles during ball milling

The layered structure cannot be refined indefinitely by deformation due to the increasing hardness with decreasing crystallite size (for details see Sect. 3.3.2). The main alloying, therefore, occurs by an interdiffusion reaction at the created clean interfaces, if a thermodynamic driving force (negative free enthalpy of mixing) exists for this diffusion couple. The required temperature rise is provided by the heat released during the ball collisions (Sect. 3.3.2).

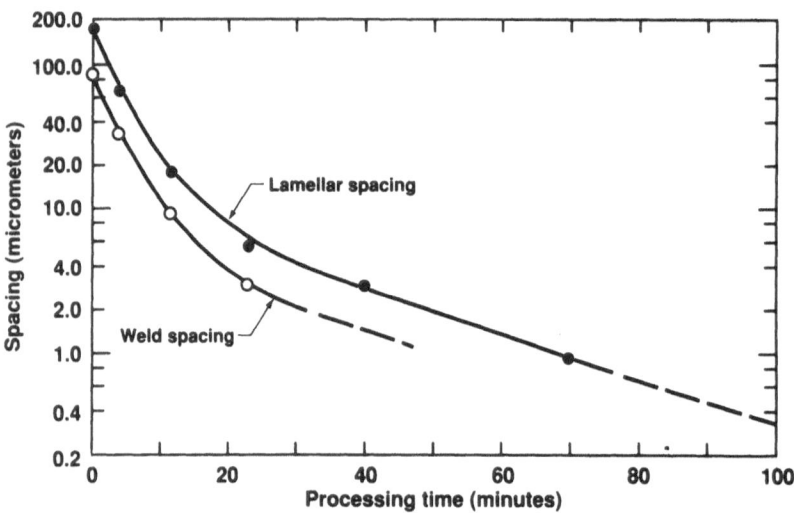

Fig. 3.3. Lamellar thickness and weld spacing vs. processing time in Fe–50 vol. % Cr powder [3.1]

3.3 Glass Formation by Mechanical Alloying

3.3.1 Basic Principles

Glass formation by mechanical alloying of elemental crystalline powders can be considered a special form of solid-state interdiffusion reaction. The basic principles of such a reaction [3.15] are described in Fig. 3.4. As is well known, the thermodynamic stable state of a system is determined by a minimum in the free enthalpy G. In metallic systems, the free enthalpy of the equilibrium crystalline state G_x is always lower than that of the amorphous state G_a below the melting temperature. The amorphous state is a metastable state, i.e., an energy barrier prevents the amorphous phase from spontaneous crystallization. To form an amorphous metal by a solid-state reaction, it is necessary to establish first a crystalline initial state with a high free enthalpy G_0 (Fig. 3.4). Depending on the formation process, this initial state can be achieved, for example, by

1) the system intermetallic phase (Zr_3Rh) plus hydrogen gas [3.16],
2) a layered system of two crystalline elemental metals [3.4, 3.17], if the alloy system has a negative free enthalpy of mixing,
3) an increase in the free enthalpy of one intermetallic phase by introducing lattice defects [3.18–20],
4) a crystalline phase far from equilibrium [3.21].

Starting from this initial state G_0, the free enthalpy of the system can be reduced either by the formation of the metastable amorphous phase or by the formation

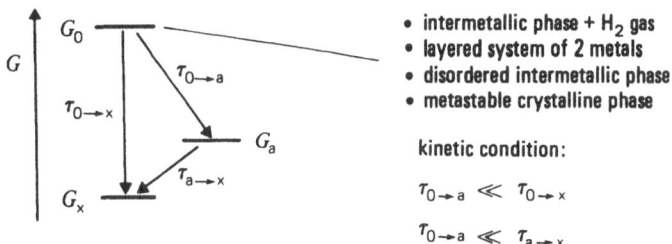

Fig. 3.4. Principle of amorphization by solid-state reaction

of the crystalline intermetallic phase (or phase mixture). Energetically favored is, of course, the crystalline equilibrium phase. The decisive factor is, however, the kinetics of phase formation. To evaluate this, the timescales of all possible reactions must be taken into consideration. The formation of the amorphous phase is then possible and likely, if the reaction to the amorphous phase proceeds substantially faster than the reaction to the crystalline phase, i.e.,

$$\tau_{0 \to a} \ll \tau_{0 \to x}$$

($\tau_{i \to j}$ is the characteristic reaction time). On the other hand, however, the amorphous phase must not crystallize under the given experimental conditions as the reaction proceeds:

$$\tau_{0 \to a} \ll \tau_{a \to x} \ ,$$

i.e., the reaction temperature must be markedly below the crystallization temperature T_x.

These conditions are fulfilled in many cases. The sequential formation of phases having different stabilities has in principle been known for a long time, in particular in the case of precipitation processes, and is described phenomenologically as the *Ostwald rule*. Typical timescales on which these solid-state reactions occur are in the range from 10^4 to 10^6 s, being consequently slower by many orders of magnitude than rapid quenching from the liquid (10^{-6}–10^{-3} s) or the vapor phase (10^{-14}–10^{-10} s).

3.3.2 The Process of Glass Formation by Mechanical Alloying

As mentioned before, mechanical alloying is performed as dry ball milling of elemental powders in a high-energy ball mill under an inert atmosphere. As a typical example, we will describe in detail the formation of glassy Ni–Zr alloys, which is, by far, the best investigated alloy system with respect to glass formation by mechanical alloying. Optical micrographs of $Ni_{50}Zr_{50}$ powder revealing the microstructure after different milling times are shown in Figs. 3.5 a–d. After 0.5 h milling, the particles have a well-aligned layered microstructure (Fig. 3.5a), and the individual layer thickness varies from 2 to 20 µm. The

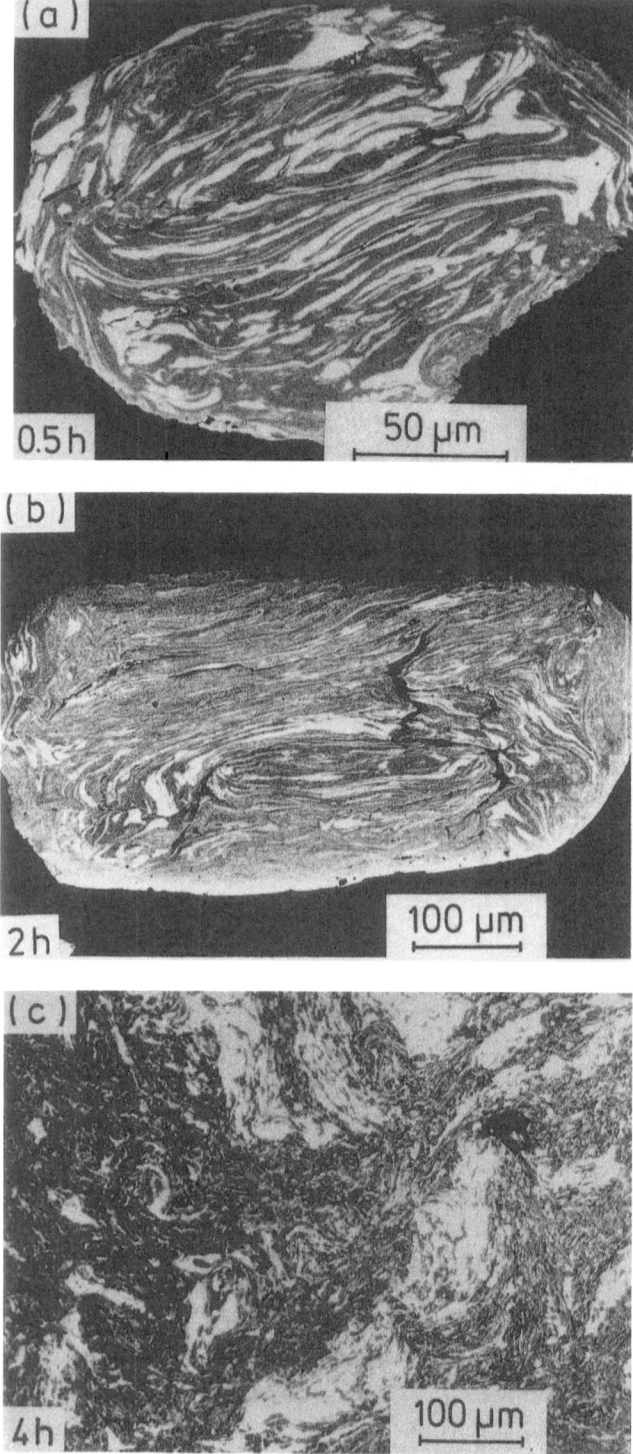

Fig. 3.5. Optical micrographs of Ni–Zr powder particles after (a) 0.5, (b) 2, (c) 4, and (d) 16 h milling time

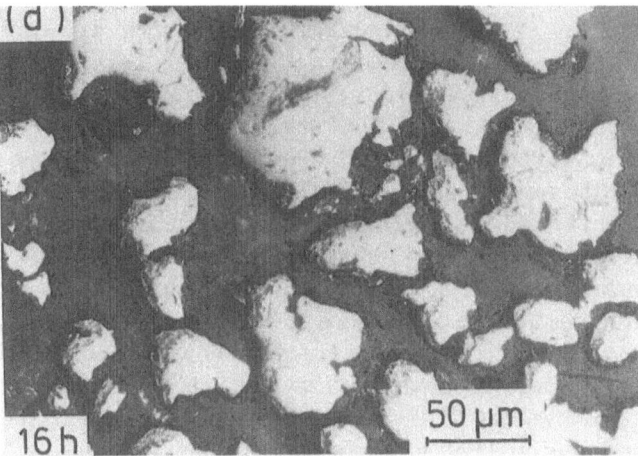

Fig. 3.5. (*Continued*)

particles have grown by repeated cold welding to about 100 μm. For longer milling times, a further growth of the particles is observed (Fig. 3.5b). After 4 h the particle size is about 500 μm (Fig. 3.5c). Due to the welding of grains with different orientation, the particles no longer exhibit a flat shape. Well-aligned parallel layers are no more visible over a larger distance. After 16 h of milling, the microstructure has completely changed (Fig. 3.5d). Most of the particles show a nearly featureless image, which is clearly distinguishable from the earlier milling states. They are mainly single-phase amorphous, as has been proved by TEM, X-ray diffraction and thermal analysis (see discussion below). Some inhomogeneities due to a weaker deformation of some particles are still present. The average diameter of the particles is about 40 μm and decreases to about 10 μm after 60 h. In this final stage, the microstructure is nearly the same when compared to the 16 h milled sample. These observations agree well with those reported for mechanically alloyed Fe–Zr [3.22–24].

During mechanical alloying the individual powder particles are not only cold welded with each other but also stick to the steel balls and the container wall. The milling tools are covered by a thin coating. The material is partially removed during further milling and alloyed with the residual loose powder. The coating of the balls and the container wall and the loose powder are deformed in the same way and, therefore, show the same microstructural changes during subsequent milling. After about 8 h, nearly all the material is deposited on the steel balls and the container wall, but after 16 h the material has broken off again and is available as loose powder. This is accompanied by a drastic change in constitution of the material, as shown above. The transformation from the crystalline layered composite to the amorphous phase mainly takes place between 4 and 16 h milling. Therefore, the transformation from coating to loose powder is interpreted as the result of a drastic change in the deformation behavior. The amorphous alloy is much more brittle than the starting material and, therefore, it no longer sticks to the milling tools. These observations are also revealed by SEM investigations. Figures 3.6a–c show the microstructure of $Ni_{50}Zr_{50}$ powder for different milling times. The decrease of the average thickness of the individual layers to about 100 nm after 8 h is clearly visible. From these images, one cannot distinguish between crystalline and amorphous regions; yet additional EDX analysis provides information on the composition of individual layers and of the whole powder particles and, therefore, gives a hint for the reaction progress. Whereas the pure elements coexist for short milling times, the Ni and the Zr content of the lamellae strongly decreases from 4 to 8 h. After 16 h, the desired average composition of $Ni_{50}Zr_{50}$ is attained, indicating a nearly complete alloying at this stage. Although relatively strong deviations from the average composition show up for different particles after short milling, the powder becomes more and more homogeneous after further milling.

Whereas the microstructural refinement and the compositional changes can be followed by SEM, we can distinguish between crystalline and amorphous material by TEM investigations. Figures 3.7a–c show bright-field images of $Ni_{50}Zr_{50}$ powder milled for 8 h and the corresponding selected area diffraction (SAD) pattern. Again the typically layered microstructure is visible. At the interface between the about 100 nm thick elemental Ni and Zr layers, a thin amorphous interlayer has formed which appears as a grey featureless band in the micrographs. The formation of the amorphous phase is confirmed by SAD (Fig. 3.7c). Besides the reflexes of unreacted Ni and Zr, the typical diffuse halo of the amorphous phase shows up in the diffraction pattern. Obviously, the intense cold working creates a multilayered composite with a layer thickness characteristic for evaporated diffusion couples [3.15] or mechanically co-deformed composite wires [3.17, 25] before solid-state amorphization. For comparison, Fig. 3.8 shows a TEM picture of mechanically co-deformed and partially reacted Ni–Zr composites obtained by Eckert et al. [3.26], which clearly demonstrates that, for amorphization by an interdiffusion reaction in these layered composites, the amorphous phase is formed at the interfaces between the two elements. The similarity of the two TEM pictures (Figs. 3.7 and 3.8) gives further evidence that the glass-forming processes are closely related. A decrease of the elemental layers to atomic dimension is not observed and, therefore, the amorphization cannot be ascribed to a mixing of the pure elements on an atomic scale. During further milling, the quantity of amorphous material increases until after 60 h the powder is fully amorphous.

X-ray diffraction patterns (Fig. 3.9) of these powders, taken after different milling times, confirm these observations. The intensity of the crystalline Ni and Zr diffraction lines is gradually reduced with increasing processing time. The crystalline lines are broadened due to the decreasing crystallite size during milling. After 8 h, a diffuse maximum has grown separately at a position between the lines of elemental Ni and Zr. After 16 h, the clearly visible maximum of the amorphous phase shows up at about $2\theta = 39°$ besides some weak crystalline diffraction peaks of unreacted Ni and Zr. The remaining crystalline reflexes are reduced in intensity due to the reduction of the amount of elemental Ni and Zr, which react to amorphous Ni–Zr. A continuous decrease of the crystallite size to

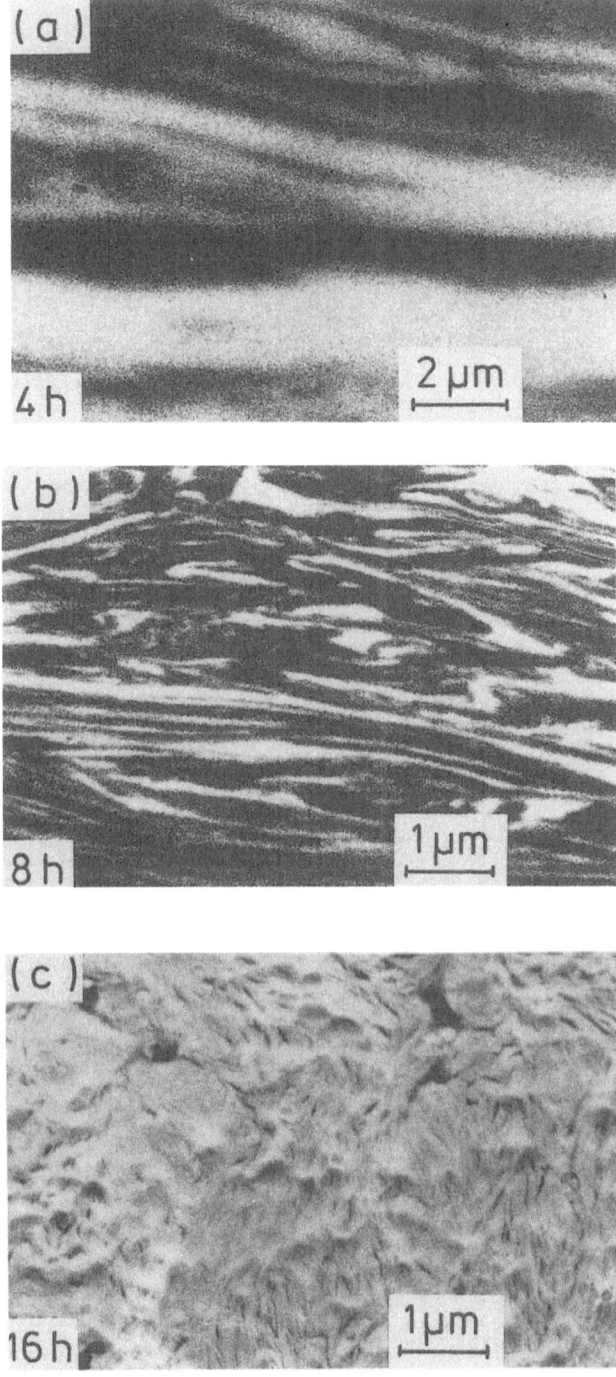

Fig. 3.6. SEM micrographs of Ni–Zr powder particles after (a) 4 h, (b) 8 h, and (c) 16 h milling time

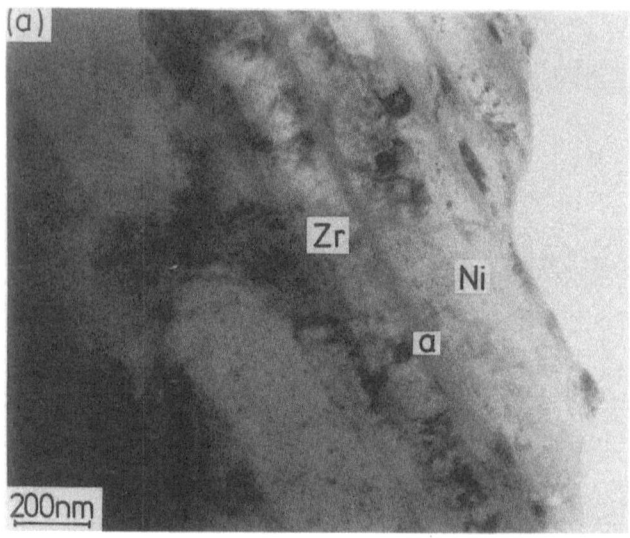

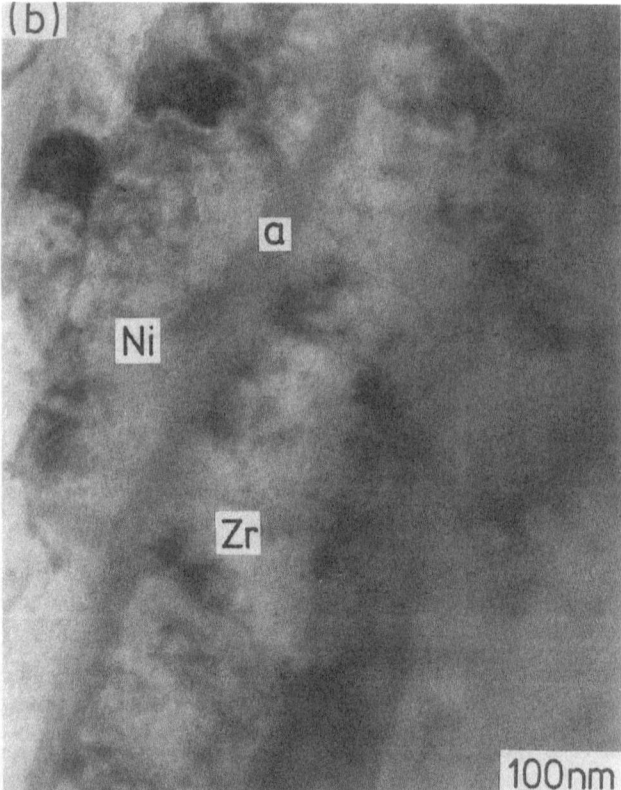

Fig. 3.7. Bright-field TEM micrographs and corresponding SAD pattern of mechanically alloyed $Ni_{50}Zr_{50}$ after 8 h milling time: (a) survey over a larger area of the sample, showing the typical layered microstructure; (b) detail of (a) showing that the amorphous phase (featureless grey band) has formed at the interface between the elemental layers.

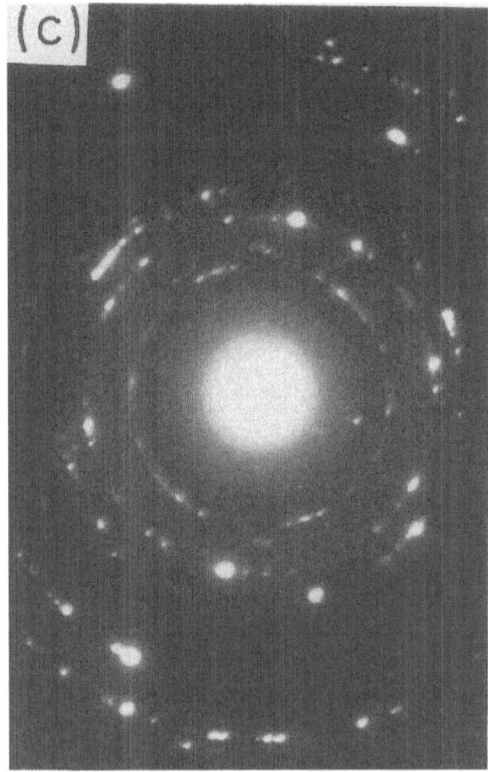

Fig. 3.7. (c) Corresponding SAD pattern showing the characteristic diffuse halo of the amorphous phase besides reflections of unreacted, crystalline Ni and Zr

the atomic scale is not observed. This agrees well with results obtained for the Fe–Zr alloy system [3.24]. After 15 h of milling, Fe–Zr powder consists of an ultrafine layered composite with an individual crystallite size of about 20 nm. Similarly, recent experiments by *Hellstern* et al. [3.27] show that for elemental Ru and for the intermetallic AlRu phase, which remain both crystalline during milling, the crystallite size saturates at 12 or 7 nm, respectively. This is explained by considering the Hall–Petch relation describing the dependence of the yield stress on the grain size. For a small grain size, a very high stress is required to maintain plastic deformation via dislocation motion. For instance, further deformation of 10 nm crystallites would require a yield stress in the order of the theoretical shear stress of the material [3.27], which cannot be supplied under the given milling conditions. Therefore, it is assumed that further deformation occurs by a grain boundary glide mechanism, which fails to refine the micro-structure. This means that ball milling does not lead to an atomic mixing, but

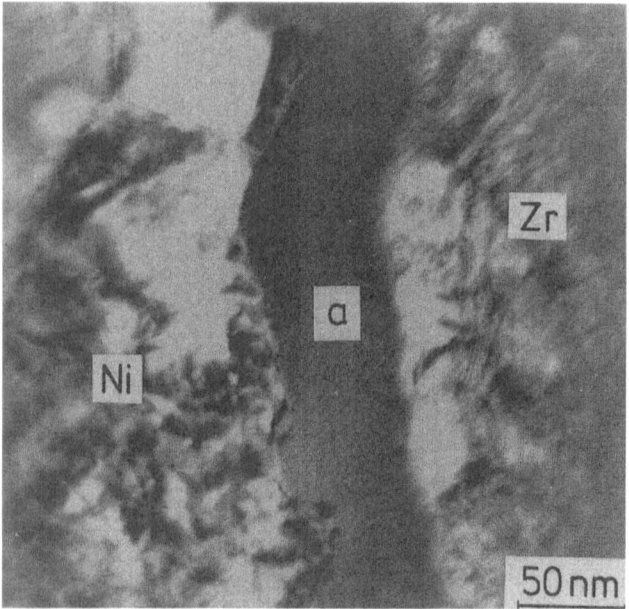

Fig. 3.8. TEM picture of a mechanically co-deformed and partially reacted Ni–Zr composite [3.26]

produces an ultrafine layered microstructure, as has been proved by the TEM investigations. Milling for 60 h finally leads to fully amorphous powder without any indication for residual crystalline material.

Quantitative results on the progress of amorphization reaction can be obtained from saturation magnetization measurements. Whereas amorphous Ni–Zr and pure Zr are not ferromagnetic at room temperature, pure Ni exhibits a saturation magnetization M_s of 54 emu/g at 20°C [3.28]. Since no intermetallic phases are detected in the X-ray diffraction patterns, measuring M_s provides direct information on the residual Ni content in the milled powders. Figure 3.10 shows the saturation magnetization versus milling time for $Ni_{30}Zr_{70}$ and $Ni_{50}Zr_{50}$. After a slight reduction during the first 2 h, the magnetization values strongly decrease for both compositions between 2 and 16 h processing time. For longer milling, the magnetization completely disappears and after 60 h no elemental Ni is left. The slight initial decrease is not caused by the amorphization reaction, but is related to the decreasing crystallite size. Because of small crystallites, the relative volume fraction of grain boundaries compared to bulk material increases. The enhanced density of grain boundaries results in a decrease of the magnetization as shown for nanocrystalline Fe [3.29]. An estimate of the fraction of Ni atoms at the grain boundaries gives a value of 3.6%, if the individual crystallites are assumed to be spherical particles with

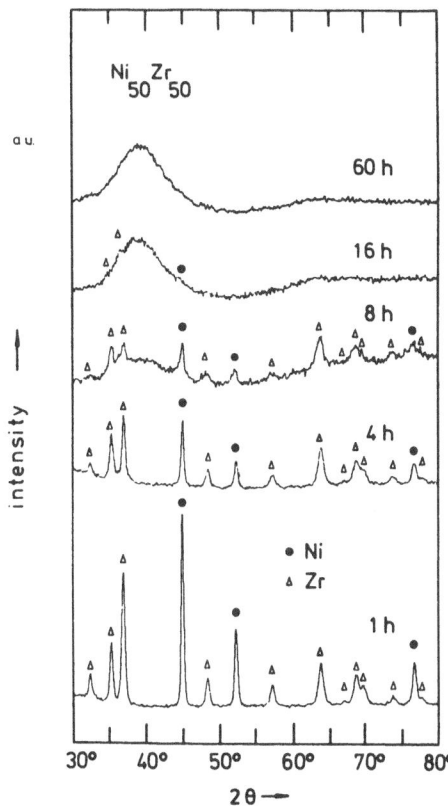

Fig. 3.9. X-ray diffraction patterns for $Ni_{50}Zr_{50}$ after different milling times

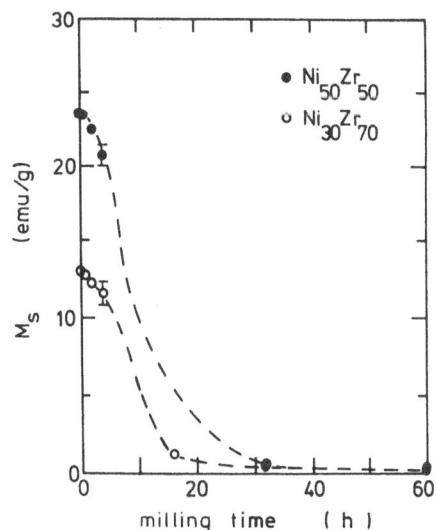

Fig. 3.10. Saturation magnetization M_s of $Ni_{50}Zr_{50}$ and $Ni_{30}Zr_{70}$ vs milling time

40 nm in diameter, a result in good agreement with the observed M_s reduction of 3.8% and 5% for $Ni_{50}Zr_{50}$ and $Ni_{30}Zr_{70}$ after 2 h of milling.

Assuming that the fraction of amorphous material X_a is proportional to the consumption of elemental Ni, X_a is given by $[M_s^{Ni}(t = 0) - M_s^{Ni}(t)]/M_s^{Ni}(t = 0)$. X_a strongly increases from 2 to 16 h and remains nearly constant after 32 h of milling (Fig. 3.11). This is in good agreement with results obtained from hydrogen storage investigations of mechanically alloyed Ni–Zr [3.30]. The transformation into the amorphous state can be described by a common curve for both compositions, indicating that the amorphization proceeds in the same way for both alloys. After 7 h, about half the material is amorphous. The dashed line is the derivative dX_a/dt of the solid line and indicates that the maximum amorphous-phase production rate (APPR) occurs at 7 h of milling.

Further information on the formation of the amorphous phase is obtained by DSC measurements. The measured crystallization enthalpy ΔH_x of the samples is proportional to the amount of amorphous phase present in the material after milling and heating to crystallization. The normalized fraction of amorphous material X_a and the APPR calculated from the crystallization enthalpy seems to be similar to what is observed for the saturation magnetization measurements shown above. But, since during heating in the DSC additional solid-state amorphization occurs at the interfaces of the partially reacted powders produced in the early stages of milling, the measured crystallization enthalpy also contains this additional contribution. This is illustrated in Fig. 3.12a, where the DSC scans (40 K/min) of partially reacted powders after 4 h milling are shown. All samples exhibit several exothermic maxima. This is nearly identical to what is observed for deformation-produced Ni–Zr composite wires with thick starting crystalline layers (Fig. 3.12b). As recently shown [3.31, 32, 26], the increase in the reaction rate at low temperatures originates from the interdiffusion of Ni into the Zr layers and the formation of an amorphous phase.

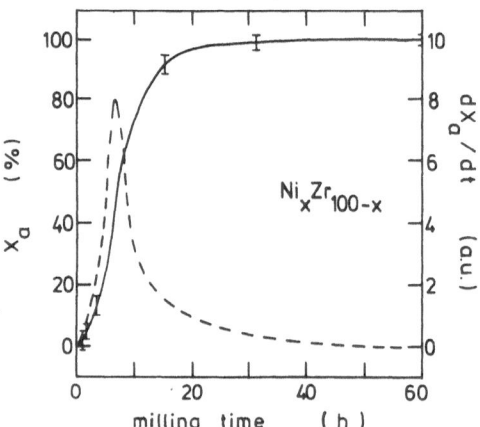

Fig. 3.11. Normalized fraction of amorphous material X_a derived from the saturation magnetization vs. milling time. The dashed line corresponds to the amorphous-phase production rate dX_a/dt

The amorphization reaction slows down as the thickness of the amorphous interlayer increases, and the intermetallic NiZr phase forms when the amorphous interlayer exceeds a critical thickness d_a^{crit}. Due to the much slower diffusion in the intermetallic phase, the reaction rate decreases. The second increase above 350°C is attributed to the reaction kinetics accelerated again at higher temperatures. Besides the growing NiZr phase, the intermetallic $NiZr_2$ phase forms. The consumption of Ni leads to the second decrease of the reaction

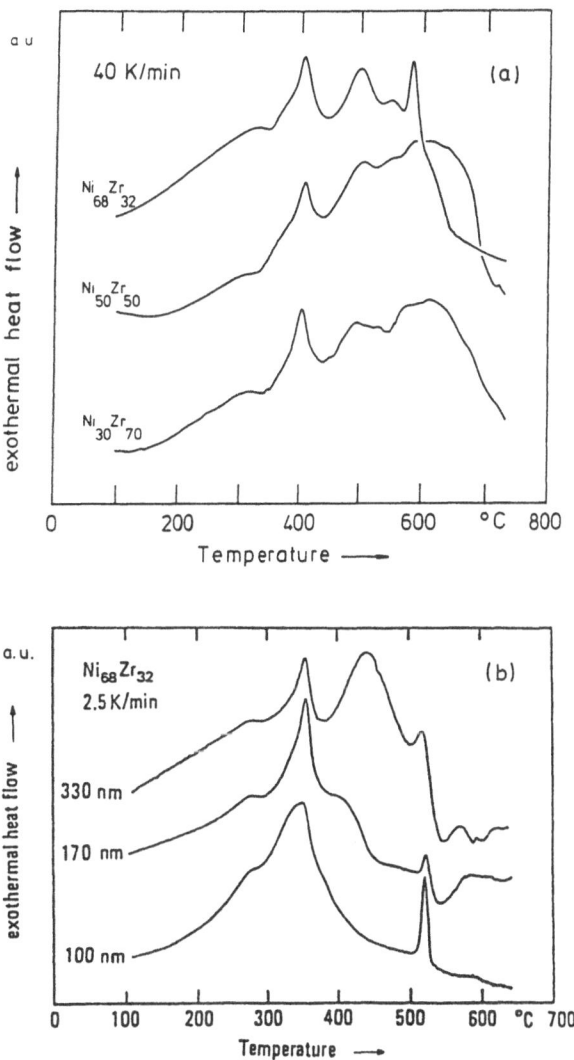

Fig. 3.12. (a) DSC traces (40 K/min) of partially reacted $Ni_{30}Zr_{70}$, $Ni_{50}Zr_{50}$ and $Ni_{68}Zr_{32}$ after 4 h milling. (b) DSC traces (2.5 K/min) of unreacted $Ni_{68}Zr_{32}$ composites with different elemental layer thickness

rate. Finally, at 520°C (2.5 K/min) the amorphous phase, that has remained unchanged up to this temperature, crystallizes.

The obvious similarity between the DSC scans of partially reacted powders and composites with thick elemental layers reveals that the interdiffusion reaction during constant-rate heating proceeds in the same way for both types of samples. Therefore, the measured crystallization enthalpy for short milling times does not give the amount of amorphous material formed by mechanical alloying (as proposed in [3.33]), but also contains the contribution of the propagating amorphization reaction during the DSC run. The fraction of amorphous phase X_a^{DSC} after milling time t_m is given by

$$X_a^{DSC} = [d_a^{crit}(dT/dt) / d(t_m)] \ ,$$

with $d = (d_{Ni} + d_{Zr})/2$. In particular, X_a^{DSC} is nearly independent of what became amorphous during milling before the DSC run. Furthermore, the critical layer thickness d_a^{crit} depends on temperature – i.e., $d_a^{crit}(T_1) > d_a^{crit}(T_2)$ for $T_1 < T_2$ – and is attained at lower temperatures, if the as-milled material exhibits an amorphous interlayer. The fraction of amorphous phase is, therefore, markedly overestimated for short milling times, and the crystallization enthalpy gives no quantitative information on the amount of amorphous material formed during milling.

Only conclusions on the layer thickness can be drawn from these results. Since the thickness of the individual layers decreases with increasing milling time, the clearly visible two-step process in the DSC scans becomes less pronounced with longer processing and finally changes to the single reaction peak characteristic for thin elemental layers (Fig. 3.12b). This is shown in Fig. 3.13 for $Ni_{68}Zr_{32}$ powder after different milling times. Accordingly, the crystallization peak at 577°C (40 K/min) becomes more and more pronounced with additional processing, indicating an increasing amount of amorphous material formed during the mechanical alloying and the subsequent constant-rate heating. The higher crystallization temperature compared to $T_x = 520°C$ for the composites results from the higher heating rate used for the mechanically alloyed powders. After 16 h milling, no reaction maximum but only a sharp crystallization peak appears in the DSC scan, revealing a nearly complete amorphization by milling only.

So far, it has been shown that mechanical alloying creates an ultrafine layered microstructure of the two crystalline elemental metals as an intermediate product. The free enthalpy of this initial state is higher than that of the amorphous state (compare Fig. 3.4) if the alloy system has a negative heat of mixing term. It can, therefore, be lowered by an interdiffusion reaction if the temperature during processing becomes high enough to allow the diffusional process. To check this, we investigated the influence of the milling intensity on glass formation for Ni–Zr [3.34]. Figure 3.14 shows the X-ray diffraction patterns of mechanically alloyed Ni–Zr powders for different compositions milled at intensities 3, 5, or 7, which correspond to velocities of 2.5, 3.6 or 4.7 m/s of the balls during the collisions. (In these experiments a Fritsch

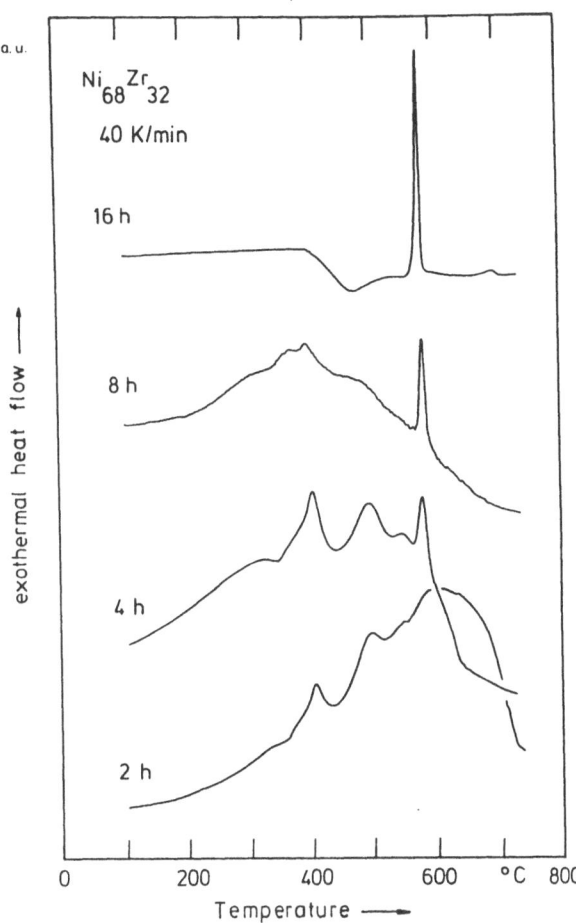

Fig. 3.13. DSC traces (40 K/min) of mechanically alloyed $Ni_{68}Zr_{32}$ after 2, 4, 8, and 16 h processing time

Pulverisette 5 planetary ball mill was used). For the low milling intensity 3, a complete amorphization could not be achieved within 60 h of milling as revealed by the residual crystalline lines. For intensity 5, complete amorphization occurs between about 30 and 82 at.% Ni, whereas for intensity 7 the crystalline intermetallic Ni_3Zr phase appears from 66 to 75 at.% Ni. It must be concluded that, at intensity 7, the individual powder particle is heated up considerably during the collision event, so that even crystallization occurs. An estimate of the maximum temperature of the powder particles resulting from shear deformation during collision of the balls, using a model proposed by *Schwarz* and *Koch* [3.19], gives temperatures of 130°C, 250°C and 407°C for milling intensities 3, 5 and 7, respectively [3.34]. Since amorphization by interdiffusion is always a faster process than crystallization for these layered composites of early and late

Ni$_x$Zr$_{100-x}$

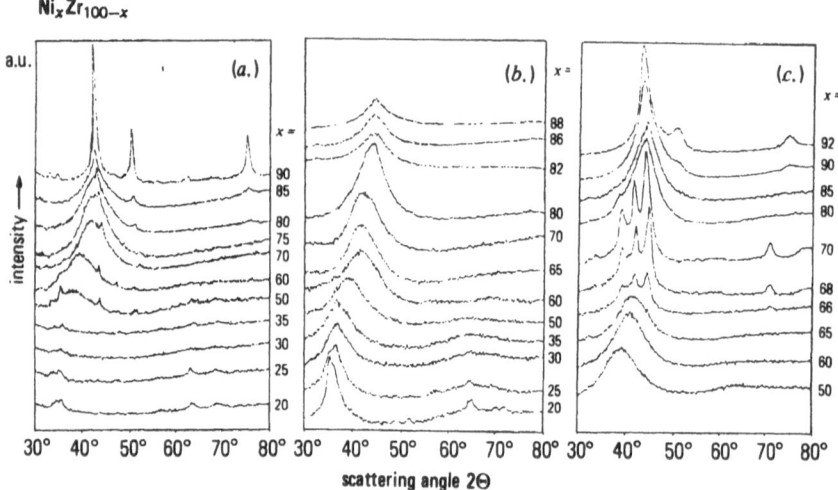

Fig. 3.14. X-ray diffraction patterns of mechanically alloyed Ni$_x$Zr$_{100-x}$ powders milled at different intensities [3.32]: (a) intensity 3; (b) intensity 5; (c) intensity 7

transition metals, the temperature reached in the powder particles during milling is sufficient for the interdiffusion reaction [3.35].

These results, i.e., the formation of an ultrafine layered microstructure, the lower limit of the crystallite size, the TEM observations, and the fact that considerably high temperatures are reached in the powder particles, indicate that glass formation by mechanical alloying is due to an interdiffusion reaction in a similar way as for amorphization by solid-state reaction in layered composites which were artificially prepared by evaporation [3.4] or mechanical co-deformation [3.17]. Further evidence is given by the investigation of the glass-forming ranges which will prove that the reaction takes place under a metastable equilibrium.

3.4 Structural Characterization

The formation of the amorphous phase is generally detected in the case of mechanically alloyed samples by X-ray diffraction (see previous Sect.). However, this method does not always produce unambiguous results since an extremely fine-crystalline material may also yield similar diffraction diagrams. Although DSC investigations suggest the initial formation of the amorphous phase as a result of the crystallization behavior, other processes such as the annealing of lattice defects can have a similar effect energetically. TEM investigations [3.36, 37] make it possible to detect the amorphous phase unambiguously. However, this does not yield any data on structural details.

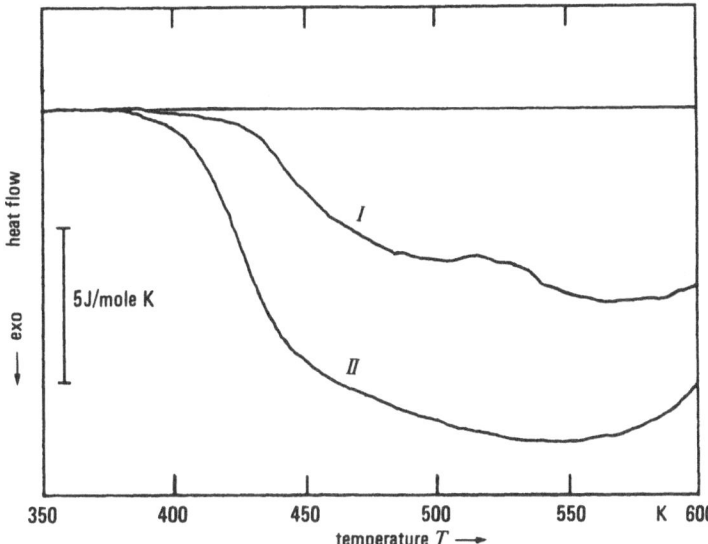

Fig. 3.15. DSC plot showing structural relaxation in mechanically alloyed and rapidly quenched Ni–Zr [3.39]

A detailed neutron–diffraction study to compare the atomic structure of amorphous Ni–Zr alloys prepared by mechanical alloying, melt spinning and sputtering was performed by *Kuschke* et al. [3.38]. From the pair correlation functions, the atomic distances, and the coordination numbers, it follows that the differently prepared amorphous alloys show nearly the identical neighborhood in the first coordination sphere, but differ slightly in the second and higher spheres. For example, a mechanically alloyed $Ni_{30}Zr_{70}$ sample gives the coordination numbers $Z_{NiZr} = 5.9$ and $Z_{ZrZr} = 11.4$, compared to respective values of 5.9 and 11.7 for melt-spun $Ni_{31}Zr_{69}$. The mechanically alloyed specimens also show a compound-forming tendency twice as large as that in the melt-spun specimens and stronger for Ni-rich than for Zr-rich alloys [3.38].

The structural relaxation on heating was investigated in detail for these powders [3.39]. In principle, the same behavior is revealed as for rapidly quenched samples (Fig. 3.15). However, the degree of thermal relaxation is two times larger in the case of the mechanically alloyed samples, and this suggests a substantially unrelaxed state. Mössbauer spectroscopy investigations carried out on Fe–Zr samples [3.40] also show that the local structure of mechanically alloyed amorphous samples is identical to that of rapidly quenched or sputtered samples within the glass-forming range from 30 to 78 at.% Fe, which can be concluded from Fig. 3.16, where the average quadrupole splitting $\overline{QS}$ is plotted versus the Fe concentration. The Mößbauer spectra also provide evidence that the Zr-solubility in α-Fe is enhanced to 5 at.% [3.40]. Similar results were obtained for Fe–Zr–B alloys [3.41] (Sect. 3.7.1).

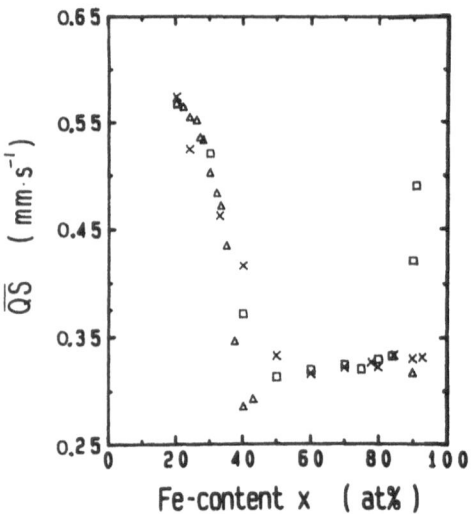

Fig. 3.16. Average quadrupole splitting $\overline{QS}$ vs. Fe concentration [3.40] (□: mechanically alloyed, △: rapidly quenched, ×: sputtered)

The atomic arrangement can also be characterized satisfactorily by investigating structure-sensitive physical properties such as superconductivity or hydrogen absorption. These properties have been investigated in detail, in particular, for mechanically alloyed amorphous Ni–Zr powders. Figure 3.17 shows the superconducting transition temperature T_c of mechanically alloyed Ni_xZr_{100-x} powder compared with that of rapidly quenched samples [3.42]. The almost constant behavior of $T_c(x)$ for $x \leq 25$ indicates the existence of two phases (Sect. 3.6). For $x \geq 30$, the two series of samples reveal the same linear dependence on the composition. The difference between the two curves can be attributed to the different methods of measurement (ac susceptibility versus resistivity) and to an oxygen content of 1–2 at.% in the mechanically alloyed samples. These results provided the first unambiguous confirmation of the amorphous structure of these samples since superconductivity is very strongly dependent on the atomic structure in the case of Ni–Zr. The transition temperature falls markedly even for amorphous samples, which are structurally relaxed by annealing at temperatures below the crystallization temperature. Crystallized samples do not show any superconductivity above 1.3 K. Specific heat measurements of mechanically alloyed $Ni_{30}Zr_{70}$ samples by *Sürgers* et al. [3.43] showed that the samples are bulk superconductors with $T_c = 2.4$ K and a rather homogeneous composition, as inferred from the width of the transition. Compared with melt-spun Ni–Zr, T_c and the electronic density of states are reduced due to the influence of oxygen impurities incorporated during the milling process. Although minor residual element inclusions preclude a definite assignment, it is most likely that all samples show a low-temperature specific heat contribution C_{TLS} of two-level tunneling states of the same magnitude as for melt-spun or vapor-quenched samples [3.44].

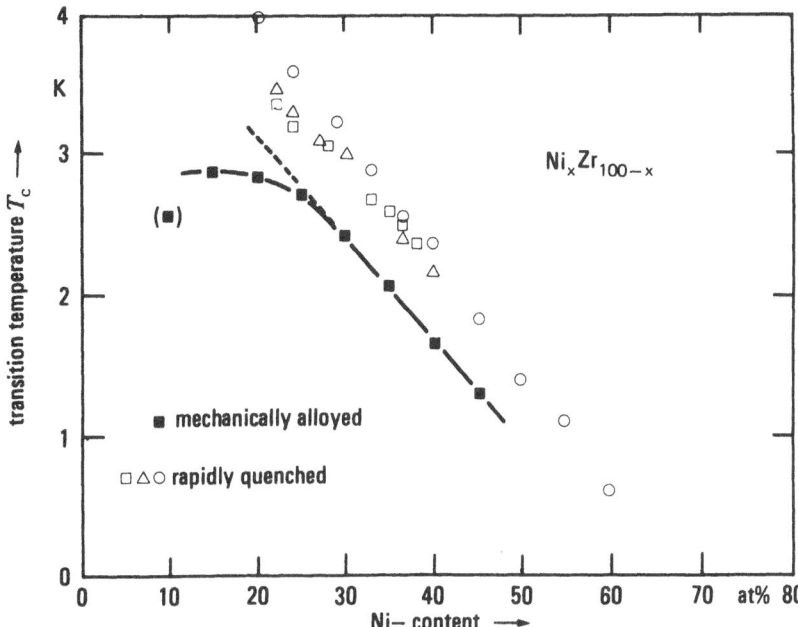

Fig. 3.17. Superconducting transition temperature T_c of mechanically alloyed and rapidly quenched Ni_xZr_{100-x} as a function of Ni content

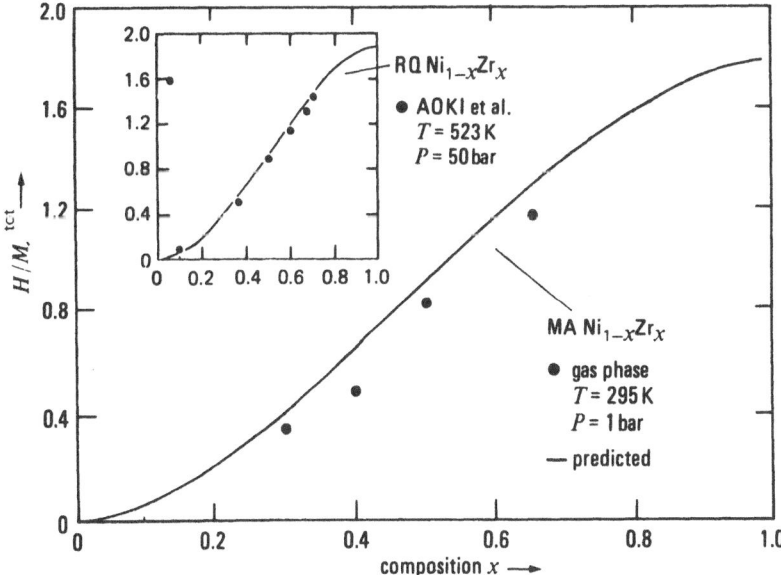

Fig. 3.18. Total hydrogen-to-metal ratio $H/M]^{tot}$ in mechanically alloyed (MA) and rapidly quenched (RQ) $Ni_{1-x}Zr_x$ as predicted (full line) and obtained experimentally (dots) [3.30]

The hydrogen absorption of amorphous metals is strongly dependent on the particular atomic arrangement, in particular on the type and number of the vacancies, the composition and any chemical short-range order. A comparison of the occupation statistics and of the thermodynamics of the hydrogen absorption of mechanically alloyed and melt-spun amorphous Ni–Zr samples by *Harris* et al. [3.30] revealed that the maximum possible loading and the number of sites occupied by hydrogen are identical (Fig. 3.18). No chemical short-range order is observed. However, the peak of the density of states (DOS) for the hydrogen occupation is markedly wider for Ni_2Zr_2 tetrahedral sites and somewhat displaced energetically (Fig. 3.19). Interestingly enough, changes in this width and in the peak position were not observed with annealing. Since hydrogen energetics are extremely sensitive to the size of interstitial sites (on the scale of less than 0.1 Å), the differences in energetics and width of the DOS could be due to intrinsic strain in the mechanically alloyed material or it may reflect slightly different (relaxed) atomic packing characteristics for the two types of alloys [3.30].

Despite of the above differences, these results again show the similarity between mechanically alloyed and melt-spun samples. Maybe more interesting is to look for differing or advantageous properties of mechanically alloyed samples. *Grütter* [3.45] found that our mechanically alloyed $Fe_{91}Zr_9$, material

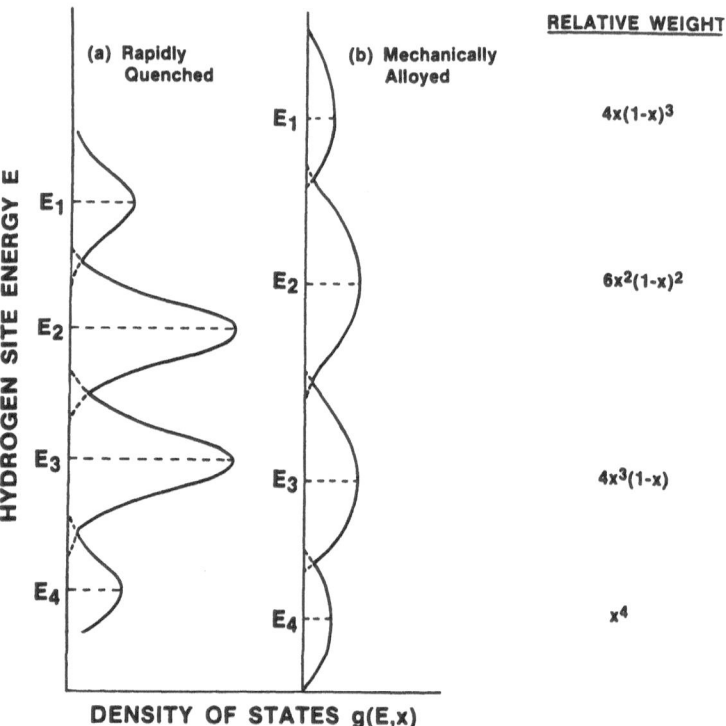

Fig. 3.19. Schematic density of states $g(E, x, T)$ for hydrogen in amorphous Ni–Zr on tetrahedral interstitial sites comparing rapidly quenched (a) and mechanically alloyed (b) materials [3.30]

representing a two-phase mixture of amorphous Fe–Zr and the primary Fe-rich phase [3.22], shows an excellent behavior as an ammonia catalyst which is by far superior to melt-spun or high-vacuum crystallized samples.

3.5 Glass-Forming Ranges

In the preceding sections, the formation of the amorphous phase and its characterization have been described. As shown above, the process of glass formation by mechanical alloying differs completely from rapid quenching; therefore, it is likely that the composition range where glass formation is possible will also be different. In what follows, the concentration range in which the formation of the amorphous phase is possible and the appearance of the resultant phase diagram of the metastable state will be discussed. Figure 3.20 [3.46] shows the phase diagram of the thermodynamically stable state (solid curves, Fig. 3.20a), the phase diagram of the metastable state (dashed curves in Fig. 3.20a) and the free enthalpy curves at the reaction temperature T_r (Fig. 3.20b) for a hypothetical A–B alloy. The equilibrium phases at the temperature T_r are the two primary phases α and β, and the intermetallic phase γ. The phase boundaries are obtained by constructing double tangents on the free enthalpy curves. For T_r, the free enthalpy curve for the amorphous phase (dashed curve) is always above the curve for the stable state. The amorphous phase therefore does not appear at the temperature T_r in the equilibrium phase diagram (to which the solid curves in Fig. 3.20a belong). Experiments have now

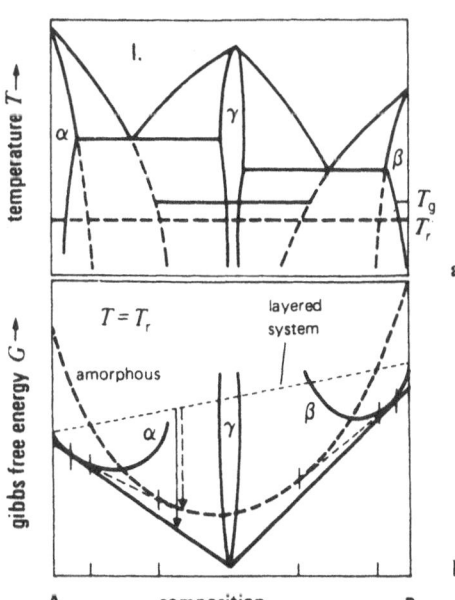

Fig. 3.20. (b) Free-enthalpy curves of a hypothetical A–B alloy (lower) and (a) the resultant phase diagram for the equilibrium (——) and non-equilibrium (– – –) states (upper) [3.46]

shown that the intermetallic phase does not form during the solid-state reaction under consideration. This means, however, that the associated free-enthalpy curves in Fig. 3.20 may be ignored. Consequently, the amorphous phase is the most stable phase in the central part of the phase diagram. The phase diagram of the metastable state (dashed curves in Fig. 3.20a) are obtained by making use of the free enthalpies of the α, amorphous and β phase and again employing the double-tangent construction.

This results in a wide glass-forming range in the center of the phase diagram, two two-phase regions on either side of the amorphous phase, and considerably extended solubilities for the primary phases α and β. In order to check these predictions, the glass-forming ranges for transformation to the amorphous state by mechanical alloying were investigated in detail.

Amorphization by mechanical alloying has been studied in a wide composition range for Ni–Ti [3.5], Cu–Ti [3.7], Fe–Zr [3.22], Ni–Zr [3.42, 34, 47, 48], Co–Zr [3.49, 50], Ni–Nb [3.51], and Nb–Al [3.52] (for a review see [3.8, 9]). In most cases, X-ray diffraction patterns were used to show the formation of an amorphous phase and to determine the glass-forming range, but most of these patterns give only a qualitative description. Again, as for the structural characterization of the amorphous state, it is much more reliable to measure physical properties as a function of composition [3.9, 42]. Especially useful are intensive physical properties such as the superconducting transition temperature (Fig. 3.17) and the crystallization temperature (Fig. 3.21), which are qualitative properties depending on composition within the homogeneity range of the amorphous phase and being constant in the two-phase region. Figure 3.21 gives an excellent example of such a composition dependence [3.42]. Whereas the crystallization temperature of rapidly quenched amorphous Ni_xZr_{100-x} samples decreases monotonically with decreasing Ni content (for $x < 45$), it stays almost constant for mechanically alloyed samples with less than 27 at. % Ni. The phase boundary between the amorphous phase and the two-phase region is therefore determined to be 27 at. % Ni. The plot of the superconducting transition temperature vs. composition (Fig. 3.17) gives the same result. In a similar way, extensive (quantitative) physical properties which depend on composition within the homogeneity range and vary linearly with the amount of phase present in the two-phase region according to the lever rule can be used.

Following these considerations, we can determine the glass-forming range of alloy systems. The results are summarized in Table 3.1. Amorphous metals can be formed by mechanical alloying in the central part of the phase diagrams. To both sides of the amorphous phase two-phase regions exist. The solubility of the primary phases is largely extended (e.g., 5 at. % Zr in Fe [3.22], 7 at. % Ni in Zr [3.42]). This demonstrates that the qualitative predictions based on the above thermodynamic considerations are, in fact, fulfilled; i.e., the results can be at least qualitatively described by a metastable phase diagram constructed from the free enthalpy curves of the two primary phases and of the amorphous phase neglecting the intermetallic phases. For a quantitative evaluation the free-enthalpy curves must be calculated. For instance, *Bormann* et al. [3.53] used the

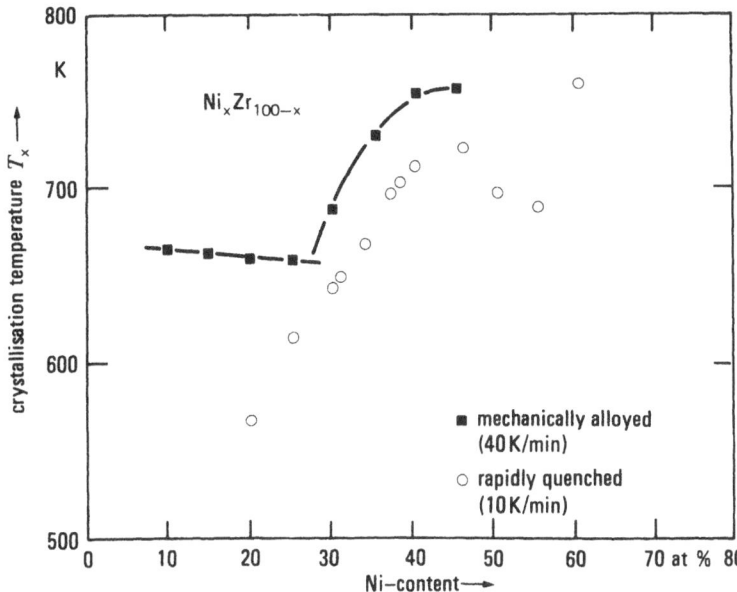

Fig. 3.21. Crystallization temperature T_x of mechanically alloyed (full squares) (40 K min^{-1}) and rapidly quenched (open circles) (10 K min^{-1}) Ni$_x$Zr$_{100-x}$ as a function of Ni content [3.42]

Table 3.1. Experimentally and theoretically obtained glass-forming ranges

	Experimental x [at. %]	Ref.	Predicted Miedema [3.8] x [at. %]	Predicted Calphad [3.53] x [at. %]
Ni$_x$Zr$_{100-x}$	27–83	[3.42, 47]	24–83	33–83
Fe$_x$Zr$_{100-x}$	30–78	[3.22]	27–79	
Co$_x$Zr$_{100-x}$	27–92	[3.50]		
Ni$_x$Ti$_{100-x}$	28–72	[3.5]	24–77	28–72
Cu$_x$Ti$_{100-x}$	10–87	[3.7]	28–15	
Pd$_x$Ti$_{100-x}$	15–58	[3.54]	20–54	
Ni$_x$Nb$_{100-x}$	20–80	[3.51, 55]	31–80	
Ni$_x$V$_{100-x}$	35–55	[3.56]		
Co$_x$V$_{100-x}$	40–67	[3.57]		
Cu$_x$Hf$_{100-x}$	30–70	[3.58]	29–79	
Ni$_x$Hf$_{100-x}$	16–65	[3.58]	25–85	

CALPHAD technique for the Ni–Zr system, predicting the glass-forming range for mechanically alloyed Ni–Zr to extend from 33 to 83 at. % Ni (Table 3.1). *Miedema* calculations [3.59] performed by *Weeber* and *Bakker* [3.8] predict 24 to 83 at. % Ni (Table 3.1). On the Ni-rich side, the value determined experimentally [3.34] hits exactly the value predicted by both theoretical techniques, whereas on the Zr-rich side, the experimental value lies between the values of the two theoretical procedures, although here also the results are very

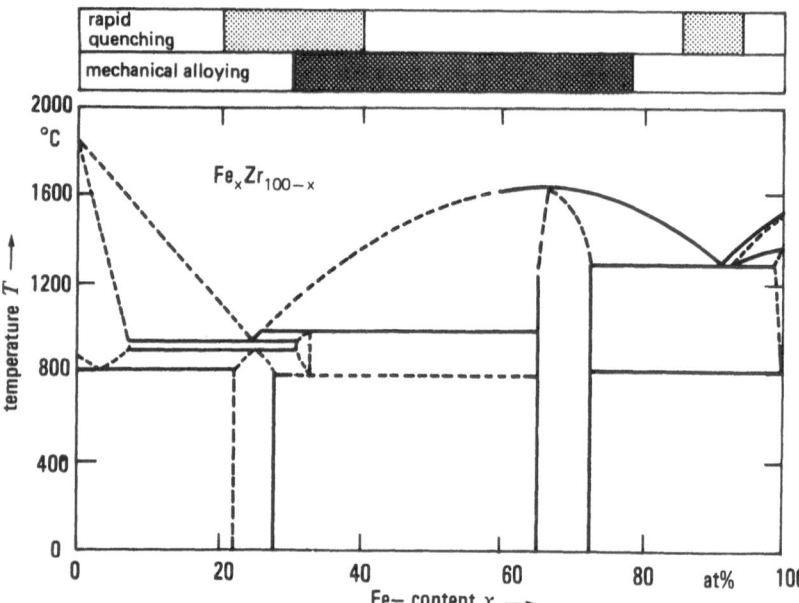

Fig. 3.22. Fe–Zr phase diagram and glass-forming ranges for mechanical alloying and rapid quenching [3.22]

close. These results show that an excellent agreement of the experimentally and theoretically obtained glass-forming ranges is achieved if these are carefully determined. The glass-forming ranges by mechanical alloying can be derived from the free enthalpy curves of the primary phases and the amorphous phase simply by neglecting the existence of intermetallic phases, which are prevented from forming by the kinetic restrictions of the mechanical alloying process.

Finally, the glass-forming range of mechanically alloyed samples should be compared with that of melt-spun samples. The most interesting example for this is the Fe–Zr system (Fig. 3.22; [3.22]), where the glass-forming ranges overlap only in a very small region, indicating again that the two processes are completely different. Whereas rapid quenching is a truly non-equilibrium process allowing glass formation preferentially close to deep eutectics, glass formation by mechanical alloying takes place under a metastable equilibrium preferentially in the central part of the phase diagram, i.e., also in the range of high-melting intermetallic phases. Eutectic compositions of the equilibrium phase diagram do not play a role.

3.6 Glass-Forming Ability

Glass formation by mechanical alloying has been found in a large number of transition metal–transition metal (TM_1–TM_2) alloys, where TM_1 is Cr, Mn, Fe, Co, Ni, Cu, Pd or Ru, and TM_2 is Zr, Ti, V, Cr, Mn, Nb or Hf [3.3, 5–8, 22, 52,

Fig. 3.23. X-ray diffraction patterns of several $TM_{60}Ti_{40}$ samples after mechanical alloying [3.61]

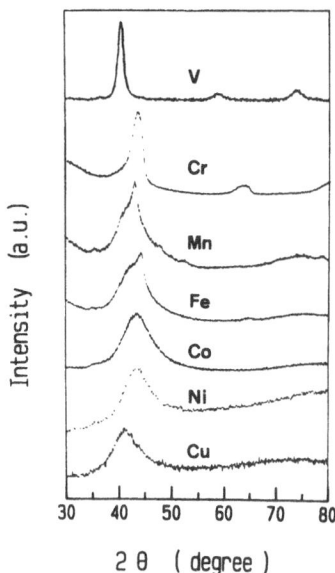

56, 57, 60–63]. A systematic study of the glass-forming ability has been performed for 3d transition metal alloys with Zr [3.8, 61], Ti [3.8, 63], Hf [3.8] and V, Mn or Cr [3.56, 57]. Figure 3.23 shows the X-ray diffraction patterns of $TM_{60}Ti_{40}$ alloys (where TM varies from Cu to V) after mechanical alloying. A complete amorphization is achieved only in the Ni–Ti, Cu–Ti and Co–Ti systems. The Fe–Ti, Mn–Ti and Cr–Ti samples exhibit a remaining crystalline portion increasing in this sequence. In the V–Ti system, the crystalline solid solution forms, which is stable over the whole composition range at higher temperatures and which, therefore, has no nucleation barrier. Within the TM–Zr alloy series with the same TM elements, only Cr–Zr and V–Zr do not become amorphous [3.61]. Similar results have been obtained for Ni–TM or Co–TM alloys [3.56, 57].

Whereas Ni–V and Co–V become fully amorphous, the Ni, Co–Cr, Mn samples do not become amorphous but form extended solid solutions. This is shown in Fig. 3.24 for the Co–Cr alloy system as a typical example. The diffraction patterns reveal the formation of the primary α-Cr phase (bcc) up to 40 at. % Cr and the primary α-Co phase (hexagonal) shows up for more than 60 at. % Co. From 40 to 60 at. % Co both phases coexist. This situation is illustrated in Fig. 3.25, where the lattice spacing d for the closest-packed lattice planes (d_{110}^{bcc} and d_{101}^{hex}) is plotted versus cobalt content. For less than 40 at. % Co the lattice spacing increases to the value of pure chromium. For more than 60 at. % Co, the lattice spacing decreases to the value of pure Co. Annealing in the DSC does not lead to structural changes for alloys below 40 at. % Co or above 60 at. % Co. However, the samples in the two-phase region from 40 to 60 at. % Co transform to the tetragonal σ-CoCr phase [3.57], as expected from the equilibrium phase diagram [3.64, 65].

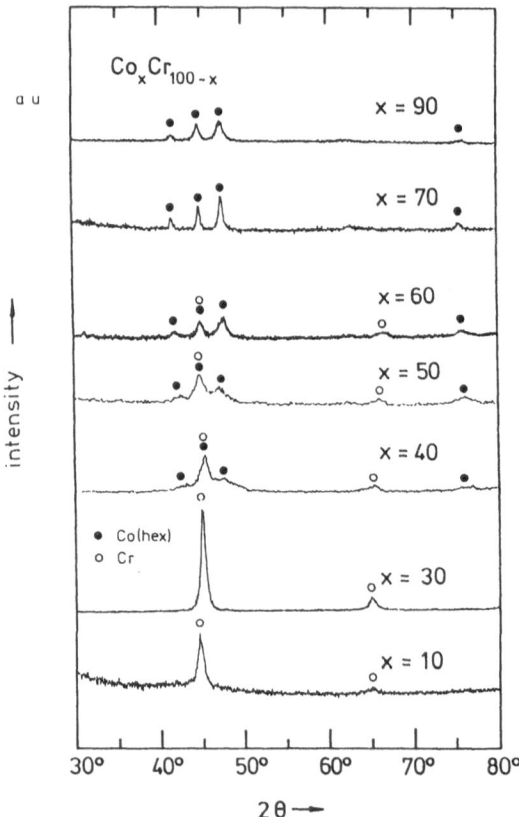

Fig. 3.24. X-ray diffraction patterns of several CO_xCr_{100-x} samples after 60 h of mechanical alloying [3.57]

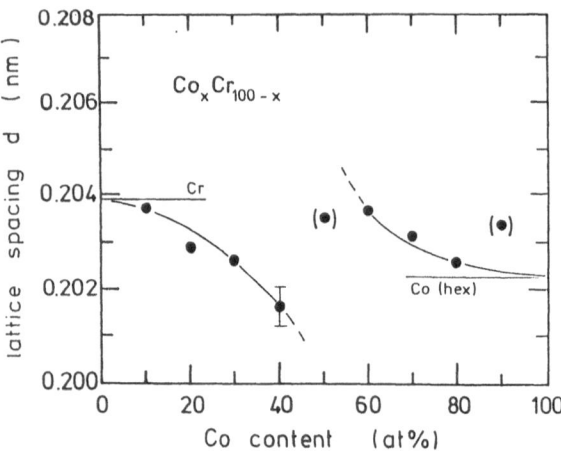

Fig. 3.25. Lattice spacing d of closest-packed lattice planes vs. Co content for mechanically alloyed Co_xCr_{100-x} [3.57]

The different alloying behavior can be explained by considering the enthalpy of mixing as estimated using the *Miedema* model [3.59]. For alloy systems with a large negative enthalpy of mixing, the glass-forming range is expected to be wider than that of systems with a relatively small one. Maximum values for the negative mixing enthalpy calculated for a composition of approximately $Co_{55}TM_{45}$ are shown in Fig. 3.26, together with the observed glass-forming range ΔGFR. For comparison, the values for the 4d element zirconium are plotted together with the data for the early 3d transition metals chromium, manganese and vanadium. Since the tendency of mixing increases from chromium to zirconium, the widest glass-forming range is expected for Co–Zr. Accordingly, the experimentally obtained glass-forming range in the Co–Zr alloy system extends from 27 to 92 at. % Co [3.47]. Obviously, the enthalpy of mixing for Co–V is also large enough for the formation of an amorphous phase from 40 to 67 at. % Co. Owing to the smaller enthalpy of mixing, the glass-forming range for Co–V is relatively small (Fig. 3.26). For Co–Cr and Co–Mn, the negative enthalpy of mixing as thermodynamic driving force for a solid-state reaction is not sufficient for the formation of an amorphous phase via inter-diffusion, but solid solutions with extended solubilities compared with the equilibrium phase diagram and intermetallic phases are produced by mechanical alloying. The easy formation of solid solutions with extended solubilities or intermetallic phases at low temperatures during milling is due to the small atomic size mismatch between cobalt and chromium or manganese. A similar example is Fe–Nd, where the difference of the free enthalpies between the amorphous phase and the layered composite is positive. Therefore, neither alloying nor amorphization occurs during milling [3.65] (Sect. 3.7.1).

Using a similar approach, *Weeber* and *Bakker* [3.8] showed, that the glass-forming ability is in general governed by two criteria: (1) the system has to have

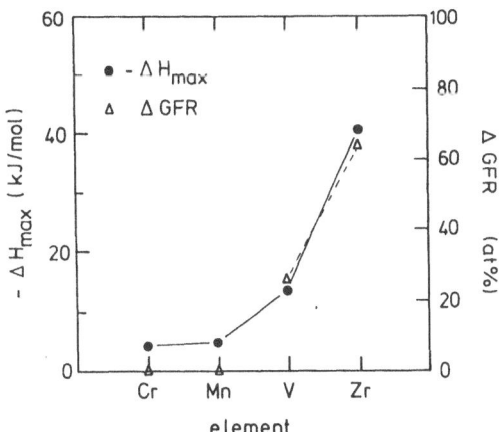

Fig. 3.26. Maximum negative enthalpies of mixing ΔH_{max} in $Co_{55}TM_{45}$ alloys calculated by *Miedema's* model [3.59] and observed glass-forming ranges ΔGFR [3.57]

a negative enthalpy of mixing; (2) the metal-to-host volume ratio must be less than a certain value, which depends on the host (e.g., 0.58 for zirconium and about 0.7 for titanium). The first criterion determines the driving force for the solid-state interdiffusion which leads to amorphization during mechanical alloying. The kinetics of the growth of the amorphous phase by diffusion depends on the second criterion given above. Obviously, there is a strong relation between diffusivity and the volume ratio of the impurity in the host lattice: the larger the volume ratio, the slower the diffusion. When the volume ratio exceeds a critical value, the characteristic time for diffusion becomes larger than the characteristic time for nucleation and growth of a crystalline phase, and crystalline instead of amorphous phases form.

Most of the TM–TM alloys mentioned here can also be amorphized by melt spinning, although the glass-forming ranges are different from those of the mechanically alloyed samples. Ti–Al is an example where an amorphous phase cannot be obtained by melt spinning. Figure 3.27 shows the X-ray diffraction patterns of various Ti_xAl_{100-x} samples after mechanical alloying [3.9]. Only the $Ti_{52}Al_{58}$ and the $Ti_{40}Al_{60}$ samples become amorphous. $Ti_{60}Al_{40}$ and other Ti-rich samples show mainly crystalline material, as the $Ti_{25}Al_{75}$ sample does. The glass-forming range, therefore, extends from about 45 to 65 at. % Al, i.e., it is in the central part of the phase diagram but shifted slightly to the Al-rich side [3.66]. The Ti–Al system also has a sufficiently high negative heat of mixing.

Further binary alloy systems in which amorphization by mechanical alloying has been found are Nb–Al [3.52], Nb–Ge [3.67, 68], Nb–Sn [3.2, 67], Co–Sn [3.69] and Co–Gd [3.70]. Recently, amorphization by mechanical alloying has even been reported for combinations of immiscible elements with $\Delta H_{mix} > 0$ like Cu–Ta [3.71], Cu–V [3.72] and Cu–W [3.73]. But, for these cases, *Yavari* et al. [3.74] stated, that, when dissolved gases such as some 5 at. % oxygen or nitrogen are present (which is typical for Ta, V or W powder) during the mechanical alloying, there is no immiscibility left and the heat of mixing of the amorphized composition becomes negative ($\Delta H_{mix} < 0$). In any case, amorphization by mechanical alloying has been established for a wide variety of alloy systems.

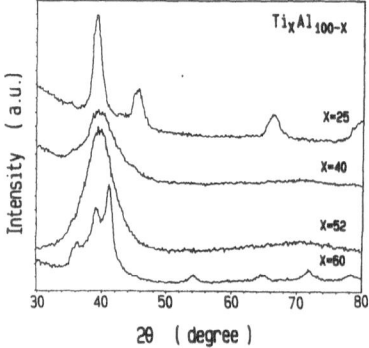

Fig. 3.27. X-ray diffraction patterns of mechanically alloyed Ti_xAl_{100-x} samples [3.9]

3.7 Special Systems and Applications

3.7.1 Boron-Containing Alloys and Nd–Fe–B

The commercially most important group of amorphous metals, the metal-metalloid systems such as Fe–Ni–B, has not been mentioned so far. Initial experiments to form amorphous Fe–B and Fe–Ni–B by mechanical alloying had been unsuccessful. We, therefore, studied the Fe–Zr–B system [3.9, 75]. Milling of elemental Fe, Zr and submicron amorphous B powder first produces a layered microstructure of Fe and Zr. The undeformed boron particles are caught by the colliding Fe and Zr particles and are imbedded in the Fe/Zr interfaces. Further milling leads to a refinement of this layered microstructure until finally the Fe and the Zr layers react to form amorphous Fe–Zr as in the case of binary Fe–Zr (Fig. 3.28). The B particles are still finely dispersed within the amorphous Fe–Zr (Fig. 3.29). An additional solid-state reaction process (2 h at 550°C) enables the boron to diffuse into the Fe–Zr , forming amorphous Fe–Zr–B (Fig. 3.29). The boron addition to amorphous Fe–Zr also increases the crystallization temperature considerably, as shown in Fig. 3.30. For a series of $(Fe_{0.75}Zr_{0.25})_{100-x}B_x$ compositions, perfectly amorphous material can be obtained up to $x = 15$. For 20 at. % B the sample crystallized, possibly because of structural changes when the TM–TM amorphous structure transforms to the TM–metalloid amorphous structure. Using the same process, also amorphous Ni–Nb–B can be formed.

The Fe–Zr–B samples have also been studied by Mößbauer spectroscopy [3.41]. Figure 3.31 shows the average quadrupole spitting $\overline{QS}$ versus the boron content for the as-milled and the reacted (2 h at 550°C) state. $\overline{QS}$ increases during the annealing because of the boron addition. However, the increase of $\overline{QS}$ with boron content for the as-milled samples shows that there is some boron (about 5 at. % B for the $(Fe_{0.75}Zr_{0.25})_{80}B_{20}$ sample) dissolved in the amorphous Fe–Zr during the mechanical alloying.

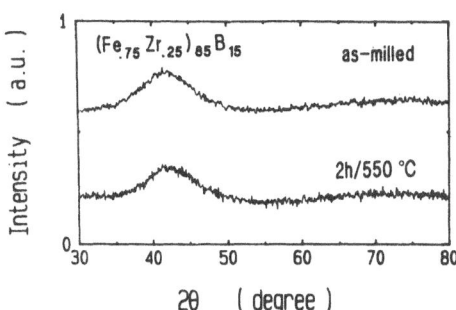

Fig. 3.28. X-ray diffraction patterns of $(Fe_{0.75}Zr_{0.25})_{85}B_{15}$ samples in the as-milled state and after 2 h annealing at 550°C [3.9]

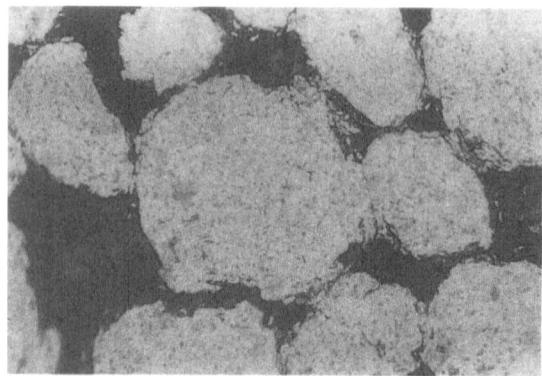

Fig. 3.29. Microstructure of Fe–Zr–B powder particles after 30 h milling time. The boron particles are finely dispersed within the amorphous Fe–Zr [3.75]

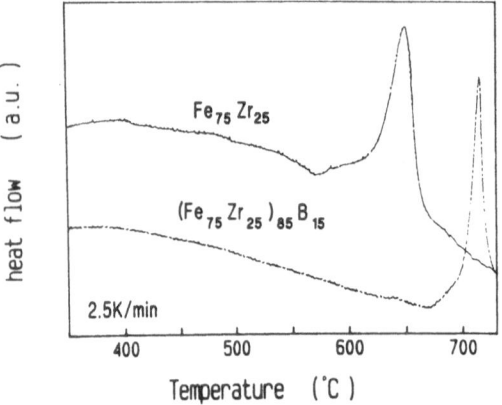

Fig. 3.30. DSC traces showing the crystallization of mechanically alloyed amorphous $Fe_{75}Zr_{25}$ and $(Fe_{0.75}Zr_{0.25})_{85}B_{15}$ at a heating rate of 2.5 K/min [3.9]

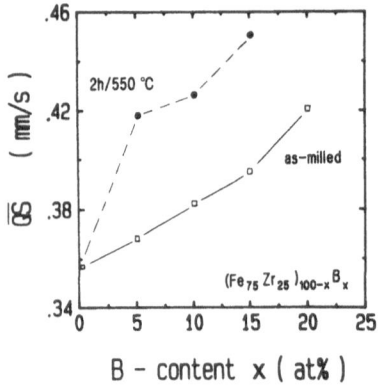

Fig. 3.31. Average quadrupole splitting $\overline{QS}$ vs. boron content for $(Fe_{0.75}Zr_{0.25})_{100-x}B_x$ in the as-milled state and after 2 h annealing at 550°C [3.41]

Milling of binary Fe–B with 20 at.% B also resulted in a fine distribution of boron particles in crystalline Fe powder [3.75]. An annealing of 4 h at 500°C, sufficient for crystallization of amorphous $Fe_{80}B_{20}$, is insufficient for the dissolution of the boron particles. The same effect occurs for Fe–Si–B, where crystalline Fe–Si alloy powders with imbedded boron particles are formed. Therefore, opposite to boron, silicon can be mechanically alloyed with iron. Recently, *Omuro* and *Miura* [3.76] even showed that amorphous TM–Si alloys (TM: Ni, Co, Mo, Mn, Cr) can be prepared by mechanical alloying using a revolution-step-like-decreasing mode. The different behavior of the two non-ductile elements is probably caused by their different hardness. It is extremely difficult to crack the very hard boron particles while imbedded in the soft Fe matrix, although a partial alloying occurs due to interdiffusion (Fig. 3.31). *Calka* et al. [3.77] confirmed our results for Fe–B. Both for $Fe_{80}B_{20}$ and $Fe_{66}B_{34}$ compositions, the milling did not produce an alloying reaction; yet for very high boron contents ($Fe_{50}B_{50}$ and $Fe_{40}B_{60}$), an amorphous structure is found after 300 h of milling. Obviously, sufficient boron can be dissolved in this case, which would correspond to an extrapolation of the $\overline{QS}$ versus boron content plot in the as-milled state in Fig. 3.31. Due to the negative heat of mixing of the Fe–B system, amorphization then takes place.

We replaced Zr by Nd [3.78] in order to investigate the possibility of forming amorphous or at least microcrystalline Nd–Fe–B by mechanical alloying and of evaluating whether this material possesses a similar potential with respect to the magnetic properties as rapidly quenched Nd–Fe–B, which is known to be an excellent hard magnetic material [3.79]. Although the microstructure develops quite similarly during milling as in the case of Fe–Zr–B, the Fe and Nd layers do not alloy due to the thermodynamic relations as reported in Sect. 3.6. Therefore, after milling, the powder particles consist of very fine Fe and Nd layers with imbedded boron particles [3.78]. No amorphous phase is formed (Fig. 3.32a). A solid-state reaction at relatively short annealing times or low temperatures leads to the formation of the hard magnetic $Nd_2Fe_{14}B$ phase (Fig. 3.32b). These powders can be either used to produce resin-bonded isotropic magnets or they can be hot pressed at about 700°C to compacted isotropic magnets. An additional hot deformation (die-upsetting) leads to a mechanical texturing and, therefore, to anisotropic magnets.

Figure 3.33 shows hysteresis loops of mechanically alloyed resin-bonded and compacted and hot die-upset anisotropic $Nd_{16}Fe_{76}B_8$-type magnets with small additions of other elements [3.80]. The anisotropic magnet was measured parallel and perpendicular to its press direction. The resin-bonded sample (its magnetization values relate only to the magnetic powder) exhibits a coercivity of 15.8 kA/cm. The compacted sample (not shown in Fig. 3.33) shows a similar coercivity (16.1 kA/cm). The hot deformation to form the anisotropic sample reduces the coercivity to 10.7 kA/cm, but improves the remanence, the squareness of the magnetization loop and the energy product considerably. The ratio of the remanences measured parallel and perpendicular to the press direction, which is a measure of the degree of alignment, is 3.3. The remanence of the

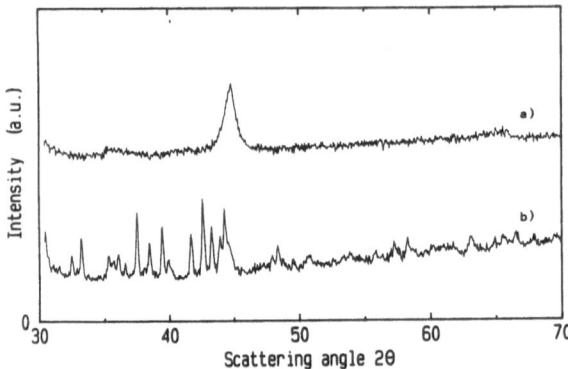

Fig. 3.32. X-ray diffraction patterns of Nd–Fe–B powders; a) after 30 h milling, b) after 30 h milling and 1 h at 600°C heat treatment

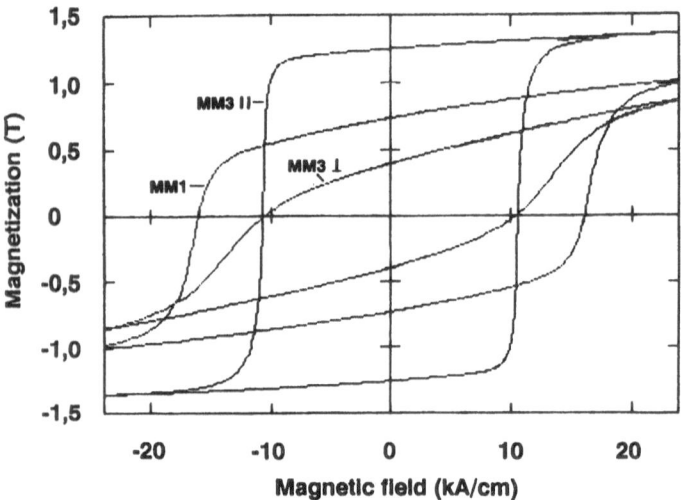

Fig. 3.33. Hysteresis loops of mechanically alloyed resin-bonded (MM1) and compacted and hot die-upset anisotropic (MM3) $Nd_{16}Fe_{76}B_8$-type magnets. The MM3 magnet was measured parallel and perpendicular to the press direction [3.80]

anisotropic sample is 1.25 T and its energy product reaches $295 \, kJ/m^3$ (Fig. 3.33). This demonstrates that the mechanical alloying process is able to provide anisotropic Nd–Fe–B magnets of a quality comparable to the best values of commercial magnets produced either by the powdermetallurgical [3.81] or the rapid quenching technique [3.79]. Therefore, mechanically alloyed Nd–Fe–B permanent magnets have a good chance to become the second large-scale commercial application of mechanical alloying besides the ODS materials.

3.7.2 Sm–Fe–X Alloys – New Permanent Magnets

Whereas mechanical alloying of Nd–Fe–B permanent magnets developed from amorphization studies in the Fe–Zr–B system without involving any amorphization, it will be shown in this section that amorphization by mechanical alloying is essential in the formation of new Sm–Fe–X phases, which show a high potential as novel generation of permanent magnets [3.80].

The Sm–Fe system has a sufficient negative heat of mixing to allow amorphization by mechanical alloying in the central part of the phase diagram. Since a high saturation magnetization is required for magnetic materials, the interesting compositions are on the Fe-rich side of the phase diagram, where a two-phase region between the amorphous phase and primary iron (with a somewhat extended solid solubility for Sm) exists (Sect. 3.5). But most essential for the formation of these new phases is the fact that crystalline intermetallic phases are not present after mechanical alloying, whereas all other competing preparation techniques (like melting or rapid quenching) cannot avoid this.

For ternary Sm–Fe–X alloys, the situation is only slightly modified. As an example, Fig. 3.34 shows X-ray diffraction patterns of as-milled Sm–Fe–Ti powders of different compositions. For $Sm_{24}Fe_{66}Ti_{10}$, complete amorphization

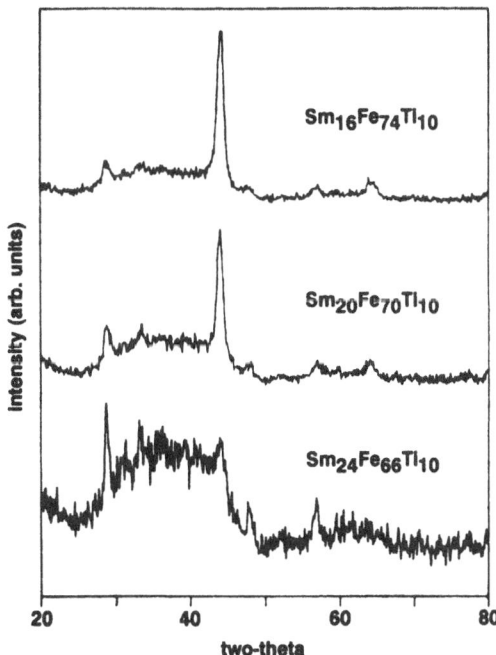

Fig. 3.34. X-ray diffraction patterns of as-milled, mechanically alloyed Sm–Fe–Ti powders of different compositions

takes place during mechanical alloying, whereas the $Sm_{20}Fe_{70}Ti_{10}$ and $Sm_{16}Fe_{74}Ti_{10}$ powders are two-phase mixtures (amorphous plus primary iron) [3.82]. (The additional small diffraction peaks in Fig. 3.34 are due to oxide formation during the X-ray investigations, since the powders were exposed to air.) The hard magnetic phases are formed during a heat treatment in the temperature range from 600°C to 900°C for annealing times between 15 min and 1 h. Depending on the alloy system, the crystallization or reaction process can either lead to equilibrium or metastable phases with a microcrystalline microstructure. For the magnetic measurements, the resulting powder particles are imbedded in epoxy resin to form magnetically isotropic resin-bonded magnets.

1:12 Magnets. Besides the Sm–Co and the RE–Fe–B systems, the best investigated RE-containing phases, which are suitable for permanent magnets, are the $1:12$ phases $(Sm(Fe,TM)_{12})$ (TM: Mo, Ti, V) with the $ThMn_{12}$ crystal structure. The anisotropy field H_A of $Sm(Fe,Mo)_{12}$, for example, is about 9 T, which is higher than that of $Nd_2Fe_{14}B$ and is, therefore, very promising for achieving high coercivities. As in the case of Nd–Fe–B, the microcrystalline $1:12$ magnets need a secondary grain-boundary phase for an effective magnetic hardening. Mechanically alloyed $Sm_{15}Fe_{70}V_{15}$ magnets of the $1:12$-type exhibit coercivities up to 9.4 kA/cm at room temperature. The saturation magnetization of the $1:12$ phase materials is about 1 T, which leads to remanences of the magnetically isotropic material of 0.4–0.49 T, resulting in an energy product of up to 41 kJ/m^3 [3.83]. As shown in general for the $1:12$ magnets [3.84], a Co substitution can rise the saturation magnetization to 1.2 or 1.3 T. The remanence increases correspondingly.

The A_2 phase in the Sm–Fe–Ti system. We tried to reproduce the beneficial effect of an increased Sm content on coercivity also for the $1:12$-type Sm–Fe–Ti system. Surprisingly, the remanence of the samples dropped rapidly to rather

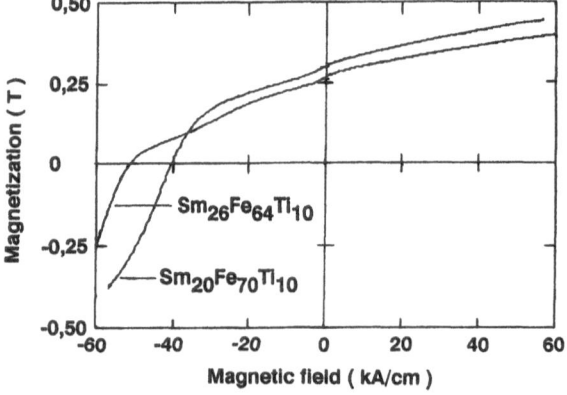

Fig. 3.35. Room-temperature demagnetization curves of resin-bonded $Sm_{20}Fe_{70}Ti_{10}$ and $Sm_{26}Fe_{64}Ti_{10}$ samples prepared by mechanical alloying

small values (0.1 T at 16 at.% Sm), but, at even higher Sm contents (20 at. % Sm), a new phase is formed [3.85], which, earlier, had only been reported for thin sputtered films [3.86]. Figure 3.35 shows the demagnetization curves of magnetically isotropic resin-bonded $Sm_{20}Fe_{70}Ti_{10}$ and $Sm_{26}Fe_{64}Ti_{10}$ samples prepared by mechanical alloying and annealing for 30 min at 725°C. The $Sm_{20}Fe_{70}Ti_{10}$ sample shows a room-temperature coercivity of 40.3 kA/cm and a remanence of 0.3 T. There is only a small step in the demagnetization curve at very low fields indicating the presence of a small amount of a soft magnetic phase. The $Sm_{26}Fe_{64}Ti_{10}$ sample even shows a coercivity of over 51.6 kA/cm, which belongs to the highest coercivities ever observed at room temperature. The crystal structure of this new phase is hexagonal with $a = 2.014$ nm and $c = 1.233$ nm. It resembles the A_2 phase recently found in the Nd–Fe system (Nd_5Fe_{17}) [3.87].

The Sm–Fe–Zr system. In the Sm–Fe–Zr system neither the 1:12 nor the A_2 phase was observed for mechanically alloyed samples with a high Sm content. Instead, a $(Sm,Zr)Fe_3$ phase forms with the rhombohedral $PuNi_3$ crystal structure. Also this phase proved to be hard magnetic when prepared as a microcrystalline material by mechanical alloying [3.82]. Its coercivity reaches 11.8 kA/cm.

Sm_2Fe_{17} nitride and carbide. Due to their limited saturation magnetization (between 0.9 T for the A_2 phase [3.85] and 1.31 T for Co-substituted 1:12 phases [3.84]), these new phases can, at present, not compete with Nd–Fe–B-based magnets. This might be different for the Sm–Fe–N and Sm–Fe–C systems, which were recently detected as new permanent magnet systems by *Coey* et al. [3.88] and *de Mooij* and *Buschow* [3.89], respectively. Whereas Sm_2Fe_{17} with the Th_2Zn_{17} crystal structure has a Curie temperature of only 116°C and an in-plane anisotropy, nitriding or carbonation of this phase to $Sm_2Fe_{17}N_x$ or $Sm_2Fe_{17}C_y$ changes the anisotropy to uniaxial and considerably improves both the Curie temperature T_c and the saturation magnetization.

For $Sm_2Fe_{17}N_x$, it results that its saturation magnetization of 1.51 T is comparable to that of Nd–Fe–B (1.6 T) and T_c (470°C compared to 314°C) and the anisotropy field H_A (22 T compared to 7.5 T) are considerably higher than those of Nd–Fe–B. For $Sm_2Fe_{17}C_y$, the improvement of the intrinsic magnetic properties is not as large as for $Sm_2Fe_{17}N_x$. Whereas both phases can be prepared by means of a gas–solid reaction at elevated temperatures, the carbides can also be produced by arc melting of Sm, Fe and Fe_3C, which results, however, in a maximum carbon content of only $x = 1$ [3.89]. Gas-phase carbonation using hydrocarbon gases leads to the formation of high-carbon compounds with $x = 2$ increasing T_c up to 400°C and the anisotropy field to over 10 T [3.90], which is a lot higher than that reported for the melted carbides.

Figure 3.36 shows X-ray diffraction patterns of the two-phase Sm–Fe powder (amorphous Sm–Fe plus primary iron) after milling (top), of the reacted Sm_2Fe_{17} phase (middle) and of the gas-phase nitrided 2:17 material with the slightly

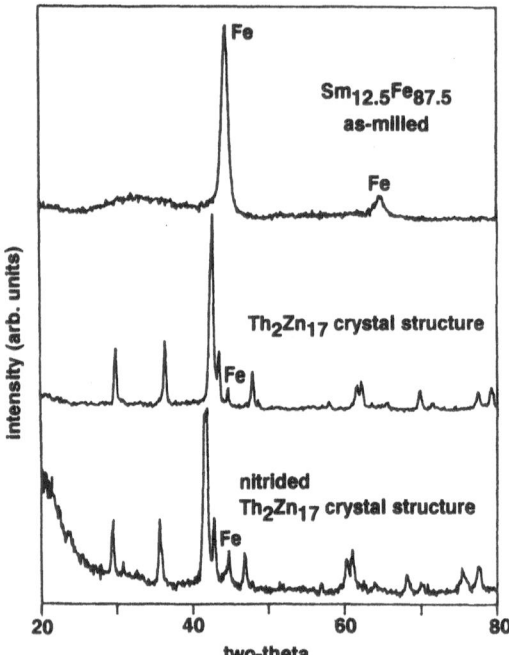

Fig. 3.36. X-ray diffraction diagrams of $Sm_{12.5}Fe_{87.5}$ after mechanical alloying (top), after the formation of Sm_2Fe_{17} by annealing in vacuum (middle), after the nitriding reaction (bottom)

shifted X-ray diffraction peaks (bottom) [3.91]. The absence of any crystalline intermetallic phases after the mechanical alloying seems to be decisive for the formation of the 2:17 phase, although both the reacted and the nitrided samples still contain a small amount of residual iron. Almost identical X-ray results as in Fig. 3.36 (bottom) were obtained for the gas-phase carbonated material.

The room-temperature hysteresis loops of magnetically isotropic, resin-bonded $(Sm_{12.5}Fe_{87.5})_{1-x}N_x$ and $(Sm_{13.5}Fe_{86.5})_{1-y}C_y$ magnets are shown in Fig. 3.37. According to the small amount of iron present in the crystallized powder (Fig. 3.36 middle), the hysteresis loops show a small soft magnetic step in the demagnetization curve. The remanence of the nitride magnet is 0.71 T, which is the same as that of mechanically alloyed isotropic $Nd_{16}Fe_{76}B_8$. The room-temperature coercivity H_c is 23.5 kA/cm and the maximum energy product $(BH)_{max}$ of the isotropic material is 87 kJ/cm³. At 150°C, the coercivity is still 14.4 kA/cm [3.91]. The values for the carbide magnet are somewhat lower: $H_c = 18.5$ kA/cm, $M_r = 0.61$ T and $(BH)_{max} = 59$ kJ/cm³. So far, comparable results have not yet been obtained neither for Sm–Fe–N nor for Sm–Fe–C material prepared by other techniques.

From DSC investigations [3.91], we conclude that the 2:17 nitride and carbide are metastable phases which form at relatively low temperatures where nitrogen or carbon can diffuse into the 2:17 phase, but both the iron and the

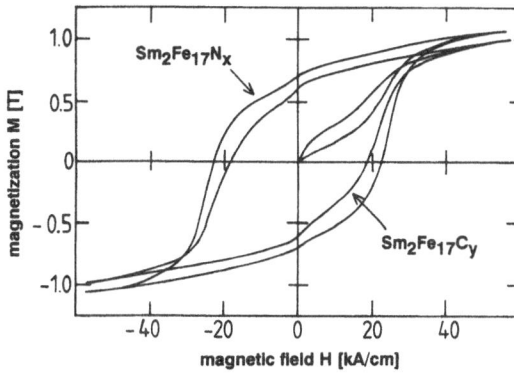

Fig. 3.37. Initial magnetization curves and hysteresis loops of $(Sm_{12.5}Fe_{87.5})_{1-x}N_x$ and $(Sm_{13.5}Fe_{86.5})_{1-y}C_y$ at room temperature

samarium atoms are immobile preventing the phase separation into the thermo-dynamically stable state – a situation similar to amorphization by a solid-state reaction in Zr_3Rh upon hydrogenation [3.16]. This instability of the $2:17$ nitride and carbide prevents the application of standard sintering techniques as used for Nd–Fe–B or Sm–Co for the production of anisotropic magnets.

3.8 Formation of Quasicrystals and Milling-Induced Phase Transitions

3.8.1 Basic Principles

Quasicrystalline phases have received much attention since first reported by *Shechtman* et al. [3.92] in 1984 for Al–Mn. Within the last few years, various preparation techniques have been applied for the preparation of alloys in the quasicrystalline state. Liquid-phase quenching [3.92, 93] and relatively slow cooling from the melt [3.94], sputter or vapor deposition [3.95, 96], ion-beam techniques [3.97–100], heat treatment of the amorphous phase [3.93, 101, 102], and solid-state reaction during interdiffusion [3.103–106]. Recently it was demonstrated that the quasicrystalline phase can be produced by mechanical alloying [3.107–112].

The structure of amorphous metals, quasicrystals, and crystalline inter-metallic compounds can be modelled by atom clusters with icosahedral arrange-ment [3.113–117]. The differences between the various phases result from a different arrangement of the individual atom clusters. Therefore, it is evident that there exists a close relation between the different states of matter, and that the different phases corresponding to minima of the free enthalpy can be quite easily transformed into each other. For example, rapid cooling from the melt results in an amorphous alloy for high quenching rates, and a quasicrystalline phase forms for lower quenching rates [3.118]. Similar results were obtained for sputtered and evaporated samples [3.95, 96, 104, 105] or electron irradiated thin films [3.102,119]. These observations are a strong hint for a close relation between the different phases on the thermodynamic scale.

The basic principles of a solid-state reaction leading to a quasicrystalline phase are very similar to what was described for an amorphization reaction during mechanical alloying (Sect. 3.3.1). As illustrated in Fig. 3.38, the free enthalpy of the equilibrium crystalline state G_x is in general lower than that of the quasicrystalline state G_q for metallic systems below the melting temperature. Although icosahedral Al–Cu–Fe and Al–Li–Cu appear to be thermodynamically stable [3.120–122], most quasicrystalline phases are known to be metastable with respect to equilibrium phases, i.e., an energy barrier exists preventing quasicrystals from spontaneous crystallization. To form a quasicrystalline alloy by a solid-state reaction, it is necessary to create an initial crystalline state G_0 with a high free enthalpy (Fig. 3.38). As mentioned above, this can be realized by the formation of a layered microstructure during the early stages of milling. Starting from this initial state G_0, the free enthalpy of the system can be lowered either by the formation of a crystalline phase (or phase mixture) or by the formation of a metastable amorphous or quasicrystalline phase. Energetically favored is, of course, the crystalline equilibrium phase, but the kinetics of the phase formation decide which phase is in fact formed. The quasicrystalline phase forms if the reaction to this phase proceeds substantially faster than the reaction to the crystalline phase. Due to their higher stability, the quasicrystalline phase forms at a higher reaction temperature compared to the amorphous phase, since the quasicrystalline structure requires a higher atomic mobility of the individual components to form a "quasilattice" exhibiting a higher degree of order than the more random atomic arrangement of the amorphous phase. This temperature is usually in the range of or above the crystallization temperature of the competing amorphous phase, thus preventing amorphous phase formation. Accordingly, the quasicrystalline phase forms if $\tau_{0\to q} \ll \tau_{0\to x}$ ($\tau_{i\to j}$ being the characteristic

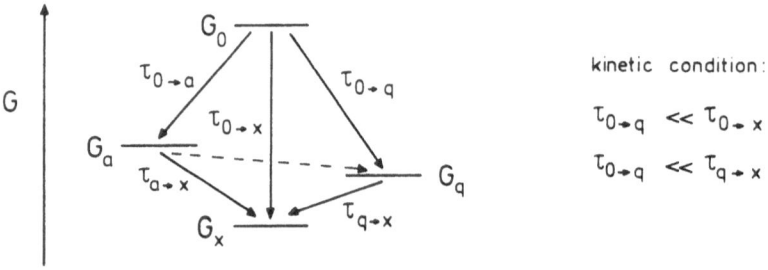

energetically favored: crystalline phase

kinetically favored: quasicrystalline phase

Fig. 3.38. Basic principles of quasicrystal formation by a solid-state reaction

timescale of the reaction). During this reaction, the quasicrystalline phase must not crystallize: $\tau_{0 \to q} \ll \tau_{q \to x}$, i.e., the reaction temperature must be well below its crystallization temperature.

3.8.2 Development of the Microstructure, Formation of the Quasicrystalline Phase, and Formation Range

Quasicrystal formation by mechanical alloying has been observed for several ternary or quaternary TM-alloy systems [3.111, 112]. In the following discussion, the progress of solid-state reaction during mechanical alloying is described in detail for Al–Cu–Mn powders. Figure 3.39 shows the X-ray diffraction patterns of $Al_{65}Cu_{20}Mn_{15}$ after different milling times at intensity setting 7 (velocity of the milling balls = 4.7 m/s). After 25 h of milling, the diffraction patterns exhibit the peaks of elemental Al, Cu, and Mn. With increasing milling time, the elemental lines in the diffraction pattern are broadened due to a reduction in crystallite size during milling and reduced in intensity. Furthermore, additional intensity appears between the elemental peaks in the range from $40° < 2\theta < 45°$. This becomes more and more pronounced until after 90 h of milling the elemental peaks have disappeared. The diffraction pattern shows an intense double peak with maxima at 43.3° and 44.3° and two weaker peaks with maxima at 63.1° and 75.4°. These peaks can be identified as the (1 0 0 0 0 0), (1 1 0 0 0 0), (1 1 1 0 0 0), and (1 0 1 0 0 0) diffraction maxima of an icosahedral phase using the indexing scheme of *Bancel* et al. [3.123].

The development of the microstructure with increasing milling time can be followed by electron microscopy. Figure 3.40 shows a SEM micrograph of an $Al_{65}Cu_{20}Mn_{15}$ powder particle after 70 h of processing. Alternating Cu-rich

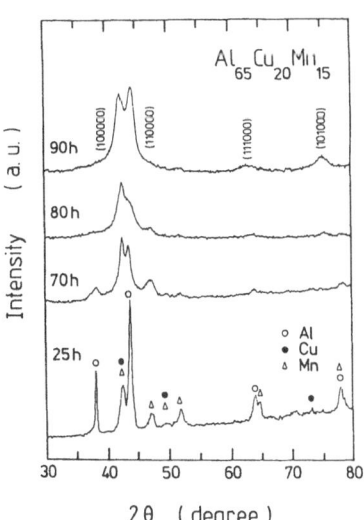

Fig. 3.39. X-ray diffraction patterns of $Al_{65}Cu_{20}Mn_{15}$ after different milling times at milling intensity 7

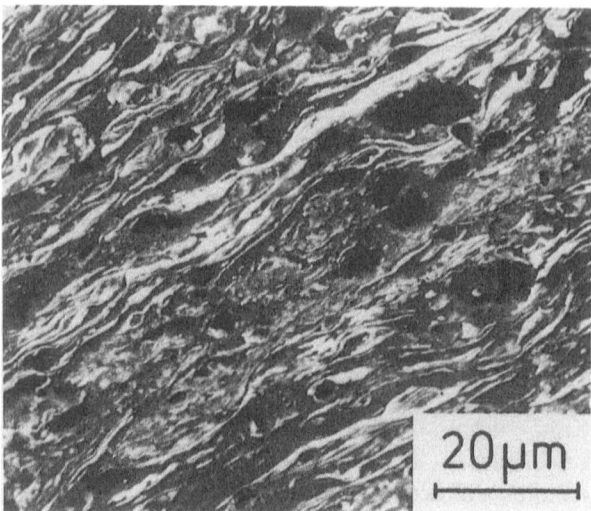

Fig. 3.40. Scanning electron micrograph of an $Al_{65}Cu_{20}Mn_{15}$ powder particle milled for 70 h at milling intensity 7

(bright areas) and Al-rich layers (dark areas) can be observed, both containing a fraction of 5–20 at.% Mn. Furthermore, nearly circular dark areas are visible, resulting from particles extracted from the sample during grinding and polishing. This layered microstructure is typical for mechanically alloyed samples during the early stages of milling. Additional mechanical alloying reduces the thickness of the individual lamellae. After 80 h of milling, the layered microstructure has disappeared, and the powder particles exhibit a nearly featureless image, which is clearly distinguishable from the earlier milling stages. Obviously, the reaction is nearly finished after 80 h of milling, a point confirmed by TEM investigations. Figure 3.41 shows SAD patterns of material milled for 100 h. The figure is a composite of patterns taken at different exposure times permitting identification of weak intensity details. The two characteristic (1 0 0 0 0 0) and (1 1 0 0 0 0) Debye–Scherrer rings are quite sharp and can be well separated. Measurements of the radii of the (1 0 1 0 0 0) and the (1 1 0 0 0 0) rings indicate that their ratio is equal to $\tau = (1 + \sqrt{5})/2$, the golden mean. These features confirm the interpretation of the X-ray results in terms of a quasicrystalline structure. Additional bright-field and dark-field electron micrographs reveal quasicrystal grain sizes ranging from some nanometers up to some ten nanometers (Fig. 3.42). Neither the diffraction pattern nor the images give indications of additional phases formed besides the quasicrystalline phase.

These results, i.e., the reduction in intensity of the elemental X-ray diffraction peaks and the formation of an ultrafine layered composite during the early stages of milling, and the coexistence of the quasicrystalline phase with crystalline material for some period, are quite similar to what is known from amorphous phase formation by mechanical alloying (Sect. 3.3.2). The local

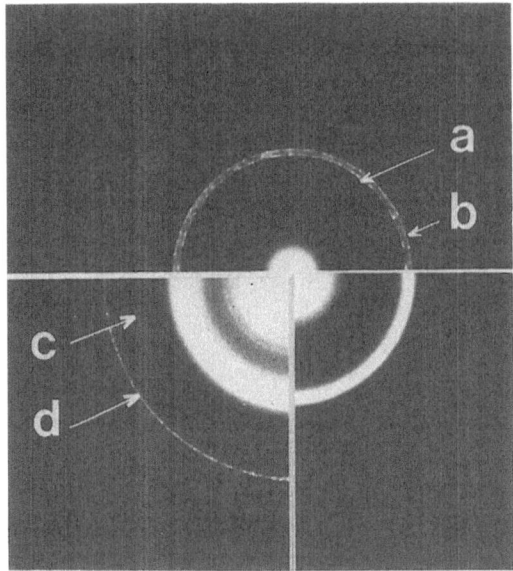

Fig. 3.41. Selected area diffraction pattern of $Al_{65}Cu_{20}Mn_{15}$ powder milled for 100 h at milling intensity 7. Three different exposure times (increasing clockwise) were used for a better identification of weak intensity details. All Debye–Scherrer rings can be indexed in icosahedral notation, e.g., (a) (1 0 0 000), (b) (1 1 0 000), (c) (1 1 1 000), and (d) (1 0 1 000)

temperature during mechanical alloying is, due to the energy dissipation of the colliding balls, high enough to allow the interdiffusion reaction. In particular, this is also the case for the experimental conditions used for the preparation of quasicrystals. The estimated temperature for milling intensity 7 ($\approx$ 407°C) is compatible with the temperature range where quasicrystal formation is observed in multilayer films of Al and transition metals [3.96, 103, 104]. From this, it is concluded that the quasicrystalline phase forms directly by an interdiffusion reaction from the crystalline starting materials. The X-ray diffraction patterns are fully compatible with those reported for melt-spun quasicrystalline material [3.120], and the SAD patterns correspond to those reported for quasicrystalline Al–Mn and Al–V alloys produced by ion beam techniques [3.97, 98], liquid-phase quenching [3.124], or heat treatment of the amorphous phase [3.102].

Figure 3.43 shows a schematic illustration of the formation range of the quasicrystalline phase in mechanically alloyed Al–Cu–Mn. Single-phase quasi-crystalline powder forms for 15–25 at.% Cu and 10–20 at.% Mn. Besides, a two-phase region of quasicrystalline phase and an unknown crystalline phase exists for 30 at.% Cu and 5–10 at.% Mn. The nature of this crystalline phase (in our discussion termed *solid solution*) can not be clearly determined from these preliminary experiments. For $Al_{70}Cu_{25}Mn_5$ a mixture of quasicrystalline material and elemental Cu forms. Samples with low Al content exhibit a single-phase *solid solution*, and Al-rich powders consist of *solid solution* and the pure

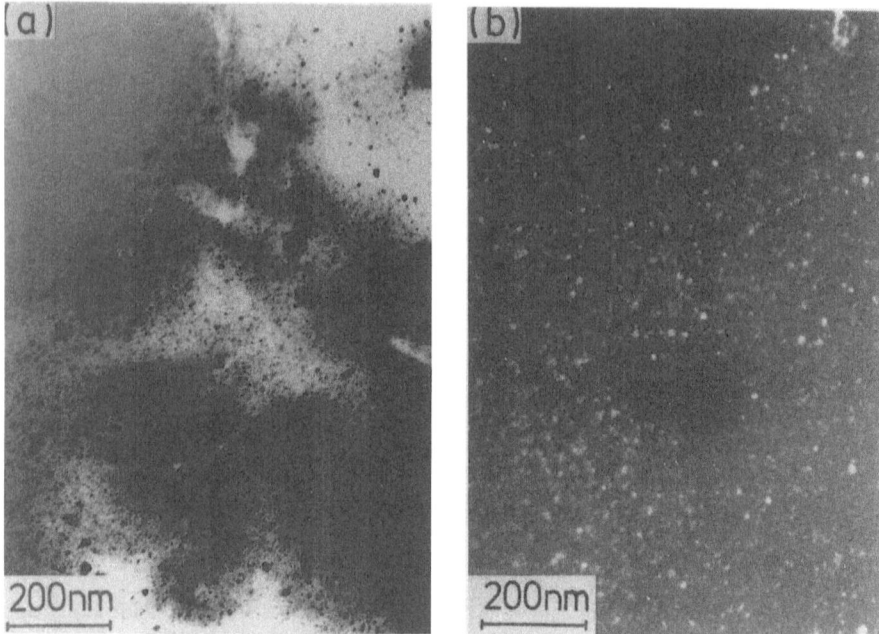

Fig. 3.42. Bright-field (a) and dark-field (b) TEM micrographs of $Al_{65}Cu_{20}Mn_{15}$ powder milled for 80 h and the corresponding SAD pattern (c). Three different exposure times (increasing clockwise) were used for a better identification of weak intensity details

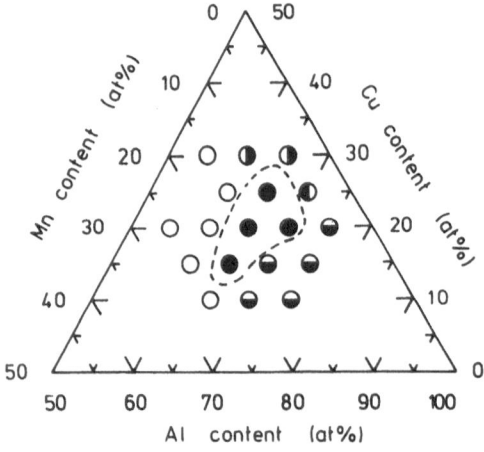

Fig. 3.43. Formation range of the quasicrystalline phase in mechanically alloyed Al–Cu–Mn: (○) *solid solution*; (◒) *elements + solid solution*; (●) quasicrystalline; (◐) quasicrystalline + *solid solution*; (◑) quasicrystalline + elements

elements. The observation of a single-phase quasicrystal composition range near $Al_{65}Cu_{20}Mn_{15}$, which is enclosed by coexisting quasicrystalline and crystalline phases, is similar to what is reported for melt-spun Al–Cu–Mn alloys [3.125, 126].

3.8.3 Influence of the Milling Intensity – Milling-Induced Phase Transitions

The X-ray diffraction patterns in Fig. 3.44 show the influence of the milling intensity on phase formation for $Al_{65}Cu_{20}Mn_{15}$ during mechanical alloying. Milling for 510 h at intensity 5 produces amorphous powder. In addition, a small irregularity at the amorphous diffraction maximum appears, which originates from a small amount of unreacted elemental α-Mn. Obviously, the relatively low reaction temperature is not sufficient for a complete solid-state reaction even after 510 h of milling. Therefore, no steady state is reached and the powder consists of a mixture of amorphous material with lower Mn content compared to the nominal composition of $Al_{65}Cu_{20}Mn_{15}$ and of unreacted microcrystalline α-Mn. Contrary to this, a quasicrystalline phase forms after 90 h of milling at intensity 7. The transformation of the quasicrystalline phase upon heating at temperatures between 50°C and 600°C was studied by DSC. The DSC scan exhibits a broad exothermic maximum from 100°C to 410°C, and a sharp peak at 500°C. Since powder heated to 600°C shows only X-ray lines of an unknown crystalline phase but no hint of icosahedral material, the sharp DSC peak at 500°C is attributed to crystallization. The crystallization enthalpy was found to be 2.2 kJ/mol. The nature of the crystalline phase cannot be determined from these preliminary experiments. Additional X-ray investigations reveal no significant structural changes for powder heated to 410°C, but only a sharpening of the individual X-ray lines. Therefore, the broad low-temperature exothermic maximum of the DSC scan is presumably caused by the relaxation

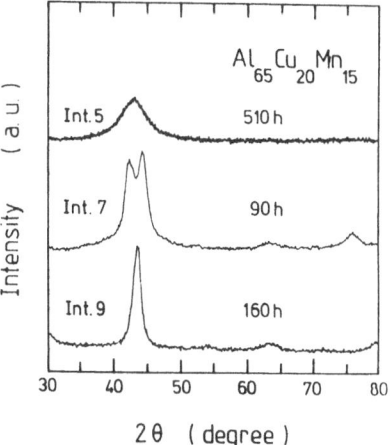

Fig. 3.44. X-ray diffraction patterns for $Al_{65}Cu_{20}Mn_{15}$ milled at intensity 5, 7, and 9 demonstrating the influence of the milling intensity on phase formation

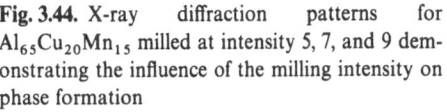

of phason strains introduced during milling [3.127–129], and the growth of the individual particles. Finally, milling for 160 h at intensity 9 results in crystalline material, which remains unchanged during further milling. This is caused by the different reaction temperatures during milling. Milling intensity 5 is thus too smooth, i.e., the reaction temperature is too low, to obtain a quasicrystalline alloy. Nevertheless, the thermal conditions are appropriate for the formation of the less stable amorphous phase. On the other hand, the high reaction temperature provided at intensity 9 leads to the formation of the crystalline phase.

It is not only that, depending on the milling intensity, mechanical alloying of the elemental powders produces the three different structural states, but these states can also be transformed into each other by additional milling at a higher or lower intensity. Figure 3.45 shows the transition of completely quasicrystalline $Al_{65}Cu_{20}Mn_{15}$ powder (90 h of milling at intensity 7) to the amorphous state by further milling (398 h) at the low intensity 3. The broad diffuse maximum of the amorphous state appears instead of the characteristic diffraction pattern of the quasicrystalline phase. Additionally, a small irregularity at $2\theta = 42.3°$ is observed originating from a small quantity of an unidentified metastable crystalline phase. Therefore, a mixture of amorphous phase and metastable crystalline phase forms during further low-intensity milling of quasicrystalline starting powder. Presumably, this can be explained by a mechanism similar to the amorphization of intermetallic phases by ball milling [3.18, 19, 130, 131]. The severe deformation during milling leads to an accumulation of lattice defects rising the free enthalpy of the faulted quasicrystal above that of the amorphous alloy. The amorphous phase represents the energetically favored state under these conditions and starts to form. The mixture of amorphous phase and metastable crystalline phase probably results from a change in composition of the powder by the formation of an amorphous phase with lower Mn-content. Therefore, the formation range of the quasicrystalline phase and the amorphous phase can be considered as not completely compatible. The inverse transition, the formation of quasicrystalline material by the transformation of an amorphous phase, was also demonstrated [3.108].

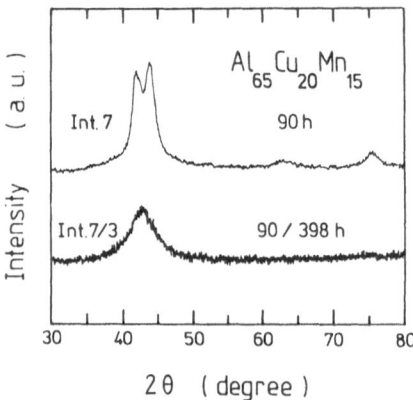

Fig. 3.45. X-ray diffraction patterns for $Al_{65}Cu_{20}Mn_{15}$ milled at intensity 7 and after further milling at intensity 3 demonstrating the quasicrystal-to-amorphous transition

Amorphous powder formed by mechanical alloying for 510 h at intensity 5 was further milled at intensity 7 for 25 h. The typical broad diffuse maximum of the amorphous state disappeared, and the characteristic intensity distribution of the quasicrystalline phase showed up in the X-ray diffraction pattern. Therefore, additional milling at higher intensity led to an amorphous-to-quasicrystal transition. Amorphization can also be achieved for crystalline starting powder mechanically alloyed for 206 h at intensity 9 by a further milling for 433 h at intensity 3 [3.108].

These results reveal the influence of the milling conditions on the phase formation during mechanical alloying. The various possible transitions between composite, amorphous phase, quasicrystalline phase, and crystalline phase are illustrated in Fig. 3.46. Depending on the chosen experimental conditions, an amorphous phase as well as a quasicrystalline or a crystalline phase can be produced directly by a solid-state reaction from the composite of the starting elements. Which phase forms depends on the milling intensity, i.e., on the actual temperature during milling. Furthermore, the various phases can be transformed into each other by additional milling at higher or lower intensity. Further milling at higher intensity leads to crystallization of the metastable amorphous or quasicrystalline phases at elevated temperatures, since milling at higher intensity increases the actual milling temperatures to values above the transformation temperatures of the metastable phases. This temperature rise is caused by the higher energy transfer between the hitting balls at high milling intensities. On the other hand, it is possible to achieve a transformation from the energetically more stable to the less stable state by ball milling at relatively low intensity. In particular, this is demonstrated for the crystal-to-quasicrystal

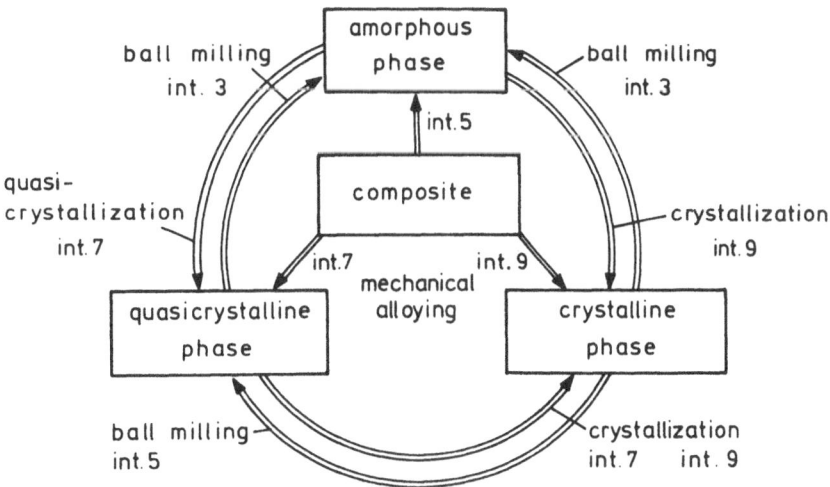

Fig. 3.46. Schematic illustration of possible transitions between composite, amorphous phase, quasicrystalline phase, and crystalline phase

transition, the crystal-to-amorphous transition, and the quasicrystal-to-amorphous transition.

3.9 Conclusions

In this overview, phenomenological results concerning the formation of glassy metals by mechanical alloying of elemental crystalline powders and the resulting properties were presented. Initially, ball-milling leads to an ultrafine composite. At the created clean interfaces, a diffusion reaction starts if the amorphous phase is thermodynamically favored compared to the crystalline composite. A large negative heat of mixing promotes the amorphization, which seems not to be possible in alloy systems with a low chemical driving force. TEM studies showed that there is a close similarity between the microstructure of mechanically alloyed powders and solid-state reacted multilayers (Chap. 2) at an intermediate amorphization state. The temperature rise during mechanical alloying is sufficiently high to enable an interdiffusion reaction, as shown by the effect of a too-intense milling, which can lead to crystallization. Especially the phenomena related to the formation of quasicrystalline material by mechanical alloying and the possible phase transitions between the amorphous, the quasicrystalline and the crystalline state, which can be controlled by the milling intensity, demonstrate the relation between milling intensity and effective temperature. The main question on the mechanism of glass formation by mechanical alloying concentrates, therefore, on why the thermodynamically stable phases do not form during the reaction. From investigation of the solid-state amorphization reaction in ultrafine layered Ni–Zr composites [3.31], we conclude that the amorphous phase grows much faster than the crystalline intermetallic phase, i.e., a diffusion related growth selection favors the amorphous phase. This can even lead to a situation in which a crystalline nucleus once formed can be destabilized and re-dissolved.

Besides the mechanical alloying of elemental powders, ball-milling of an intermetallic compound can also lead to amorphization, as demonstrated for several alloys [3.18, 19, 130, 131] (for more details see Chap. 2). This cannot be explained by the above statements, since in this case no composition-induced destabilization provides the driving force for an interdiffusion reaction. Amorphization by milling starting from powders of crystalline intermetallics is attributed instead to the accumulation of lattice defects – mainly the creation of antiphase boundaries – which raise the free enthalpy of the faulted intermetallic above that of the amorphous alloy. Therefore, there exists some similarity with irradiation-induced amorphization [3.20].

Mechanically alloyed and rapidly quenched glassy metals are structurally very similar, both topologically and chemically, as proved not only by X-ray and neutron diffraction, TEM and Mößbauer investigations and structural relaxation studies in the DSC, but also by measuring structure-sensitive physical properties as superconductivity or hydrogen absorption. Despite the observed minor differences, these results are basically consistent with the idea that the *glassy* state of a material is a unique metastable phase that may be formed by a variety of rather different procedures (mechanical alloying, rapid quenching, sputtering) starting from different initial phases as elemental crystalline powders, the liquid alloy, or the vapor phase.

The glass-forming ranges for mechanical alloying are exactly determined also by measuring physical properties as a function of composition. These experimentally obtained glass-forming ranges agree quite well with those derived theoretically from the free enthalpy curves of the primary phases and the amorphous phase by simply neglecting the existence of intermetallic phases prevented from forming by the kinetic restrictions of the mechanical alloying process. This indicates that glass formation by mechanical alloying can be described as a metastable equilibrium process. Amorphous metallic powders can be formed by mechanical alloying in a large number of alloy systems. Also, compositions that are not accessible by melt spinning can be made amorphous.

Mechanical alloying, therefore, offers an attractive technique for forming amorphous metals with possible technological applications, such as coatings with excellent wear resistance or corrosion resistance, new catalytic material or precursor material for permanent magnets. Especially with regard to new magnetic materials as nitrides or carbides, mechanical alloying is able to provide precursor powder excluding any intermetallic phases, from which magnets with superior properties can be formed by crystallization. Since these magnets will be able to compete with Co–Sm and Nd–Fe–B magnets, we predict that the mechanical alloying process will get its share on the growing rare earth permanent magnet market.

Acknowledgements. The authors acknowledge effective cooperation and stimulating discussions during the preparation of this manuscript with E. Hellstern, W. L. Johnson, C. Kuhrt, K. Samwer, K. Schnitzke, and J. Wecker. This work has been supported by the German Ministry for Research and Technology and by the Brite/Euram program of the European Community.

References

3.1   J.S. Benjamin: Am. Sci. **234**, May 40 (1976)
3.2   R.L. White: "The Use of Mechanical Alloying in the Manufacture of Multifilamentary Superconductor Wire" Ph.D. Thesis, Stanford University (1979)
3.3   C.C. Koch, O.B. Cavin, C.G. McKamey, J.O. Scarbrough: Appl. Phys. Lett. **43**, 1017 (1983)
3.4   R.B. Schwarz, W.L. Johnson: Phys. Rev. Lett. **51**, 415 (1983)
3.5   R.B. Schwarz, R.R. Petrich, C.K. Saw: J. Non-Cryst. Solids **76**, 281 (1985)
3.6   E. Hellstern, L. Schultz: Appl. Phys. Lett. **48**, 124 (1986)
3.7   C. Politis, W.L. Johnson: J. Appl. Phys. **60**, 1147 (1986)
3.8   A.W. Weeber, H. Bakker: Physica B **153**, 93 (1988)
3.9   L. Schultz: Mater. Sci. Eng. **91**, 15 (1988)
3.10   C.C. Koch: In *Materials Science and Technology*, Vol. 15, Processing of Metals and Alloys, ed. by R.W. Cahn (VCH, Weinheim 1991)

3.11   N. Burgio, A. Iasonna, M. Magini, S. Martelli, F. Padella: Il Nuovo Cimento **13D**, 459 (1991)

3.12   D.R. Maurice, T.H. Courtney: Metall. Trans. A **21**, 289 (1990)

3.13   R.M. Davis, B.T. McDermott, C.C. Koch: Metall. Trans. A **19**, 2867 (1988)

3.14   H. Hashimoto, R. Watanabe: Proc. Int'l Symp. on Mechanical Alloying, Kyoto, May 1991 In: Mater. Sci. Forum 88–90, 89 (1992)

3.15   W.L. Johnson: Mater. Sci. Eng. **97**, 1 (1988)

3.16   X.L. Yeh, K. Samwer, W.L. Johnson: Phys. Rev. Lett. **51**, 415 (1983)

3.17   L. Schultz: In: *Amorphous Metals and Non-Equilibrium Processing*, ed. by M. von Allmen (Les Editions de Physique, Les Ulis 1984) p. 135
        L. Schultz: In: *Proc. 5th int'l Conf. on Rapidly Quenched Metals*, ed. by S. Steeb, H. Warlimont (Elsevier, Amsterdam 1985) p. 1585

3.18   A.E. Ermakov, E.E. Yurchikov, V.A. Barinov: Fiz. Metal. Metalloved. **52**, 1184 (1981)

3.19   R.B. Schwarz, C.C. Koch: Appl. Phys. Lett. **49**, 146 (1986)

3.20   L.E. Rehn, P.R. Okamoto, J. Pearson, R. Bhadra, M. Grimsditch: Phys. Rev. Lett. **59**, 2987 (1987)

3.21   A. Blatter, M. von Allmen: Phys. Rev. Lett. **54**, 2103 (1985)

3.22   E. Hellstern, L. Schultz: Appl. Phys. Lett. **49**, 1163 (1986)

3.23   E. Hellstern, L. Schultz: Mater. Sci. Eng. **93**, 213 (1987)

3.24   E. Hellstern, L. Schultz: J. Appl. Phys. **63**, 1408 (1988)

3.25   M. Atzmon, J.D. Verhoeven, J.D. Gibson, W.L. Johnson: In: *Proc. 5th Int'l Conf. on Rapidly Quenched Metals*, Würzburg, ed. by S. Steeb, H. Warlimont (Elsevier, Amsterdam 1985) p. 1561

3.26   J. Eckert, L. Schultz, K. Urban: J. Mater. Res. **6**, 1874 (1991)

3.27   E. Hellstern, H.J. Fecht, Z. Fu, W.L. Johnson: J. Appl. Phys. **65**, 305 (1989)

3.28   R. Weast: *CRC Handbook of Chemistry and Physics* 57th ed, (Chemical Rubber, Cleveland 1976)

3.29   H. Gleiter, P. Marquardt: Z. Metallkde. **75**, 263 (1984)

3.30   J.H. Harris, W.A. Curtin, L. Schultz: J. Mater. Res. **3**, 872 (1988)

3.31   L. Schultz: Z. Phys. Chem. **157**, 257 (1988)

3.32   J. Eckert, L. Schultz, K. Urban In: *Proc. Europ. Conf. on Advanced Materials and Processes*, Aachen 1989 ed. by H.E. Exuer, V. Schumacher (DGM, Oberursel 1990) p. 1043

3.33   R. Schulz, M.L. Trudeau, J.Y. Huot, A. Van Neste: Phys. Rev. Lett. **62**, 2849 (1989)

3.34   J. Eckert, L. Schultz, E. Hellstern, K. Urban: J. Appl. Phys. **64**, 3224 (1988)

3.35   J. Eckert, L. Schultz, K. Urban: J. Mater. Sci. **26**, 441 (1991)

3.36   F. Petzoldt, B. Scholz, H.D. Kunze: Mater. Lett. **5**, 280 (1987)

3.37   J. Eckert, L. Schultz, K. Urban: J. Non-Cryst. Solids **130**, 273 (1991)

3.38   W.M. Kuschke, L. Schultz, P. Lamparter, S. Steeb: Z. Naturforsch. **46a**, 491 (1991)

3.39   R. Brüning, Z. Altounian, J.O. Strom-Olsen, L. Schultz: Mater. Sci. Eng. **97**, 317 (1988)

3.40   C. Michaelsen, E. Hellstern: J. Appl. Phys. **62**, 117 (1987)

3.41   C. Michaelsen, L. Schultz: Acta Metall. Mater. **39**, 987 (1991)

3.42   L. Schultz, E. Hellstern, A. Thomä: Europhys. Lett. **3**, 921 (1987)

3.43   C. Sürgers, H. von Löhneysen, L. Schultz: Phys. Rev. B **40**, 8787 (1989)

3.44   H. von Löhneysen: Mater. Sci. Eng. A **133**, 51 (1991)

3.45   P. Grütter: Diploma Thesis, University Basel (1986)

3.46   R.B. Schwarz: Mater. Res. Soc. Bulletin 55 (May/June 1986) p. 55

3.47   J. Eckert, L. Schultz, K. Urban: J. Less-Common Met. **145**, 283 (1988)

3.48   F. Petzold, B. Scholz, H.D. Kunze: Mater. Sci. Eng. **97**, 25 (1988)

3.49   A. Thomä, G. Saemann-Ischenko, L. Schultz, E. Hellstern: Jpn. J. Appl. Phys. **26**, 977 (1987)

3.50   E. Hellstern, L. Schultz, J. Eckert: J. Less-Common Met. **140**, 93 (1988)

3.51   P.Y. Lee, C.C. Koch: J. Non-Cryst. Solids **94**, 88 (1988)

3.52   E. Hellstern, L. Schultz, R. Bormann, D. Lee: Appl. Phys. Lett. **53**, 1399 (1988)

3.53   R. Bormann, F. Gärtner, K. Zöltzer, R. Busch: J. Less-Common Met. **145**, 19 (1988)

3.54   J.R. Thompson, C. Politis: Europhys. Lett. **3**, 199 (1985)

3.55   F. Petzold: J. Less-Common Met. **140**, 85 (1988)

3.56   J. Eckert, L. Schultz, K. Urban: In: *Proc. DGM Conf. on New Materials by Mechanical*

Alloying Techniques, Calw-Hirsau 1988 ed. by E. Arzt, L. Schultz (DGM, Oberursel 1989) p. 85

3.57   J. Eckert, L. Schultz, K. Urban: J. Less-Common Met. **166**, 293 (1990)

3.58   J.R. Thompson, C. Politis, Y.C. Kim: Mater. Sci. Eng. **97**, 31 (1988)

3.59   A.R. Miedema, P.F. de Châtel, F.R. de Boer: Physica B **100**, 1 (1980)
        A.K. Niessen, F.R. de Boer, P.F. de Châtel, W.C.M. Mattens, A.R. Miedema: CALPHAD **7**, 51 (1983)

3.60   C. Politis: Z. Phys. Chem. **157**, 209 (1988)

3.61   E. Hellstern, L. Schultz: Phil. Mag. B **56**, 443 (1987)

3.62   B.P. Dolgin, M.A. Vanek, T. McGory, D.J. Ham: J. Non-Cryst. Solids **87**, 281 (1986)

3.63   E. Hellstern, L. Schultz: Mater. Sci. Eng. **93**, 213 (1987)

3.64   M. Hansen, K. Anderko: *Constitution of Binary Alloys*, (McGraw-Hill, New York 1958)
        T.B. Massalski (ed.): *Binary Alloy Phase Diagrams* (Amer. Soc. for Metals. Metals Park, Ohio 1986) Vol. 1

3.65   L. Schultz, J. Wecker: Meter. Sci. Eng. **99**, 127 (1988)

3.66   G. Cocco, S. Enzo, L. Schiffini, L. Battezzati: In: *Proc. DGM Conf. on New Materials by Mechanical Alloying Techniques*, Calw-Hirsau 1988, ed. by E. Arzt, L. Schultz (DGM, Oberursel 1989) p. 343

3.67   C.C. Koch, M.S. Kim: J. Physique Co. **46**, 573 (1985)

3.68   C. Politis: Physica B **135**, 286 (1985)

3.69   A. Hikata, M.J. McKenna, C. Elbaum: Appl. Phys. Lett. **50**, 478 (1987)

3.70   D. Girardin, M. Maurer: In: *Proc. DGM Conf. on New Materials by Mechanical Alloying Techniques*, Calw-Hirsau 1988, ed. by E. Arzt, L. Schultz (DGM, Oberursel 1989) p. 91

3.71   G. Veltl, B. Scholz, H.D. Kunze: Mater. Sci. Eng. A **134**, 1410 (1991)

3.72   T. Fukunaga, M. Mori, K. Inou, U. Mitzutani: Mater. Sci. Eng. A **134**, 863 (1991)

3.73   E. Gaffet, C. Louison, M. Harmelin, F. Faudot: Mater. Sci. Eng. A **134**, 1380 (1991)

3.74   A.R. Yavari, P.J. Desré: Proc. Int'l Symp. on Mechanical Alloying, Kyoto (1991) In: Mater. Sci. Forum 88–90, 43 (1992)

3.75   L. Schultz, E. Hellstern, G. Zorn: Z. Phys. Chem. **157**, 203 (1988)

3.76   K. Omuro, H. Miura: Appl. Phys. Lett. **60**, 1433 (1992)

3.77   A. Calka, A.P. Radlinski: Appl. Phys. Lett. **58**, 119 (1991)

3.78   L. Schultz, J. Wecker, E. Hellstern: J. Appl. Phys. **61**, 3583 (1987)

3.79   J.J. Croat, J.F. Herbst, R.W. Lee, F.E. Pinkerton: Appl. Phys. Lett. **44**, 148 (1984)

3.80   L. Schultz, K. Schnitzke, J. Wecker, M. Katter, C. Kuhrt: Proc. MMM-Intermag Conf., Pittsburgh PA (1991) In J. Appl. Phys. **70**, (1991)

3.81   M. Sagawa, S. Fujimura, N. Togawa, H. Yamamoto, Y. Matsuura: J. Appl. Phys. **55**, 2083 (1984)

3.82   L. Schultz, K. Schnitzke, J. Wecker, M. Katter: IEEE Trans. **MAG-26**, 1373 (1990)

3.83   L. Schultz, K. Schnitzke, J. Wecker: Appl. Phys. Lett. **56**, 868 (1990)

3.84   A. Müller: J. Appl. Phys. **64**, 249 (1988)

3.85   K. Schnitzke, L. Schultz, J. Wecker, M. Katter: Appl. Phys. Lett. **56**, 587 (1990)

3.86   N. Kamprath, N.C. Jiu, H. Hegde, F.J. Cadieu: J. Appl. Phys. **64**, 5720 (1988)

3.87   J.M. Moreau, L. Paccard, J.P. Nozieres, F.P. Missell, G. Schneider, V. Villas-Boas J. Less-Common Met. **163**, 245 (1990)

3.88   J.M.D. Coey, H. Sun: J. Magn. Magn. Mater. **87**, L251 (1990)

3.89   D.B. de Mooij K.H.J. Buschow: J. Less-Common Met. **142**, 349 (1988)

3.90   J.M.D. Coey, H. Sun, Y. Otani, D.P.F. Hurley: J. Magn. Magn. Mater. **98**, 176 (1991)

3.91   K. Schnitzke, L. Schultz, J. Wecker, M. Katter: Appl. Phys. Lett. **57**, 2853 (1990)

3.92   D. Shechtman, I. Blech, D. Gratias, J.W. Cahn: Phys. Rev. Lett. **53**, 1951 (1984)

3.93   C. Politis, W. Krauss, H. Leitz, W. Schommers: Mod. Phys. Lett. B **3**, 615 (1989)

3.94   Z. Zhang, K. Urban: Scr. Metall. **23**, 767 (1989)

3.95   K.G. Kreider, F.S. Biancaniello, H.J. Kaufman: Scr. Metall. **21**, 657 (1987)

3.96   A. Csanady, P.B. Barna, J. Mayer, K. Urban: Scr. Metall. **21**, 1535 (1987)

3.97   D. A. Lilienfeld, M. Nastasi, H.H. Johnson, D.G. Ast, J.W. Mayer: Phys. Rev. Lett. **55**, 1587 (1985)

3.98   J.A. Knapp, D.M. Follstaedt: Phys. Rev. Lett. **55**, 1591 (1985)

3.99   D.M. Follstaedt, J.A. Knapp: J. Appl. Phys. **59**, 1756 (1986)

3.100   J.D. Budai, M.J. Aziz: Phys. Rev. B **33**, 2876 (1988)

3.101   S.J. Poon, A.J. Drehmann, K.R. Lawless: Phys. Rev. Lett. **55**, 2324 (1986)

3.102   K. Urban, N. Moser, H. Kronmüller: phys. stat. sol. (a) **91**, 411 (1985)

3.103   D.M. Follstaedt, J.A. Knapp: Phys. Rev. Lett. **56**, 1827 (1986)

3.104   A. Csanady, K. Urban, J. Mayer, P.B. Barna: J. Vac. Sci. Technol. A **5**, 1733 (1987)

3.105   P.B. Barna, G, Rodnoczi, A. Csanady, K. Urban: Scr. Metall. **22**, 373 (1988)

3.106   W.A. Cassada, G.J. Shiflet, S.J. Poon: Phys. Rev. Lett. **56**, 2276 (1986)

3.107   J. Eckert, L. Schultz, K. Urban: Appl. Phys. Lett. **55**, 117 (1989)

3.108   J. Eckert, L. Schultz, K. Urban: Europhys. Lett. **13**, 349 (1990)

3.109   J. Eckert, L. Schultz, K. Urban: Z. Metallkde. **81**, 862 (1990)

3.110   J. Eckert, L. Schultz, K. Urban: J. Less-Common Met. **167**, 143 (1990)

3.111   J. Eckert, L. Schultz, K. Urban: Acta Metall. Mater. **39**, 1497 (1990)

3.112   E.Y. Ivanov, I.G. Konstanchuk, B.D. Bokhonov, V.V. Boldyrev: Reactivity of Solids **7**, 167 (1989)

3.113   G. Venkataraman, D. Sahoo: Contemp. Phys. **26**, 579 (1985); Contemp. Phys. **27**, 3 (1986)

3.114   P.W. Stephens, A.I. Goldman: Phys. Rev. Lett. **56**, 1168 (1986); Phys. Rev. Lett. **57**, 2331 (1986)

3.115   D. Levine, P.J. Steinhardt: Phys. Rev. Lett. **53**, 2447 (1984)

3.116   V. Elser, C.L. Henley: Phys. Rev. Lett. **55**, 2883 (1985)

3.117   P. Guyot, M. Audier: Phil. Mag. B **52**, L15 (1985)

3.118   S. Garcon, P. Sainfort, G. Regazzoni, J.M. Dubois: Scr. Metall. **21**, 1493 (1987)

3.119   K. Urban, M. Bauer, A. Csanaday, J. Mayer: Mater. Sci. Forum **22–24**, 517 (1987)

3.120   A.P. Tsai, A. Inoue, T. Masumoto: Jpn. J. Appl. Phys. **26**, L1505 (1987)

3.121   A.P. Tsai, A. Inoue, T. Masumoto: J. Mater. Sci. Lett. **7**, 322 (1988)

3.122   M.A. Marcus, V. Elser: Phil. Mag. B **54**, L101 (1986)

3.123   P.A. Bancel, P.A. Heiney, P.W. Stephens, A.I. Goldman, P.M. Horn: Phys. Rev. Lett. **54**, 2422 (1985)

3.124   J. Mayer, K. Urban, J. Fidler: phys. stat. sol. (a) **99**, 467 (1987)

3.125   A.P. Tsai, A. Inoue, T. Masumoto: J. Mater. Sci. Lett. **6**, 1403 (1987)

3.126   A.P. Tsai, A. Inoue, T. Masumoto: J. Mater. Sci. Lett. **8**, 253 (1987)

3.127   R. Lück, H. Hess, F. Sommer, B. Predel: Scr. Metall. **20**, 677 (1986)

3.128   A. Inoue, Y. Bizen, T. Masumoto: Met. Trans. A **19**, 383 (1988)

3.129   D. Bahadur, Y. Srinivas, R.A. Dunlap: J. Phys.: Condensed Matter **1**, 256 (1989)

3.130   A.W. Weeber, H. Bakker, F.R. de Boer: Europhys. Lett. **2**, 445 (1986)

3.131   C.C. Koch, J.S.C. Jang, P.Y. Lee: In: *Proc. DGM Conf. on New Materials* by Mechanical Alloying Techniques, Calw/Hirsau 1988, ed. by E. Arzt, L. Schultz (DGM, Oberursel 1989) p. 101

General Reading

Arzt, E., Schultz, L. (eds): *New Materials by Mechanical Alloying Techniques* (DGM, Oberursel 1989)

Gilman, P.S., Benjamin J.S.: *Mechanical Alloying* Ann. Rev. Mater. Sci. **13**, 279–300 (1983)

Koch C.C., Mechanical milling and alloying, in *Materials Science and Technology*, Vol. 15, Processing of Metals and Alloys, ed. by R.W. Cahn (VCH, Weinheim 1991) pp. 193–244

Schultz, L. *Formation of Amorphous Metals by Solid-State Reactions*. Phil. Mag. B **61**, 453 (1990)

Shingu P.H. (ed.): Proc. Int'l Symp. on Mechanical Alloying, Kyoto (1991) Mater. Sci. Forum 88–90 (1992)

Weeber, A.W. H. Bakker: *Amorphization by Ball Milling – A Review*. Physica B **153**, 93–135 (1988)

4. Glassy Metals in Catalysis

A. Baiker

With 21 Figures

About a decade has elapsed since the use of metallic glasses in catalysis research was first reported [4.1–3]. Since then, numerous publications have been devoted to the use of metallic glasses in catalysis. Some of these activities have been covered in reviews [4.4–8]. Unfortunately, in many of the early investigations, relatively little attention was paid to the structural and chemical characterization of the materials used consequently yielding results of a rather phenomenological nature focusing mainly on catalytic behavior. In fact, many glassy metals were reported to exhibit higher activities and selectivities than their crystalline counterparts.

A general feature of more recent investigations is that more emphasis has been placed on the understanding of the relationship between the structural and chemical properties of the alloys and their behavior in catalysis. These investigations showed that, presumably, in most applications of metallic glasses in catalysis, the surface of the metastable amorphous alloy undergoes chemical and structural changes under reaction conditions. This observation, coupled with the fact that as-quenched alloys exhibit very low surface areas and consequently low activity, led several investigators to use metallic glasses as catalyst precursors rather than as catalysts. In several investigations it has been demonstrated that glassy metals may constitute interesting precursor materials for the preparation of supported metal catalysts. The aim of this chapter is to illustrate the progress made in this field and to draw some conclusions with regard to the potential and limitations of these materials for catalytic applications.

4.1 Preparation of Glassy Metals for Catalytic Studies

The various metallic glasses reported in the literature [4.9] fall into a few well-defined categories: (i) late transition metal + metalloid; (ii) early transition metal + late transition metal or group IB metal; (iii) earth alkali metal + group IB metal; (iv) early transition metal + alkali metal; and, (v) Actinide + early transition metal. In catalysis research, exclusively metallic glasses of categories (i) and (ii) have been used so far. Table 4.1 lists glassy metals which have been used in catalytic studies. Note that metal–zirconium alloys and Ni, Fe, and mixed Ni–Fe alloys with P and/or B as metalloid have been used most frequently.

Topics in Applied Physics, Vol. 72
Beck/Güntherodt (Eds.)
© Springer-Verlag Berlin Heidelberg 1994

Table 4.1. Glassy metal alloys used in catalytic applications

Alloy	*Reaction* (major products)	References
	Hydrogenation of CO	
$Fe_{40}Ni_{40}P_{16}B_4$	(C_1–C_3 hydrocarbons)	[4.1]
Fe–Ni with P and/or B	(C_1–C_3 hydrocarbons)	[4.3]
$Ni_{60}Fe_{20}P_{20}$	(C_1–C_5 hydrocarbons)	[4.10]
$Fe_{20}Ni_{60}P_{20}$, $Fe_{90}Zr_{10}$	(C_1–C_5 hydrocarbons)	[4.11]
$Fe_{81}B_{13.5}Si_{3.5}C_2$	(C_1–C_3 hydrocarbons)	[4.12]
$Ni_{67}Zr_{33}$	*Methanation*	[4.13]
$Pd_{35}Zr_{65}$	*Methanation*	[4.14, 15]
$Au_{25}Zr_{75}$, $Rh_{25}Zr_{75}$, $Pt_{25}Zr_{75}$	*Methanation*	[4.16]
$Pd_{25}Zr_{75}$, $Os_{25}Zr_{75}$, $Ir_{25}Zr_{75}$		
$Fe_{82.2}B_{17.8}$	(C_1–C_4 hydrocarbons)	[4.17, 18]
$Fe_{80}B_{80}$, $Fe_{40}Ni_{40}B_{20}$	(C_1–C_4 hydrocarbons)	[4.19]
$Ni_{78}P_{19}La_3$, $Ni_{81}P_{19}$	(C_1, C_2 hydrocarbons)	[4.20]
$Cu_{70}Zr_{30}$, $Cu_{60}Ti_{40}$, $Cu_{64}Hf_{36}$	*Methanol synthesis*	[4.21]
Cu–Ce–Al	*Methanol synthesis*	[4.22]
	Hydrogenation of CO_2	
$Cu_{70}Zr_{30}$	*Methanol synthesis*	[4.23]
$Pd_{33}Zr_{67}$	*Methanation*	[4.24]
	Hydrogenation of hydrocarbons	
$Fe_{44-x}Ni_{37}Cr_xP_{15}B_4$ ($x < 10$)	(Acetylene)	[4.25]
$Pd_{80}Si_{20}$, $Pd_{77}Ge_{23}$	((+)-apopinene)	[4.26, 27]
$Ni_{81}P_{19}$, $Ni_{62}B_{38}$	(Buta-1,3-diene)	[4.28]
$Ni_{81}P_{19}$	(Ethene, buta-1,3-diene)	[4.29]
$Ni_{64}Zr_{36}$	(Buta-1,3-diene)	[4.30]
$Cu_{70}Zr_{30}$	(Buta-1,3-diene)	[4.31]
$Fe_{90}Zr_{10}$	(Buta-1,3-diene)	[4.32]
$Pd_{80}Si_{20}$, $Pd_{77}Ge_{23}$	(*Cis*-cyclododecene)	[4.2, 33]
$Pd_{80}Si_{20}$, $(Ni_{50}Fe_{50})_{80}B_{20}$	(Cyclohexene, *n*-hexene Phenylethyne, α-pinene)	[4.34]
$Pd_{35}Zr_{65}$, $Pd_{20}Zr_{80}$	(Cyclohexene, benzene)	[4.35]
$Ni_{81}P_{19}$, $Ni_{62}B_{38}$	(Ethene, propene, isoprene *cis*-but-2-ene, buta-1,3-diene)	[4.36]
$Ni_{62}B_{38}$	(Ethene)	[4.37, 38]
$Ni_{36}Zr_{64}$	(Ethene)	[4.39]
$Cu_{70}Zr_{30}$	(Ethene)	[4.40]
$Cu_{62}Zr_{38}$	(Ethene, isoprene)	[4.41]
$Pd_{80}Si_{20}$, $Pd_{77}Ge_{23}$	(1-octyne, 4-octyne, phenyl-acetylene)	[4.42]
$Pd_{80}Zr_{20}$, $Pd_{35}Zr_{65}$, $Ni_{30}Zr_{30}$	(1-hexene)	[4.43]
$Ni_{40}Fe_{40}B_{20}$, $Fe_{85}B_{15}$, $Cu_{50}Zr_{50}$	(Phenylethyne)	
$Fe_{91}Zr_9$	*Ammonia synthesis*	[4.44]
$Ni_{91}Zr_9$, $Ni_{64}Zr_{36}$	*Reduction of NO*	[4.45]
$Pd_{25}Zr_{75}$, $Rh_{25}Zr_{75}$		[4.46]
$Cu_{70}Zr_{30}$, $Fe_{91}Zr_9$, $Ni_{64}Zr_{36}$		
$Pd_{33}Zr_{67}$	*Oxidation of CO*	[4.47]
$Cu_{67}Ti_{33}$	*Dehydrogenation of methanol* (Methyl formate)	[4.48]
$Fe_{80}B_{20}$	*Decomposition of formic acid* (Carbon monoxide)	[4.49]
	Hydrogenolysis	[4.37]
$Ni_{62}B_{38}$	(Ethane, cyclopropane)	
	Cycloamination	[4.50]
$Cu_{70}Zr_{30}$	(5-amino-pentanol)	

Although most of the known preparation methods [4.5, 6, 51] can be used for the production of metallic glasses for catalytic studies, the most frequently used technique is *melt-spinning*, which produces thin ribbons. In several studies, the glassy metals were used in the form of powders or flakes prepared from the ribbons by milling or chopping them under liquid nitrogen. Powders can easily be packed in the reactor tube and possess a larger geometrical surface area than the ribbons. However, special care has to be taken that no crystallization occurs during the grinding process.

4.2 Motivation for Using Glassy Metals in Catalysis

The initial motivation for using glassy metals in catalysis originated from some of their unique properties [4.5, 52, 53], which make them interesting materials for catalytic research. Some of these properties are described as follows.

(i) Ideally, the surface of amorphous materials should be devoid of any long-range ordering of the constituents and exhibit a high density of low coordination sites and defects. The important role of low coordination sites in catalysis, such as terraces, steps, and kinks has been demonstrated by *Somorjai* [4.54] on crystalline materials.

(ii) Glassy metals should possess high flexibility with regard to fine tuning of the electronic properties [4.55], mainly because thermodynamic constraints are less stringent in supercooled liquids than in crystalline materials.

(iii) Glassy metals are ideally chemically homogeneous and structurally isotropic.

(iv) Glassy metals are highly reactive due to their metastable structure and undergo solid-state reactions frequently more easily than their crystalline counterparts. This property is of importance in their use as catalyst precursors.

(v) Glassy metals exhibit good conductivity for electricity and heat. These properties paired with excellent corrosion resistance [4.56] make them particularly interesting for application in electrocatalysis.

(vi) Glassy metals prepared by melt-quenching exhibit a planar morphology ideal for investigation with modern characterization tools such as electron spectroscopy and scanning tunneling microscopy.

(vii) Their manufacturing as foils or small ribbons may facilitate the design of new reactor conceptions.

Limitations with regard to the application of glassy metals in catalysis originate from their metastable structure and the intrinsically low surface area, which corresponds to the geometrical area of as-quenched materials. Furthermore, we should note that ideal glassy metal surfaces being chemically and structurally isotropic and showing uniform short range ordering of the constituents are difficult to produce and even more difficult to maintain under conditions where catalytic reactions are carried out.

4.3 Glassy Metals as Catalytically Active Materials

Surveying the research efforts made in the past decade two main directions are discernable, namely, investigations carried out on as-quenched glassy metals and those where the glassy metals were subjected to different pretreatments and served merely as precursors to catalytically active materials.

4.3.1 Studies of Glassy Metals in As-quenched State

a) Surface Structure and Electronic Properties

Direct insight into the surface structure of glassy metals has been obtained by using scanning tunneling microscopy (STM). *Wiesendanger* et al. [4.57] performed STM measurements on glassy $Rh_{25}Zr_{75}$ prepared by melt-spinning. They found that the surface is made up of flat areas identified as disordered regions, and hill structures, which they attributed to nanocrystals imbedded in the amorphous matrix. Figure 4.1A shows a typical topographic profile plot. One can distinguish two parts: the flat area in the lower left part, which could be identified as an amorphous region of the sputtered $Rh_{25}Zr_{75}$ surface by comparison with other STM images, and the hill structure in the upper right part, which was attributed to a crystallite with a diameter of about 20 nm within the amorphous sample. This surface structure is further supported by the simultaneously recorded image of the local tunneling barrier height (Fig. 4.1B). The lower left part of the image yields a higher local tunneling barrier height than the upper right part with the hill structure. Based on their careful STM studies, the authors concluded that the appearance of small crystalline domains in as-prepared glassy metals is likely and that the cooling rate used for the rapid quenching process influences the number and size of these crystalline domains.

Similar surface morphology was found by *Walz* et al. [4.58] for amorphous $Fe_{91}Zr_9$ prepared by melt-spinning. They concluded from their STM investigations that the surface of the Fe–Zr samples was at least partially crystallized.

Schlögl et al. [4.59] investigated the surface morphology of amorphous $Fe_{91}Zr_9$ alloys using both scanning electron microscopy (SEM) and STM. Figure 4.2 shows low-resolution STM images of the shiny side of the freshly prepared amorphous $Fe_{91}Zr_9$ alloy. Note the pronounced anisotropy of the surface structure of the melt-quenched material. The three STM images show that the surface is not flat, but consists of flat disk-shaped structures several hundred angstroms in size with clear valleys separating them. The valleys were

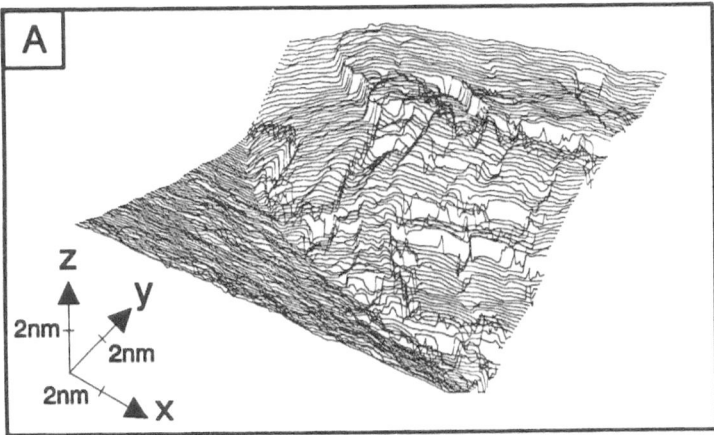

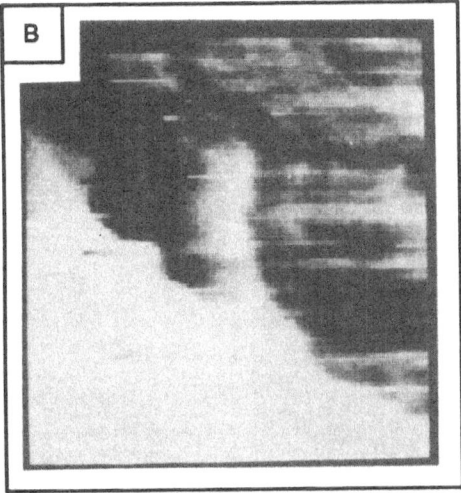

Fig. 4.1. A) STM profile plot of a 2×2 nm surface area of $Rh_{25}Zr_{75}$ after 45 min sputtering with 5 keV Ar ions [4.57]. B) Image of the local tunneling barrier height, which corresponds to the topographic image in Fig. 4.1A

attributed to inhomogeneous cooling of microdroplets of the liquid alloy. It could be clearly shown that the smooth surface of the droplets consists of irregular corrugations with an average distance of the maxima of several nanometers and an amplitude of below 1 nm. The latter structure has been suggested to originate from frozen waves excited on the surface of the formerly liquid microdroplets and preserved by the rapid quenching rate of ca. 10^6 deg/s.

The experimental investigations mentioned above indicate that the surfaces of glassy metals prepared by melt-quenching are by far not as homogeneous as advocated earlier [4.5]. In most cases, they will be rather inhomogeneous, containing disordered regions and crystalline-like regions, which are probably

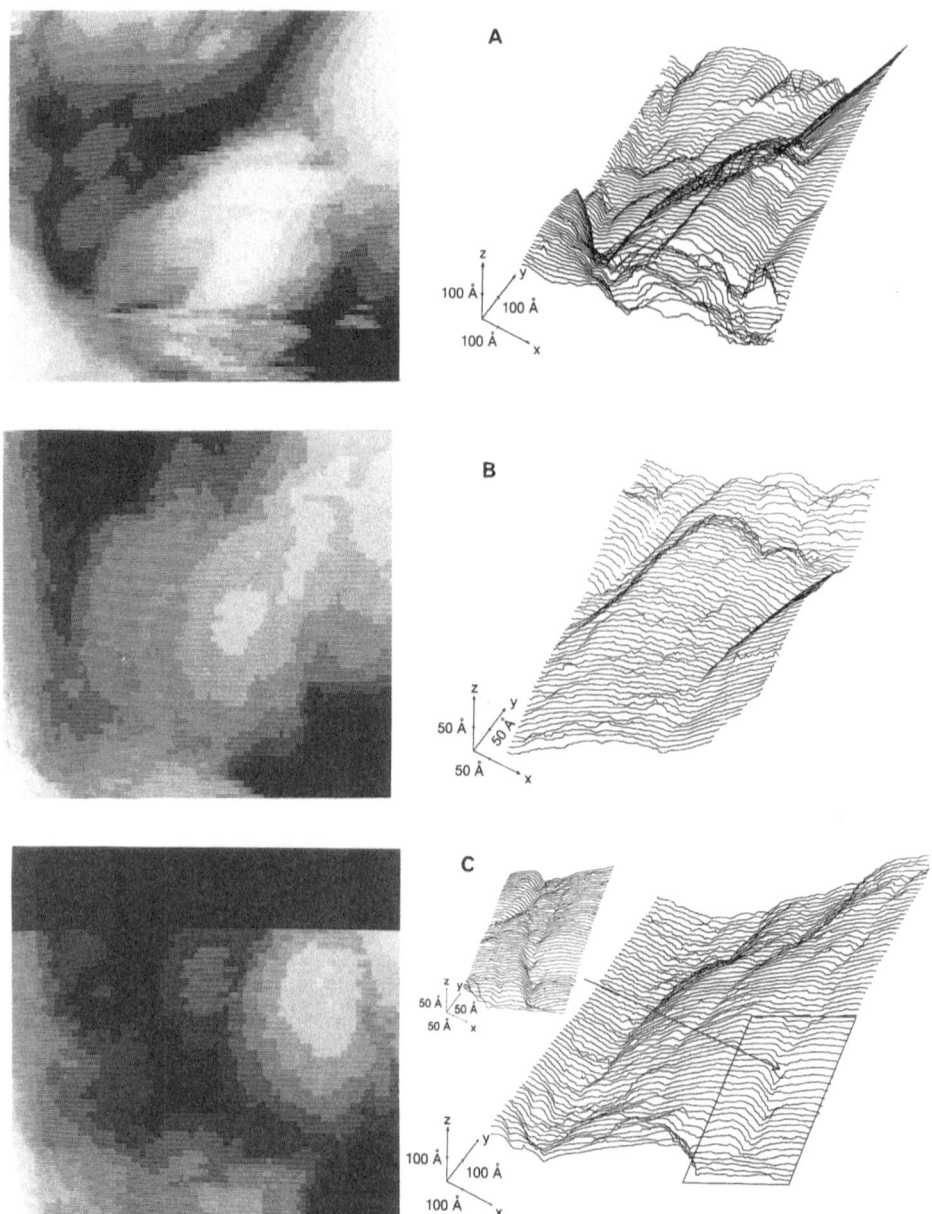

Fig. 4.2. Low-resolution STM images of the shiny side of freshly prepared amorphous $Fe_{91}Zr_9$ [4.59]. The scale in the three main figures is 100 Å/unit, and 50 Å/unit in the inset of (C). The computer-processed images always represent a slightly smaller area than the direct recordings. The height difference between successive gray tones is 5 Å

created already during the quenching process or after contamination with air. Thus the ideal amorphous metal surface being isotropic and showing uniform short-range ordering may in reality be difficult to produce, and even more difficult to maintain, particularly under conditions where catalytic reactions are performed. This phenomenon imposes severe limitations to all applications of glassy metals, where a stable disordered surface structure is derived. The final answer concerning the short-range ordering of amorphous surfaces cannot yet be provided. High resolution STM of the topography and the local tunneling barrier height of the amorphous areas could be promising to provide information about short-range order of metallic glass surfaces if the electronic surface structure and the geometric surface structure can be correlated [4.57].

Knowledge about the electronic structure of metallic glasses has emerged mainly from electron-spectroscopy studies. However, it should be noted that other types of spectroscopy such as optical absorption, X-ray emission, bremsstrahl-isochromate spectroscopy, and soft X-ray appearance potential spectroscopy also yield information about the electronic structure of solids, but have been rarely used so far to study glassy metals. In many cases, the results of these techniques are complementary to those obtained by electron spectroscopy. Metallic glasses offer the opportunity to study alloys in continuous ranges of concentrations and at concentrations at which no crystalline phase exists. Although electron spectroscopy measurements can yield information on both surface and bulk properties of a solid, in most investigations carried out until now on glassy metals, mostly the bulk properties have received attention. A comprehensive review gathering the knowledge gained in this field has been produced by *Oelhafen* [4.55].

Insight into the electronic structure of glassy metal surfaces has been obtained by studying their interaction with probe molecules [4.60–62] using mainly photoelectron spectroscopy (UPS and XPS). *Hauert* et al. [4.60] studied the chemisorption of CO on Ni–Zr glassy metals using UPS. Chemisorption of CO was investigated as a function of exposure, temperature, and alloy composition. Molecular and/or dissociative CO chemisorption was observed depending on the alloy composition, as illustrated in Fig. 4.3, which depicts the amount of molecularly adsorbed CO as a function of temperature for polycrystalline Ni and various glassy Ni–Zr alloys. Each data point was obtained by exposing a clean substrate at a given temperature to 8L CO. A general characteristic of all glassy alloys investigated is that the amount of molecularly adsorbed CO decreases at the cost of dissociatively adsorbed CO with increasing temperature. In addition, the temperature for a given amount of molecular chemisorption decreases with increasing Zr content. In contrast to glassy alloys, on polycrystalline Ni the decrease in molecularly adsorbed CO with increasing temperature was not accompanied by a corresponding increase of dissociatively adsorbed CO, but by desorption of CO. No dissociatively adsorbed CO could be detected on the pure Ni samples in the temperature range investigated. In contrast, on pure polycrystalline zirconium, CO was found to be dissociatively adsorbed in the whole temperature range. An important result emerging from this study is

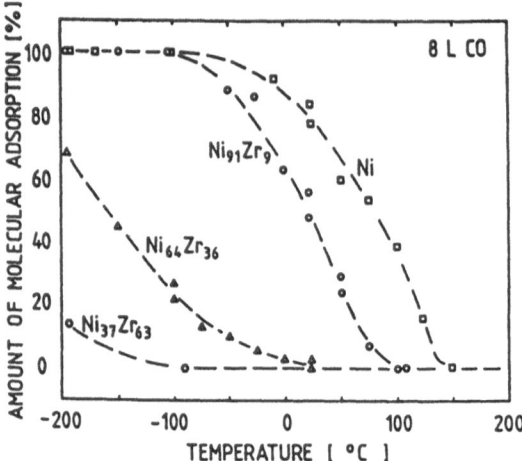

Fig. 4.3. Amount of molecularly adsorbed CO (100% = maximum observed coverage) as a function of temperature for polycrystalline Ni and various Ni–Zr glassy alloys [4.60]. Each data point was obtained by exposing a clean substrate at a given temperature to 8 Langmuirs CO

that the ratio of molecular to dissociative CO chemisorption is not directly related to the surface composition of the Ni–Zr alloys. This phenomenon was attributed to Zr modifying the local electronic structure at the Ni atom sites in such a way that the chemisorption behavior of these sites is profoundly different from elemental nickel.

Guczi and coworkers [4.61] studied CO chemisorption on glassy and crystalline FeNiB alloys in the presence and in the absence of hydrogen using XPS and UPS. They found that CO chemisorption at 300 K is characteristic of the surface structure. At 570 K, no difference could be observed in the mode of chemisorption because only dissociative carbon was present. However, the reactivity differences observed in the CO + H$_2$ reaction could be ascribed to the difference in the surface transformation of the carbidic species. The authors suggested that this species can be stabilized by the small ensemble size characteristic for glassy and partially crystallized samples, whereas the main route of the dissociative carbon on crystallized samples is the inactive bulk carbide formation. This phenomenon was found to be influenced by the alloy composition and by the presence of hydrogen.

Baiker et al. [4.62] investigated the adsorption of nitrogen and CO on amorphous Fe$_{91}$Zr$_9$ using UPS and XPS. The studies showed that both molecular and dissociated nitrogen were present on the surface after exposure to dinitrogen at 79 K. Figure 4.4 shows the N 1s X-ray photoelectron spectra measured after exposure of the sputter-cleaned Fe$_{91}$Zr$_9$ to a nitrogen atmosphere. The top trace shows two structures, the wide one is due to molecular adsorption of dinitrogen, while the narrow structure is attributed to dissociated nitrogen. The dissociation of dinitrogen at this low temperature is remarkable. In order to study the reactivity of these surface species, eight

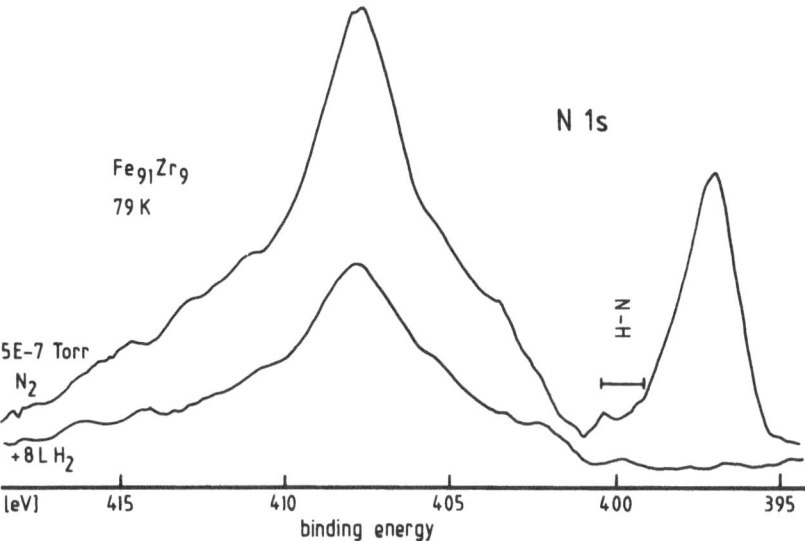

Fig. 4.4. N 1s X-ray photoelectron spectra of nitrogen adsorbed on sputter cleaned $Fe_{91}Zr_9$ alloy surface at 79 K [4.62]. Top trace was measured under static nitrogen pressure of $5 \cdot 10^{-7}$ Torr. Bottom trace was measured after admission of 8 L hydrogen to the static nitrogen atmosphere. The instrumental resolution had to be limited to 1.5 eV in order to collect each spectrum within 6 h

Langmuir of hydrogen were admitted still at low temperature with the nitrogen supply terminated. The resulting spectrum is shown in the bottom trace of Fig. 4.4. Note that upon addition of hydrogen to the surface, the dissociated nitrogen species desorbed completely, presumably as ammonia. Based on their experiments, the authors suggested that dissociatively bound nitrogen is the reactive precursor for ammonia synthesis on the investigated glassy materials. The results of the UPS and XPS studies, including binding energies and line profiles, agreed well with the results of similar investigations carried out on single-crystal surfaces of iron. Similarly, no differences were observed in the CO adsorption normalized to the number of iron sites in the surface (amount adsorbed, kinetics, chemical shift) when comparing CO adsorption on poly-crystalline iron and amorphous $Fe_{91}Zr_9$. This indicated that the local electronic structure of the adsorption sites on the amorphous alloy surface are similar to those of elemental iron. The similarity of the nitrogen adsorption, which is known to be structure sensitive, may be taken as an indication that both the polycrystalline and the *amorphous* surfaces exhibit similar microstructures. The similarities in the microstructure of the surfaces may have been produced by the extensive sputter-cleaning of both samples before adsorption experiments.

Unfortunately, only relatively little studies have been carried out so far where the influence of the electronic structure of a glassy metal substrate on the chemisorption behavior of probe molecules such as CO and N_2 has been investigated. However, in view of the high flexibility in the chemical composition of these materials, where elements with different electronic and hence chemical properties can be combined within a large concentration range, such efforts seem rewarding. The well-understood electronic alloying effects are strong enough to allow detection of chemisorption effects not only on the probe molecules but also on the substrate. Better knowledge of the interaction of carefully selected probe molecules with glassy metal surfaces would help to clarify what possibilities with regard to catalytic applications these materials can offer. Without this knowledge, no tailoring of glassy metals for catalytic applications will be possible.

b) Catalytic Studies

Relatively little catalytic work has been carried out so far under conditions where the surface of the metal alloys can be regarded as unreconstructed, i.e., where the chemical composition and structure of the surface can be assumed to be in the state characteristic for the freshly quenched material. In principle, such investigations can only be performed at temperatures far below the crystallization temperature of the alloy and require special precaution to eliminate possible contamination of the alloy during its transfer from the fabrication (melt spinning) to the catalytic reactor.

In their initial work, *Smith* and coworkers [4.2] showed that Pd–Si glasses, made by splat cooling [4.51] are active catalysts for the hydrogenation with deuterium of cis-cyclododecene at room temperature. Pd–Si glasses produced more trans-isomerization, more dideutero-saturate, and less extensive exchange than crystalline Pd. Later, these studies were extended to include Pd–Ge glasses [4.33] and compared the catalytic behavior of the glassy metals with their crystalline counterparts.

The different selectivities found for the above hydrogenation over amorphous and crystalline Pd systems were attributed to the different surface topography of the glassy metals and the crystalline alloys. The glassy surfaces were suggested to be free of atomically flat terraces and to be highly populated with protuberances approximating kinks and ledges, as compared with the surface of large crystallites, where the terraces predominate. The kinks and ledges on crystalline surfaces are of discrete dimensions [4.54], whereas the glassy surface is assumed to present protuberances with a continuum of coordination numbers.

In contrast to the results mentioned above, *Giessen* et al. [4.34] found no significant differences in the selectivity behavior between glassy and crystalline phases of $Pd_{80}Si_{20}$ during hydrogenation reactions of n-hexene, phenylethyne, α-pinene, and cyclododecene. Catalytic selectivities with regard to *cis-trans* isomerization, double-bond migration, and the stereochemistry of addition were about the same regardless of whether a glassy or a crystalline catalyst was used. Minor differences were observed in hydrogen–deuterium exchange. The reason for these contrasting results is not clear.

More recently, it has been suggested that the hydrogenation of *cis*-cyclododecene is not very suitable to characterize differences in the structure of these surfaces, since an isomer (*trans*-cyclododecene) is produced in the reaction which has a rate of hydrogenation different from the parent compound. These differences in rates of hydrogenation mask the actual amounts of isomerization occurring and, therefore, conceal the kinds of catalytic sites available on the surface. With this in mind, *Smith* and coworkers [4.26, 27] tested the molecule (+)-apopinene (6, 6-dimethyl-1R,5R-bicyclo[a]) hept-2-ene) as a surface probe to distinguish the relative percentages of terraces, ledges, and kinks available on the metallic surface. This probe molecule is more useful than the earlier used *cis*-cyclododecene because its isomerization product (−)-apopinene has the identical rate of hydrogenation on a symmetrical surface. The crystallized alloys showed a higher ratio of isomerization–to–deuteration than the parent amorphous alloys. It was suggested that the isomerization of (+)-apopinene reflects the total number of ledge, kink and terrace sites, whereas hydrogenation reflects only the number of kinks. Based on these arguments, *Smith* and coworkers concluded that the surface structure of an amorphous alloy is not two-dimensionally random (flat), but is three-dimensionally random (hilly or rolling).

Molnar et al. [4.42] studied the selective half-hydrogenation of phenylacetylene, 1-octyne, and 4-octyne over Pd–Si and Pd–Ge glassy and crystalline catalysts, and for comparison, over splat cooled Pd, reduced PdO_2 and Pd foil. They found that terminal alkynes comminute Pd structures and expose new active sites. These sites are different on the rapidly cooled catalysts compared to the regularly crystallized catalysts. Although no significant changes were detected in alkyne hydrogenation selectivities after several hydrogenations, marked changes were revealed by (+)-apopinene. On the terminal acetylene-treated foils and on the reduced PdO_2, the rates of hydrogenation and isomerization of (+)-apopinene increased, but the ratio of the two rates remained almost the same. In contrast, the splat-cooled catalysts showed a higher rate increase for isomerization than for hydrogenation. *Molnar* et al. [4.42] suggested that the effect of the terminal alkynes is to expose sites of lower coordinative unsaturation, especially ledges and kinks by comminution of the Pd structures. These newly exposed sites are assumed not to influence alkyne reactions. It is interesting to note that heat treatment of the amorphous alloys was previously [4.26] reported to have a similar effect on the surface structure of the amorphous alloys.

4.3.2 Glassy Metals as Precursors of Catalytically Active Materials

Presently available reports on catalysis over glassy metals are listed in Table 4.1. In most catalytic applications of glassy metals, pretreatment of the as-quenched materials (e.g., in a reducing gas atmosphere) was found to be crucial to obtain high catalytic activities. Several factors may contribute to this behavior, the most important being: (i) the surfaces of metallic alloys exposed to air are likely to be covered with a superficial layer of inactive metal oxides; (ii) the surface area of as-quenched materials is very small (usually less than 0.1 m^2/g) and is therefore easily deactivated in the presence of contaminants.

Several different procedures have been applied to improve the catalytic properties of as-quenched materials, including the reduction in hydrogen [4.1, 3, 11, 12, 17, 25, 31, 40] or in other reducing gas atmospheres (e.g., H_2/CO [4.10, 14–16, 21, 23], H_2/CO_2 [4.23, 24], H_2/N_2 [4.44, 45]), as well as etching in acid solutions (HCl [4.19], HNO_3 [4.20, 28, 36, 37], HF [4.39, 41]) followed by oxidation and reduction. It seems likely that, in most cases where such pretreatments were applied, the original surface structure of the amorphous alloy was altered. Thus comparative studies of the catalytic behavior of the pretreated glassy metals and their crystalline counterparts do not generally provide a reasonable basis for answering the question of whether or not the amorphous surface is more active than the corresponding crystalline surface. In any case, it is interesting to note that in most studies (exceptions are reported in [4.22, 34]) the amorphous samples were found to exhibit improved catalytic behavior, i.e., either higher activity [4.1–3, 10–12, 14, 15, 17, 20, 24, 31–33, 40, 43, 45–48] or better selectivity [4.2, 11, 25, 30, 31, 33, 42, 48], compared to their crystalline counterparts. The reason for this behavior is in many cases not clear, since too little effort has been expended on characterizing the chemical and structural properties of the amorphous and crystalline alloy surfaces. Several factors such as degree of ordering and dispersion of the active component, electronic properties, formation of new phases, nucleation and growth of crystalline domains, segregation phenomena, and textural properties will be differently influenced during pretreatment, depending on whether an amorphous or a crystalline alloy is used as starting material [4.50].

More recently, highly active catalysts were prepared from glassy metals by selectively oxidizing the more electropositive constituent of the materials. After reduction, finely dispersed transition metal particles imbedded in an amorphous or partially crystalline oxidic matrix of the more electropositive constituent were obtained. In principle, this method for the preparation of supported metal catalysts from metal alloys is not new; it had been applied earlier by *Shamsi* and *Wallace* [4.63] to crystalline intermetallic alloys. The use of glassy metals as precursors may offer several advantages, such as higher flexibility in composition, homogeneous distribution of constituents on molecular scale, and higher reactivity. These advantages emerge from the intrinsic properties of the materials outlined above.

There are several examples reported in the literature which demonstrate the potential of glassy metals as catalyst precursors [4.13, 16, 21, 23, 24, 35, 38, 39, 41, 44, 45, 47, 50]. We will discuss next some typical properties of such catalysts using the preparation of metal on zirconia catalysts from metal zirconium alloys as examples. These systems have been extensively studied and are therefore ideal for illustrating the potential of glassy metals as catalyst precursors.

a) Oxidation/Reduction Treatment of Precursor

Various supported metal catalysts [4.20, 44, 64, 65] were prepared by oxidation/reduction treatment of the metallic glass precursor. The oxidation behavior, which will be addressed later in this review, depends largely on the chemical composition of the glassy metal. Generally, the more electropositive constituent oxidizes more easily and plays a dominant role. Oxidation is in most cases accompanied by partial crystallization of the solid at temperatures far below the usual crystallization temperature. The oxidation causes in general large structural and morphological changes of the precursor, which result in a drastic increase of the surface area of the material. Surface areas in the range of 50–100 m^2/g are achievable by oxidation/reduction treatment of glassy metal precursors. The reduction step is mainly necessary to reduce the transition metal oxides formed to the corresponding metals. The result of this pretreatment is a material containing metal particles imbedded in an oxidic matrix of the more electropositive element (e.g., Zr, B, P). For illustration of this procedure, we may consider the preparation of metal/zirconia catalysts by oxidation/reduction treatment of corresponding metal–zirconium alloys.

Nickel/Zirconia from $Ni_{64}Zr_{36}$. Nickel on zirconia can be prepared by controlled oxidation of the amorphous Ni–Zr alloy in air and subsequent reduction in a hydrogen atmosphere [4.64, 65]. The oxidation in air in the temperature range 570–750 K results in solids containing ZrO_2 and metallic nickel, besides unreacted amorphous metal alloy. Significant oxidation of Ni to NiO is only observed after almost complete oxidation of the zirconium in the alloy. Figure 4.5 depicts the XRD patterns of the amorphous $Ni_{64}Zr_{36}$ alloy corresponding to different degrees of oxidation (α) of the amorphous metal alloy; α was measured gravimetrically and denotes the fraction oxygen consumed divided by the amount of oxygen required to convert Zr to ZrO_2. The XRD patterns indicate the build-up of small crystalline particles of tetragonal and monoclinic ZrO_2 and metallic nickel upon oxidation.

The crystallization behavior of as-prepared samples investigated by DSC measurements under an inert gas atmosphere is shown in Fig. 4.6. Note that the temperature range of crystallization does not depend significantly on the degree of oxidation of the alloy. This behavior is further supported by the observation that the specific heat of crystallization referred to the unreacted core of the alloy is constant ca. 40 J/g, regardless of the degree of oxidation of the alloy sample [4.64]. The presence of zirconia in the oxidized alloy was found to have only little influence on the crystallization behavior of the unreacted part of the alloy. The chemical and structural changes of the bulk are accompanied by similar drastic changes in the textural properties of the alloy. The BET surface area of the precursor material (0.02 m^2/g) increases to 10–25 m^2/g depending on the oxidation conditions used.

The surface-oxidation behavior of the amorphous precursor alloy was investigated by means of XPS and UPS [4.66, 67]. Oxygen doses up to 2000 L were used to study the initial stages of the oxidation of the clean surfaces in the temperature range from room temperature to 570 K. Figure 4.7 compares the XPS Zr 3d spectra of the fresh amorphous $Ni_{64}Zr_{36}$ alloy, the sample after exposure to 80 L O_2, and a ZrO_2 reference sample. The Zr 3d levels of the alloy

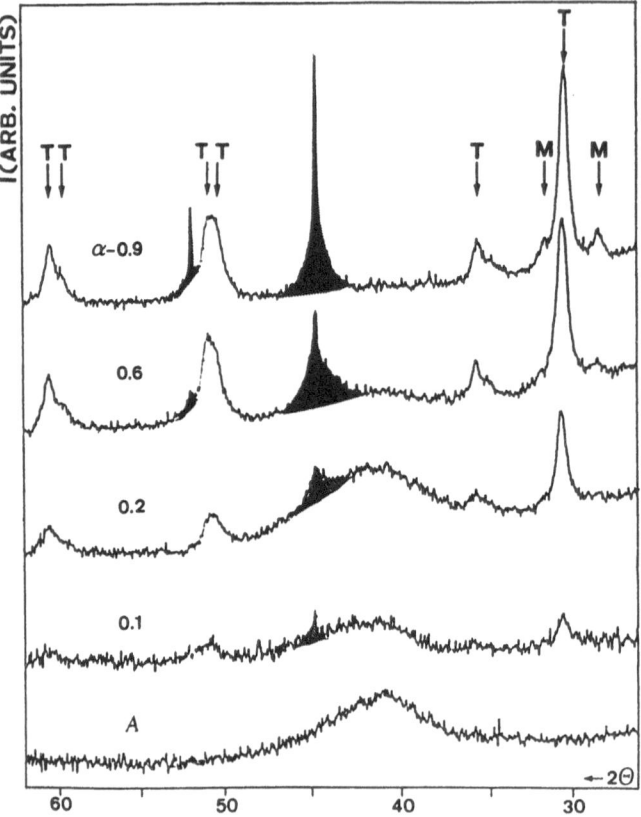

Fig. 4.5. XRD patterns of amorphous $Ni_{64}Zr_{36}$ alloys of different degree of oxidation α [4.64]. Trace A belongs to as-quenched alloy. Reflections of Ni are shaded and arrows indicate positions of main reflections of tetragonal (T) and monoclinic (M) ZrO_2

cleaned by argon ion bombardment are located at $E_b = 179.4$ eV. The shift compared to clean metallic Zr ($E_b = 179.0$ eV) is due to alloying [4.55]. After exposure to 80 L, an additional doublet can be seen, which is attributed to Zr in an oxidized state shifted by 3.1 eV with respect to pure Zr. The different shifts in the Zr 3d core levels of the ZrO_2 reference sample and the sample obtained by exposure to 80 L O_2 reveal a different stoichiometry of these zirconium oxides. Comparison of the observed shifts with literature data indicates that the zirconia formed upon oxygen exposure is deficient in oxygen $ZrO_{1 < x < 2}$. It appears that oxygen-deficient zirconia is formed predominantly in the initial stage of the oxidation, i.e., when metallic zirconium is still present in the sample. With higher degree of oxidation (i.e., when the bulk of the alloy is oxidized) stoichiometric zirconia (ZrO_2) becomes prevalent, as the XRD patterns (Fig. 4.5) indicate.

As-prepared catalysts were found to be suitable for the liquid-phase hydrogenation of *trans*-2-hexene-1-al to hexan-1-ol, which occurs in two subsequent

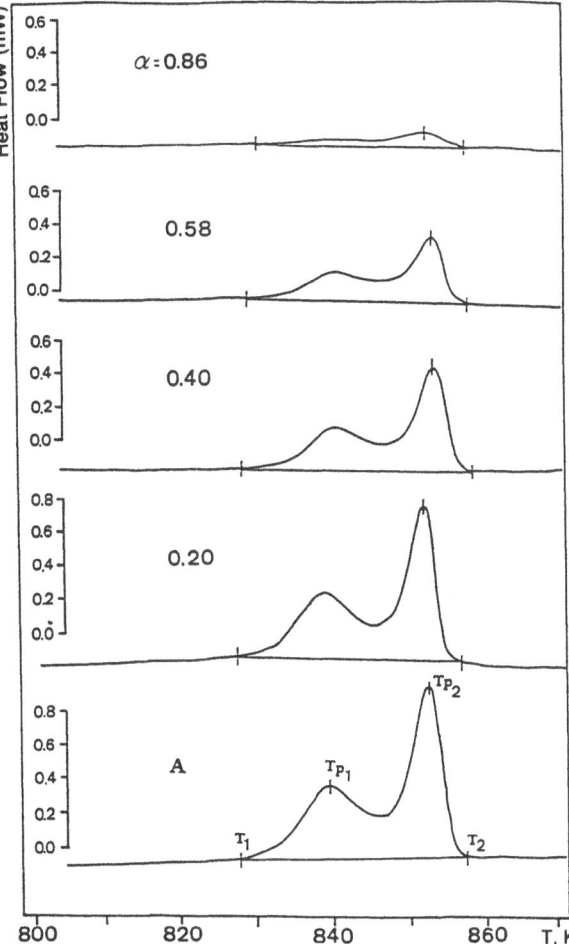

Fig. 4.6. Crystallization behavior of amorphous and partially oxidized $Ni_{64}Zr_{36}$ alloy investigated by DSC measurements. Curve A was measured for as-quenched amorphous alloy. α denotes the degree of oxidation of the sample. Heating rate 5 K/min

steps [4.65]:

$$trans\text{-}2\text{-hexene-1-al} \rightarrow \text{hexen-1-ol} \rightarrow \text{hexan-1-ol} . \tag{4.1}$$

Palladium/Zirconia from $Pd_{33}Zr_{67}$. $Pd_{33}Zr_{67}$ was found to exhibit a significantly different oxidation behavior than $Ni_{64}Zr_{36}$ [4.65]. Figure 4.8 shows the XRD patterns of the amorphous $Pd_{33}Zr_{67}$ alloy after oxidation in air at 590 K for different times. In contrast to the oxidation behavior observed with the Ni–Zr alloy, significant oxidation of the group VIII transition metal occurs already at relatively low degree of oxidation α. Note that the degree of oxidation α is defined here as the fraction oxygen consumed divided by the amount of

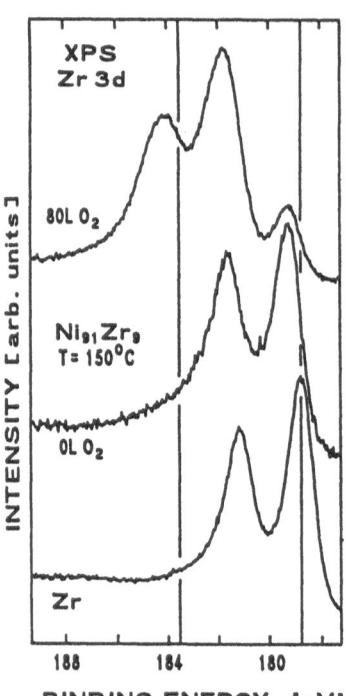

Fig. 4.7. XPS core-level spectra of Zr 3d levels of $Ni_{64}Zr_{36}$ after exposure to 80 L O_2 at 420 K [4.66, 67]. Reference spectra of ZrO_2 (obtained after exposure to 1000 L O_2) and clean metallic alloy (0 L O_2) are shown for comparison. Zr $3d_{5/2}$ core-level positions of Zr and ZrO_2 are indicated by vertical lines

oxygen required to convert the alloy to PdO and ZrO_2. The bulk concentration of metallic Pd first increases and then decreases with increasing degree of oxidation α. The fully oxidized sample contained ZrO_2, PdO and a little Pd. Although monoclinic and tetragonal ZrO_2 exist in the oxidized samples, the monoclinic phase is dominant independent of the degree of oxidation.

The crystallization behavior of the oxidized samples was investigated using DSC. The DSC curves shown in Fig. 4.9 indicate that the as-quenched sample under the conditions used starts to crystallize around 700 K. Most interesting is that the partially oxidized samples show an additional exothermal process at significantly lower temperature. Simultaneous TG measurements revealed that, during both thermal events, the sample weight does not change. The exothermal process occurring at lower temperature is due to the solid-state reduction of PdO by metallic Zr: $2PdO + Zr \rightarrow 2Pd + ZrO_2$, which occurred in the partially oxidized samples. For the sample with $\alpha = 0.3$, this reaction is not complete before the crystallization of the unreacted amorphous alloy. The superposition of the two thermal events is even more pronounced for the sample with $\alpha = 0.5$. Note that the larger the degree of oxidation (α) of the sample, the higher the temperature of the reduction of PdO by metallic Zr. This solid-state reduction is particularly interesting in practical catalysis when the contamination of the active noble metal species has to be avoided. The metallic zirconium present in as-prepared catalysts may act as oxygen scavenger.

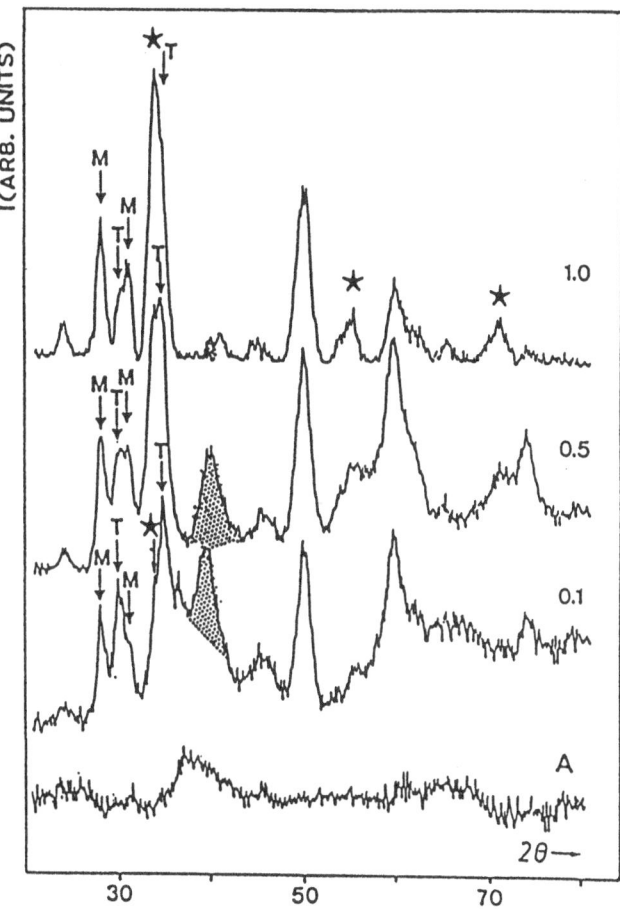

Fig. 4.8. XRD patterns of amorphous $Pd_{33}Zr_{67}$ after different degree of oxidation α indicated on right side [4.65]. Oxidation carried out in air at 590 K. Trace A belongs to as-quenched alloy. Reflections of Pd are shaded, asterisks indicate reflections due to PdO, arrows mark positions of main reflections of monoclinic (M) and tetragonal (T) ZrO_2

As a result of the drastic chemical and structural changes of the bulk material, the BET surface area of the amorphous precursor increases from 0.025 to about 60 m²/g depending on the oxidation conditions. Pore-size distribution measurements using nitrogen capillary condensation indicated that the material contained mainly pores of 2–4 nm size besides some larger pores.

XPS core level spectra of Zr 3d levels of $Pd_{33}Zr_{67}$ measured after different exposures of the alloy to oxygen at 420 K showed similarly to the Ni–Zr alloy the formation of non-stoichiometric $ZrO_{1 < x < 2}$, i.e., zirconia deficient in oxygen. Again the XRD and XPS investigations indicate that oxygen deficient ZrO_2 is prevalent in the initial stage of the oxidation (surface and subsurface region), whereas in completely oxidized samples (bulk oxidation), stoichiometric ZrO_2 prevails.

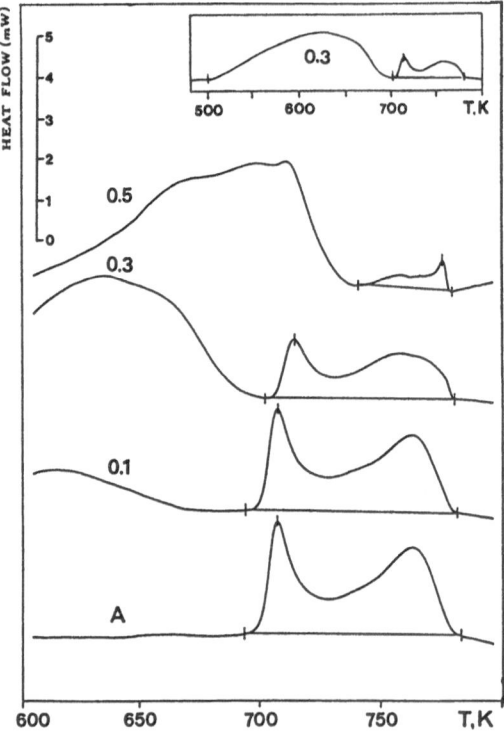

Fig. 4.9. Crystallization behavior of glassy $Pd_{33}Zr_{67}$ alloy after partial oxidation in air at 590 K investigated by DSC under inert gas atmosphere. The degree of oxidation is marked on the curves. Trace (A) corresponds to as-quenched alloy. Inset in upper right corner represents overview of DSC curve for $\alpha = 0.3$ and illustrates the occurrence of the solid-state reduction $2\,PdO + Zr \rightarrow 2Pd + ZrO_2$ (broad signal at 500–700 K) previous to crystallization of the unreacted alloy at higher temperature. Heating rate 5 K/min

As-prepared catalysts showed most interesting behavior in the liquid phase hydrogenation of *trans*-2-hexene-1-al. In contrast to the similarly prepared Ni/ZrO_2 catalyst, which produced selectively hexan-1-ol, with the Pd/ZrO_2 catalyst only the first step of reaction (4.1) is catalyzed and hexen-1-ol can be obtained selectively [4.65].

Copper/Zirconia from $Cu_{70}Zr_{30}$. Copper/zirconia catalysts, which exhibit excellent activity and selectivity for the hydrogenation of buta-1,3-diene to butenes and for the cycloamination of 5-amino-pentanol to piperidine, were prepared by pretreatment of glassy $Cu_{70}Zr_{30}$ in flowing hydrogen at 473 K [4.50]. Without this pretreatment, the glassy alloy showed only poor activity. The pretreatment led to significant changes of the bulk and surface properties of the precursor. X-ray analysis of the glassy alloy before and after the hydrogen pretreatment indicated partial crystallization of the material. After pretreatment, the X-ray patterns exhibited copper reflections superimposed on the broad intensity maximum characteristic for the glassy starting material. DSC measurements carried out with the precursor alloy before and after hydrogen pretreatment confirmed the partial loss of the amorphous character upon the hydrogen pretreatment. The heat released during crystallization of the pretreated sample was about 50% of the one measured for the original glassy alloy. For both samples crystallization started at about 740 K, a temperature which was far above the pretreatment temperature of 473 K. Similarly pretreated crystalline $Cu_{70}Zr_{30}$ exhibited the reflections of $Cu_{10}Zr_7$ together with an unidentified phase, in good agreement with other studies on the crystallization of copper–zirconium alloys [4.68]. Comparison of the X-ray diffraction pattern of the crystalline $Cu_{70}Zr_{30}$ and of glassy and crystalline $Cu_{23}Zr_{77}$ before and after pretreatment indicated that the bulk structure of these materials did not change significantly upon hydrogen pretreatment.

The bulk structural changes occurring with the amorphous $Cu_{70}Zr_{30}$ precursor during the hydrogen pretreatment were accompanied by drastic changes in the surface morphology. The initially flat surface of the amorphous precursor became very rough and exhibited the meander-like pattern shown in Fig. 4.10. This change in the surface morphology caused an increase of the BET surface area from 0.015 to 56 m^2/g. Conversely, SEM and BET surface area

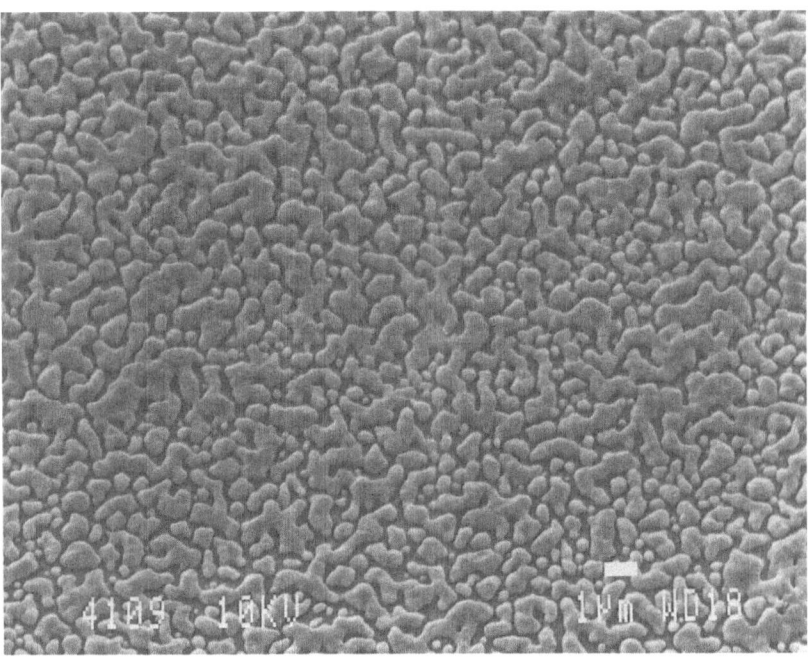

Fig. 4.10. Scanning electron micrograph showing the surface morphology of $Cu_{70}Zr_{30}$ after pretreatment in hydrogen at 473 K for 16 h [4.50]

measurements did not reveal any marked change in the surface morphology of the crystalline $Cu_{70}Zr_{30}$.

Figure 4.11 depicts the effect of the hydrogen pretreatment on the concentration profiles of the constituents for the glassy $Cu_{70}Zr_{30}$ precursor as measured by Auger electron spectroscopy (AES) combined with Argon ion sputtering [4.69, 50]. Most remarkable is that the glassy $Cu_{70}Zr_{30}$ precursor contained already a considerable amount of oxygen due to contamination with air (Fig. 4.11A). For this sample, a markedly lowered Cu concentration, a slight reduction of Zr, and relatively large oxygen contents were measured in a zone near the surface to a depth of about 15 nm. In this region, the ratio Cu/Zr was significantly lower than in the deeper bulk indicating an enrichment of zirconium in the surface region. At a depth exceeding 70 nm, the concentration ratio Cu/Zr reached the bulk value, but the oxygen concentration remained relatively high. Copper was largely metallic as evidenced from the energy of the L(3)M(4.5)M(4.5) Auger transition and from the absence of the satellite of the 2p(3/2) level. The Zr 3d core level spectrum of sputter-cleaned samples indicated the coexistence of metallic and oxidized Zr in the bulk.

The effect of the hydrogen pretreatment on the concentration profiles of the constituents is presented in Fig. 4.11B. The depth profile measurements indicate a drastic enrichment of copper in the surface and subsurface region. The copper

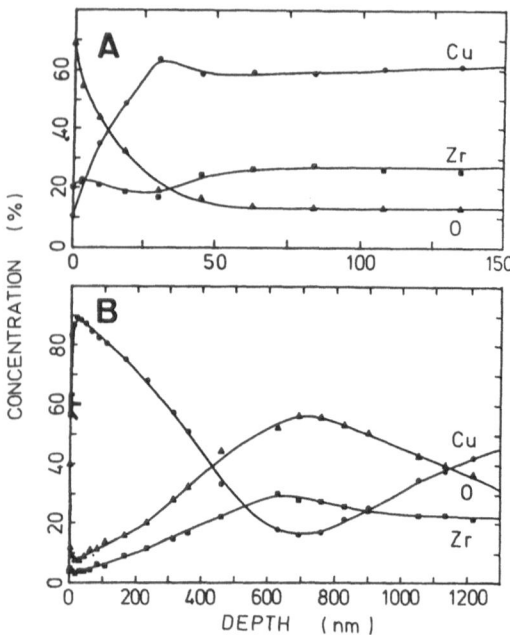

Fig. 4.11. AES depth profiles of amorphous $Cu_{70}Zr_{30}$ precursor before (A) and after (B) pretreatment in a hydrogen atmosphere at 473 K [4.50]. The concentrations were determined from the intensities of the Cu L(2,3)M(4,5)M(4,5), Zr M(2,3)N(1)N(2,3) and O KL(2,3)L(2,3) transitions

concentration reached a maximum of about 90% at a depth of 200 nm and decreased to a minimum at around 700 nm. This indicated that Cu had diffused from the bulk towards the surface. Nominal bulk concentrations of Cu and Zr were only measured at a depth of more than 700 nm. XPS measurement showed strong Zr 3d levels at 183.7 and 186.0 eV indicating that most if not all oxygen was bound to zirconium forming zirconium oxide. The fact that ZrO_2 reflections were seen only very weakly in the XRD pattern of the pretreated sample indicated that this phase was predominantly amorphous.

Interesting information about the genesis of the active catalyst was also obtained by temperature programmed reduction (TPR) studies of the precursor materials. The reduction profiles of glassy and crystalline $Cu_{70}Zr_{30}$ were found to be significantly different, particularly in the lower temperature region, i.e., at temperatures lower than 673 K. The glassy precursor showed marked hydrogen adsorption already at room temperature. Upon heating, first the reduction of superficial copper oxides and the formation of copper hydride occurred. At higher temperatures the copper hydride decomposed and the formation of zirconium hydride became prevalent. The latter phase was also not stable and decomposed at higher temperature. Similar behavior was not observed for the glassy and crystalline $Cu_{27}Zr_{73}$ and the crystalline $Cu_{70}Zr_{30}$ precursor.

Table 4.2 summarizes the butadiene hydrogenation activity and product distributions measured for the catalysts derived from the amorphous and crystalline copper–zirconium precursors. The initial butadiene hydrogenation rate of the catalyst prepared from glassy $Cu_{70}Zr_{30}$ was about an order of magnitude higher than the corresponding rates obtained with the other Cu–Zr precursors and a polycrystalline copper foil, which was used as a reference. Note that the catalyst derived from glassy $Cu_{70}Zr_{30}$ exhibited not only much higher activity but also a largely different product distribution than the other catalysts, a result reflected by the markedly lower value for the but-1-ene/but-2-ene ratio of this catalyst.

Table 4.2. Activities and selectivities for the hydrogenation of buta-1,3-diene to butenes of catalysts derived from amorphous and crystalline copper-zirconium alloys [4.50]. The experiments were performed under the following conditions: 403 K, total surface area of catalyst in reactor 0.066 ±0.003 m², total pressure 180 kPa; partial pressure ratio butadiene: hydrogen = 1

Precursor	Initial butadiene hydrogenation rate [μmol/m² s]	Butene distribution	
		Ratio but-1-ene / but-2-ene	Ratio trans-but-2-ene / cis-but-2-ene
$Cu_{70}Zr_{30}$ (amorph.)	29.9	0.8–1.6[a]	0.4–0.6
$Cu_{70}Zr_{30}$ (cryst.)	3.2	4.0–7.0	0.6–0.8
$Cu_{27}Zr_{73}$ (amorph.)	2.3	4.0–7.0	0.5–0.8
$Cu_{27}Zr_{73}$ (cryst.)	2.6	4.0–8.0	0.6–0.8
Cu (foil)	3.3	4.0–6.0	6.0–0.8

[a] depending on conversion

The above example illustrates how important the knowledge of the solid-state reactions and segregational phenomena is for successful preparation of efficient catalysts from glassy precursors. The occurrence of the copper segregation upon hydrogen exposure at elevated temperature was found to be crucial for successful preparation of copper/zirconia catalysts from Cu–Zr precursors; this segregation depends on various factors such as the structure of the precursor material, the oxygen content, and the chemical composition of the alloy.

Iron on Zirconia from $Fe_{91}Zr_9$. Walz and *Güntherodt* [4.70] prepared small α-iron particles stabilized in a zirconia matrix by oxidation/reduction treatment of an amorphous $Fe_{91}Zr_9$ alloy. The oxidation of the precursor was carried out at 570 K in a pure oxygen atmosphere. Figure 4.12 shows the development of

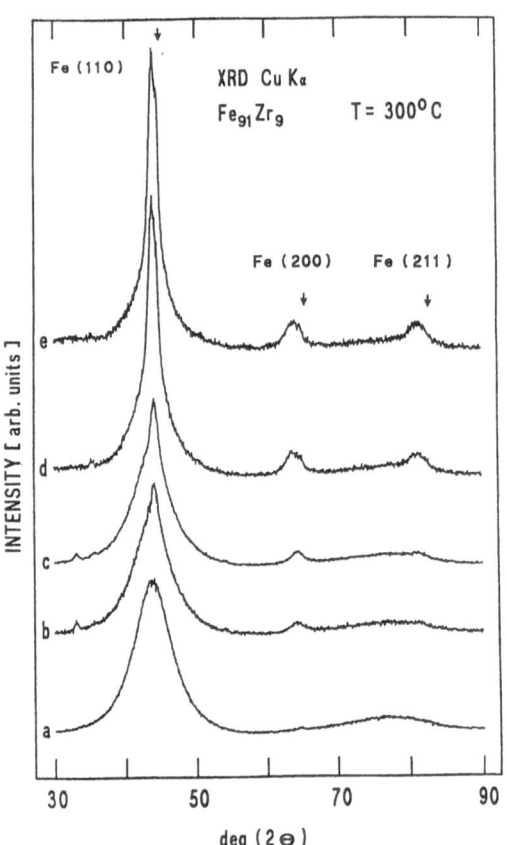

Fig. 4.12. *In situ* XRD patterns of amorphous $Fe_{91}Zr_9$ after different times of oxidation in an atmosphere of 0.9 bar oxygen at 670 K [4.70]. a) as-quenched sample; b) after 69 h; c) after 320 h; d) after 320 h and subsequent exposure to hydrogen (0.8 bar) for 76 h. Positions of main reflections of α-iron are indicated by arrows

the structure of the precursor material as a function of the time of exposure as seen by *in situ* XRD. XPS studies of the surface oxidation [4.66] in an oxygen atmosphere indicated that the surface is very reactive and transforms rapidly in a Fe_2O_3/ZrO_x layer. Note that upon exposure of the oxidized samples to a hydrogen atmosphere, the α-iron X-ray reflections grow markedly, finally reaching a steady-state in which the amorphous $Fe_{91}Zr_9$ has partially transformed into α-iron and a small amount of ZrO_2. Similarly prepared catalysts have been shown to exhibit high ammonia synthesis activity [4.62].

b) Transformation of Precursor Under Reaction Conditions

In several investigations, the active catalysts were derived from the glassy metal precursors by exposing them to reaction conditions [4.1, 13, 14, 15, 23, 24, 47, 71].

For illustration, we may consider the preparation of a palladium/zirconia catalyst highly active for the oxidation of CO [4.47, 71], the preparation of a copper/zirconia catalyst for the hydrogenation of CO_2 [4.23], and the preparation of iron/zirconia for ammonia synthesis [4.44].

Palladium/Zirconia Catalyst for CO Oxidation. Amorphous $Pd_{33}Zr_{67}$ initially does not exhibit significant activity for CO-oxidation when packed into a fixed-bed reactor and exposed to CO-oxidation conditions. Figure 4.13A shows how the activity of amorphous $Pd_{33}Zr_{67}$ develops when the sample is exposed to CO-oxidation conditions at 553 K. The initially inactive material transforms into a highly active catalyst, a transformation accompanied by large structural and textural changes of the material. The X-ray diffraction pattern of the active catalyst (Fig. 4.13B) indicates that the precursor is transformed into a solid containing monoclinic and tetragonal ZrO_2, PdO, metallic Pd and, as dominant palladium phase, a solid solution of oxygen in Pd. The latter phase is characterized by a shift of the main reflections of Pd towards lower 2θ. X-ray photoelectron spectroscopy [4.72] revealed that, in the surface and subsurface region, zirconia existed as non-stoichiometric $ZrO_{1 < x < 2}$.

The BET surface area as well as the palladium metal surface area of the precursor increases by more than two orders of magnitude during the *in situ* activation. The solid-state reactions occurring in the metallic glass during *in situ* activation result in a large increase of the BET surface area from 0.02 to 45.5 m^2/g. The palladium metal surface area of the as-prepared catalyst determined by CO chemisorption is 6.9 m^2/g, which corresponds to a palladium dispersion of about 6%.

Figure 4.14 compares the intrinsic activity of palladium in the catalyst prepared from the metallic glass with the corresponding activity of palladium in a palladium-on-zirconium catalyst prepared by conventional impregnation (incipient wetness) of zirconium dioxide with a palladium salt $((NH_4)_2PdCl_4)$. Note the markedly higher turnover frequency measured for the Pd/ZrO_2 catalyst prepared from the metallic glass as compared to the conventionally prepared catalyst.

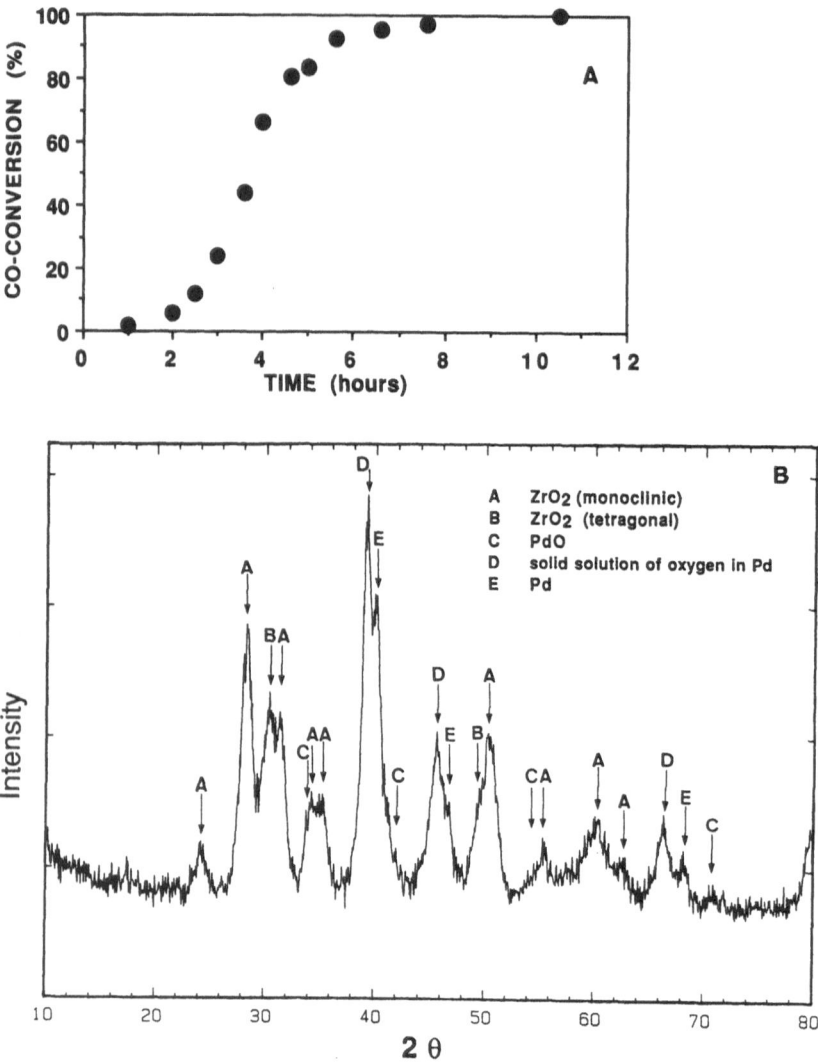

Fig. 4.13. Behavior of glassy $Pd_{33}Zr_{67}$ under CO oxidation conditions [4.71]. A) Development of CO oxidation activity of glassy alloy with time on stream during exposure to reaction conditions. Conditions: fixed-bed reactor; temperature, 553 K; feed gas mixture, 17% CO, 17% O_2, 66% N_2; flow rate 2.5 cm^3/s (STP), 1 bar; 0.3 g of alloy. B) X-ray diffraction pattern of final active catalyst after 6 h on stream (Cu K_α)

Scanning electron microscopy and electron dispersive X-ray analysis showed that the initially flat surface of the precursor alloy changes to a rough surface which is built up of small agglomerates containing zirconia and palladium. High-resolution electron microscopy as well as electron diffraction (Fig. 4.15) showed that the agglomerates are made up of intimately mixed intergrown small crystallites of zirconia and palladium. Due to this particular structural property,

Fig. 4.14. Comparison of CO oxidation activities of palladium-on-zirconia catalysts prepared by *in situ* activation from amorphous $Pd_{33}Zr_{67}$ (●) and by conventional impregnation of zirconia with a palladium salt (○). Arrhenius plots of the turnover frequencies are plotted. Conditions: reactant gas mixture, 1700 ppm of CO, and 1700 ppm O_2 in nitrogen; flow rate, 150 ml (STP) min^{-1}; amount of catalyst, ●: 0.37 g; ○: 1.24 g

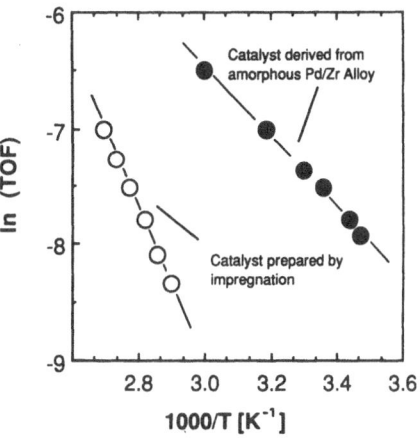

Catalyst derived from amorphous Pd/Zr Alloy

Catalyst prepared by impregnation

ln (TOF)

1000/T [K^{-1}]

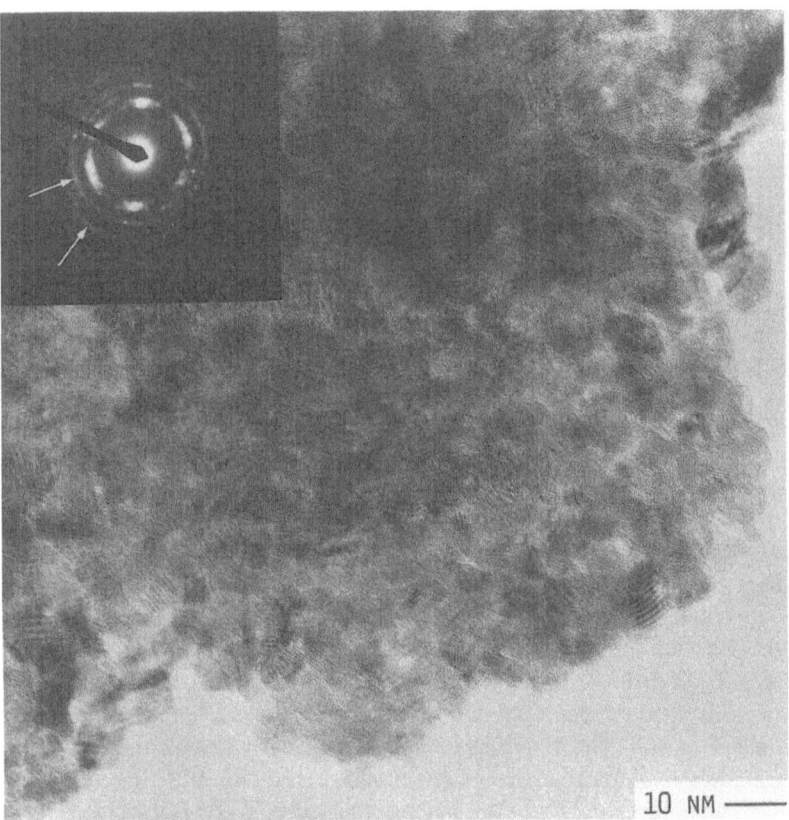

10 NM

Fig. 4.15. Morphology and structure of catalyst prepared by *in situ* activation under CO oxidation conditions as seen by high-resolution electron microscopy and electron diffraction [4.71]

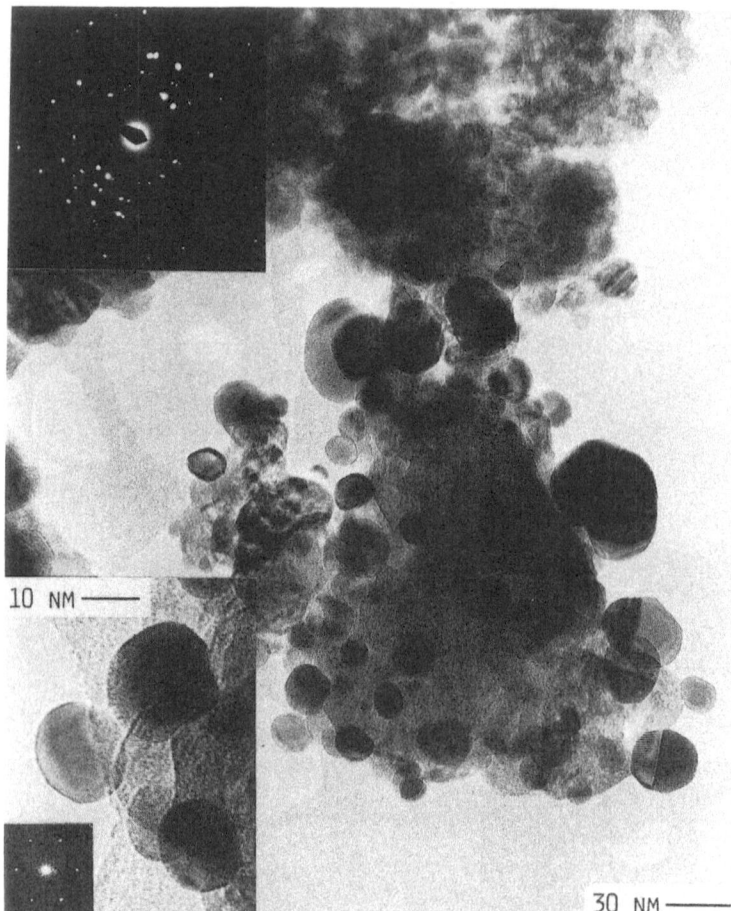

Fig. 4.16. Morphology of conventionally prepared palladium on zirconia catalyst as seen by high-resolution electron microscopy and electron diffraction [4.71]. The electron diffraction pattern give evidence for the presence of only well-crystallized species. As inset a typical palladium *single crystallite* with the corresponding optical diffraction pattern is shown

as-prepared catalysts exhibit an extremely large interfacial area between the active metal species and the oxidic support material. It is interesting to compare the electron micrograph of the catalyst derived from the amorphous alloy (Fig. 4.15) with the micrograph of a palladium-on-zirconium catalyst prepared by conventional impregnation (incipient wetness) of zirconium dioxide with a palladium salt (Fig. 4.16). This comparison clearly reveals the large difference in the interfacial area between the palladium and zirconia phases.

The higher activity of the palladium species in the catalyst derived from the glassy metal has been attributed to the unique structural and morphological properties of this catalyst. It has been suggested that the large interfacial area,

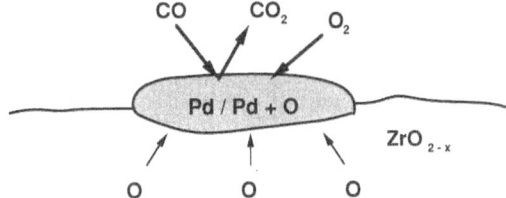

Fig. 4.17. Schematic illustration of the structural properties and the CO oxidation mechanism of palladium/zirconia catalyst derived by *in situ* activation from glassy $Pd_{33}Zr_{67}$ precursor

the build up of a solid solution of oxygen in the palladium, and the presence of the non-stoichiometric zirconia facilitate the transfer of oxygen through the ion conducting zirconia to the active palladium phase and thereby enhance the CO-oxidation rate. This behavior is illustrated schematically in Fig. 4.17. A similar oxygen transfer is unlikely to play a role in the structure of the conventionally prepared Pd/ZrO_2 catalyst due to the very low interfacial area and the lack of oxygen deficient zirconia surrounding the metallic phase.

Copper/Zirconia Catalyst for the Hydrogenation of CO_2 to Methanol. When exposed to CO_2 hydrogenation conditions, amorphous $Cu_{70}Zr_{30}$ transforms into a highly active copper/zirconia catalyst [4.23]. Figure 4.18A depicts the change of the activity and product distribution of the glassy precursor alloy during *in situ* activation at 493 K. Note that the initial CO_2 conversion is very low. The activity develops with time on stream and passes through a maximum reaching finally a steady-state after about five hours on stream. The steady-state activity is slightly lower than the calculated equilibrium conversion (dashed line in Fig. 4.18A). In the initial period, the selectivity to methanol passes through a maximum at about 80% and then decreases reaching a steady-state value of about 50%. The products formed under steady-state conditions are methanol, CO, and water. As a by-product, some ethanol is formed during the initial period of activation.

The XRD pattern of the final catalyst (Fig. 4.18B) show mainly the reflections of larger crystalline (mean size about 25 nm as estimated by X-ray diffraction line broadening) and smaller disordered particles of copper. The latter gave rise to a significant broadening of the footings of the copper reflections in the XRD pattern. AES depth profiling and XPS measurements of the active catalyst revealed that zirconium was almost quantitatively transformed into zirconia under reaction conditions, the oxidizing agents are CO_2 and H_2O. Simultaneously to the oxidation of zirconium, copper segregates onto the surface resulting in a drastic increase of the copper concentration on the surface. This behavior was confirmed by measurement of the copper surface area using N_2O chemisorption. The copper surface areas of the final catalysts were between 4 and 8 m^2/g depending on the reaction conditions. Similarly, the BET surface area increased to 46–79 m^2/g due to the solid-state reactions.

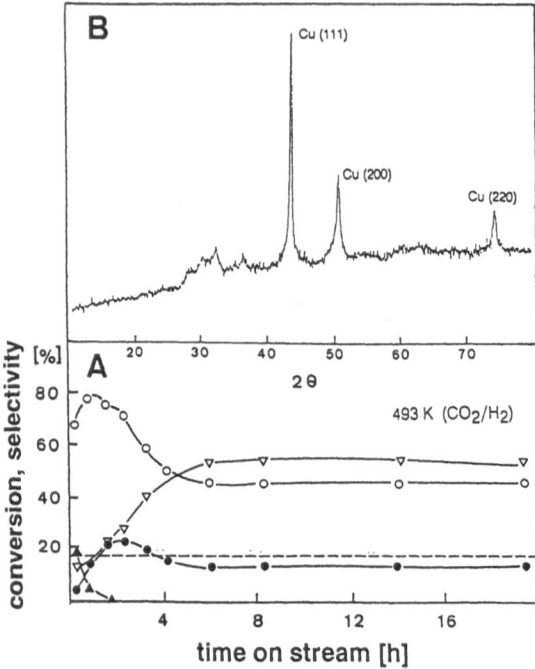

Fig. 4.18. Catalytic behavior and structural changes of glassy $Cu_{70}Zr_{30}$ alloy during exposure to CO_2 hydrogenation conditions [4.23]. A) Change of CO_2 hydrogenation activity and product distribution as a function of time-on-stream. Dashed line indicates the calculated equilibrium conversion. Symbols: CO_2 conversion ●; selectivities to methanol ○, carbon monoxide ▽, and ethanol ▲. Hydrogenation conditions: 1.2 g of sample, feed rates of reactants: CO_2, 2.3 mmol/s; H_2, 7.6 mmol/s; total pressure 15 bar. B) X-ray diffraction patterns of active sample after steady-state conversion was reached (Cu K_α)

As-prepared catalysts exhibited selectivities up to more than 80% with regard to methanol, when used at temperatures lower than about 470 K. It should be mentioned that, using a similar activation procedure, palladium/zirconia catalysts were prepared from glassy $Pd_{33}Zr_{67}$, which were highly active for the hydrogenation of CO_2 to methane [4.24].

Iron/Zirconia Catalyst for Ammonia Synthesis. Highly active ammonia synthesis catalysts were prepared successfully from amorphous $Fe_{91}Zr_9$ by exposing the alloy to ammonia synthesis conditions (690 K, 9 bar, stoichiometric mixture of H_2/N_2) [4.44, 59, 62]. Under these conditions, the initially almost-inactive glassy alloy starts to crystallize and undergoes a sequence of structural and chemical changes, which finally lead to a highly active and stable ammonia synthesis catalyst. Figure 4.19A depicts the change in the ammonia synthesis activity of the glassy metal during the *in situ* activation. After about 500 h on stream, the catalyst derived from the glassy metal alloy reaches an intrinsic activity which is more than an order of magnitude higher than the one observed

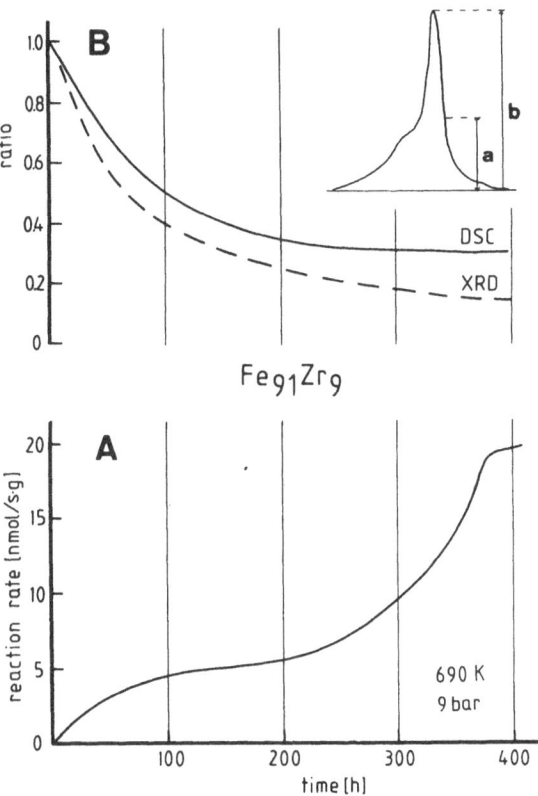

Fig. 4.19. Catalytic behavior and bulk structural changes of glassy Fe₉₁Zr₉ alloy during exposure to ammonia synthesis conditions [4.44]. A) Change of activity of glassy Fe₉₁Zr₉ alloy as a function of time on stream during *in situ* activation under ammonia synthesis conditions (690 K, 9 bar, stoichiometric feed of 40 ml (STP)/min). B) Crystallization of glassy alloy during *in situ* activation as determined by DSC and XRD analyses. The integral amount of heat evolved in the DSC of the residually glassy material is plotted in the upper curve. The inset defines the ratio a:b which was used to quantify the crystallization from the XRD data. The connection between crystallization and the evolution of the a:b ratio in XRD is given by the growth of the main peak for α-iron (increase in the absolute scattering power) as a consequence of crystallization

for polycrystalline iron. The time necessary for the activation can be drastically reduced (to a few hours) by exposing the glassy alloy to oxygen pulses prior to the *in situ* activation [4.44].

Figure 4.19B shows the progress of crystallization during the *in situ* activation as determined by differential scanning calorimetry (DSC) and X-ray diffraction (XRD). The bulk structural changes of the glassy metal alloy during the *in situ* activation are reflected by the XRD traces shown in Fig. 4.20, which depicts the results of two analyses from samples taken at typical points in the activation curve (Fig. 19A); namely, in the low activity region (top trace in Fig. 4.20) and in the final stage of activation (bottom trace in Fig. 4.20). Comparison of the two wide scans in Fig. 4.20 shows the significant increase in

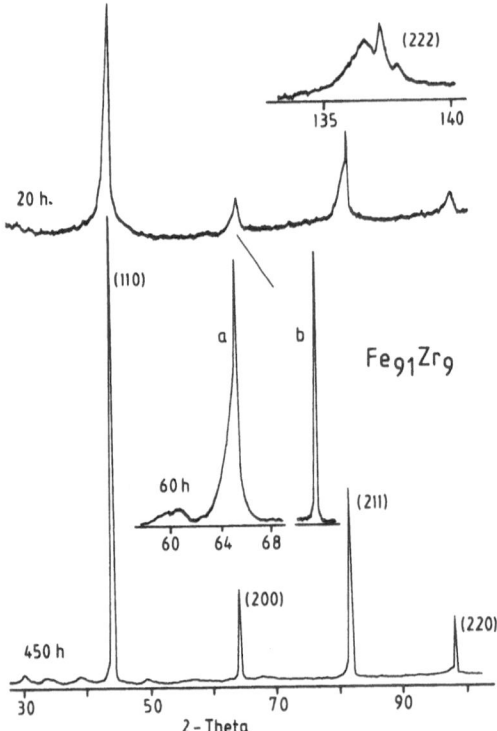

Fig. 4.20. XRD traces of active sample derived from milled glassy iron–zirconium alloy after 20 h (top) and 450 h of activation under 9 bar of a 3:1 mixture of hydrogen and nitrogen at 700 K [4.44]. The top inset shows a high-resolution scan over the Fe (2 2 2) reflection, which consists of two doublets arising from the $CuK_{\alpha 1}/CuK_{\alpha 2}$ splitting. The central inset compares the Fe (2 0 0) reflection of a sample exposed to *in situ* activation for 60 h (a) and of a sample crystallized thermally in an inert gas atmosphere (b)

the scattering power for the activated sample, the iron (1 1 0) peak is nearly eightfold more intense than in the top trace. The zirconium oxide phase is clearly visible in the bottom trace; even in the final catalyst, it is however still poorly crystallized. The final stable catalyst consists of iron particles stabilized by poorly crystalline zirconia. Two forms of iron can be distinguished: larger particles of well-crystalline α-iron and, as a minority phase, small particles of disordered iron with a considerably larger lattice constant, which has been designated δ iron.

Figure 4.21 shows the XPS core level spectra of the Fe 2p and Zr 3d electrons measured for the stable active catalyst. The outer surface is covered with iron oxide (in hematite-like forms) and zirconia exists as non-stoichiometric ZrO_{2-x}. In the iron 2p spectrum, a contribution of metallic iron is visible indicating that the surface oxide film is thin within the information depth of XPS (ca. 2.5 nm). It has been suggested that the surface oxide stabilizes the iron

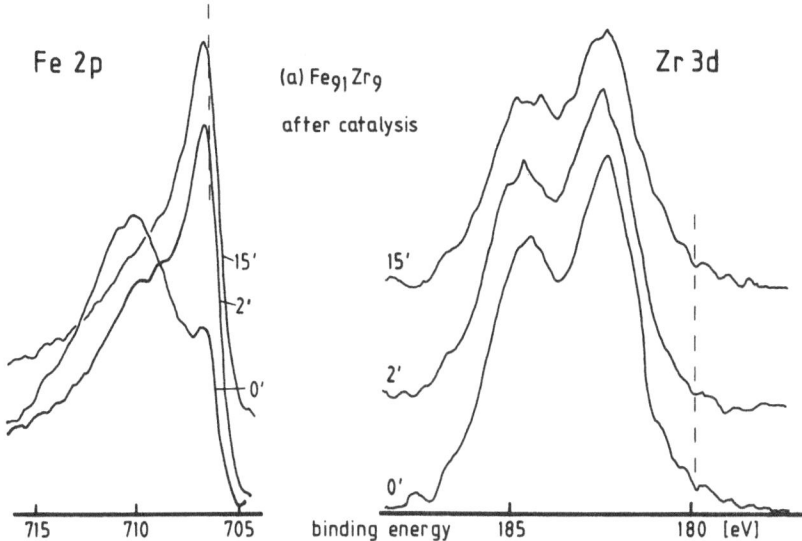

Fig. 4.21. XPS core-level spectra of final stable active catalyst derived by *in situ* activation of glassy $Fe_{91}Zr_9$ [4.44]. Times in minutes refer to Ar sputtering (scanning ion source at 5 kV with 10 mA emission at 2×10^{-6} Torr). The dotted lines denote centroid positions of elemental Fe (2p-3/2) and Zr (3d-5/2)

particles in a similar fashion as inclusions of iron aluminium oxides stabilize the technical ammonia synthesis catalyst.

It is noteworthy that structural investigations by means of *in situ* XRD [4.73] showed the presence of α-iron in catalysts prepared from unpromoted Fe_3O_4 after treatment with hydrogen at 723 K. In contrast, however, promoted industrial (BASF and ICI) ammonia synthesis catalyst precursors produce almost fully reduced catalysts with very broad diffraction peaks as a result of the action of either hydrogen or the reacting gas mixture at about 700 K. The XRD patterns were very different from those of finely divided and paracrystalline α-iron, but showed remarkable similarity to those of certain iron-containing glassy metals. These observations indicated that the catalytically active phase in the reduced industrial ammonia synthesis catalysts is probably amorphous iron, interdispersed with microcrystalline promoter phases (e.g., $FeAlO_3$).

c) Chemical Etching of Precursor

Chemical etching can be most effective in modifying the catalytic behavior of glassy metals by changing the chemical and structural properties of the surface of the materials. The removal of oxidic surface layers, which may originate from the fabrication (melt-spinning) or from exposure of the alloy to atmospheric conditions has been shown to be crucial in many catalytic applications. However, etching may also cause drastic changes in the catalytic behavior due to alterations of the structural and chemical properties of the surface. The etching is mostly carried out using acids such as HCl [19], HNO$_3$ [4.20, 28, 36, 37], HClO$_4$ [4.39, 41] and HF [4.39, 41]. However, other chemical agents such as bases or complex-forming compounds may also be useful for this procedure. Particularly promising is the use of selective chemical etching processes where a specific component of the alloy is extracted by the etching agent. Such selective etching can be used to prepare porous catalysts with a skeletal type structure, known as Raney-type catalysts [4.74]. Glassy metals may offer considerable advantages for such catalyst preparations due to their structural and chemical isotropy and their high reactivity. This has been demonstrated by *Yamashita* and co-workers [4.41], who prepared Raney-type copper catalysts from amorphous Cu$_{62}$Zr$_{38}$ alloys.

Extraction of zirconium species from the amorphous pulverized Cu–Zr alloy by treatment with HF gave a highly porous Raney-type copper catalyst. The catalytic activity of this Raney-type copper catalyst for the hydrogenation of ethene and isoprene was much higher than that of the catalyst prepared from the corresponding pulverized crystalline alloy. The great enhancement of the activity of the catalyst derived from the amorphous alloy was attributed to its porous structure, the high surface concentration of copper, the high dispersion, and the electronic state of the copper sites, which was found to be influenced by the unextracted surface zirconium species.

A considerable advantage of the etching method is that relatively large surfaces can be prepared probably without complete loss of the disordered structure of the amorphous alloy surface. However, so far only very little is known about the structural changes amorphous alloys undergo during etching and further work will be necessary to explore this potentially interesting route of catalyst preparation.

4.4 Factors Influencing Structural and Chemical Properties of Catalysts Derived from Glassy Metals

Important properties of glassy metals influencing the structural and chemical properties of the catalyst derived from them are: (i) chemical composition; (ii) chemical and structural homogeneity; (iii) thermal stability and crystallization behavior; (iv) oxidation behavior; (v) dissolution of gases; and (vi) segregation phenomena. These factors together with the conditions used for the chemical transformation of the precursor are crucial to obtain catalysts with the desired properties.

4.4.1 Chemical Composition

The chemical composition influences virtually all properties discussed subsequently and is therefore a controlling factor in the preparation of catalysts from glassy metals. We should note that up to now only a limited range of glassy metals have been used in catalysis research (Table 4.1). This has so far seriously limited the possible potential of these materials for catalysis. Future success in applying glassy metals in catalysis will largely depend on the possibility to prepare glassy metals with desired chemical compositions. In view of the large number of possible binary or multicomponent metallic systems, the number of alloy melts which can be quenched into the glassy state is still relatively small and by far not as large as one would expect from the fact that thermodynamical constraints are less stringent for metastable solids. Rigorous limitations with regard to the constituents and their concentration ranges have still to be dealt with. Very different cooling rates are necessary (10^2–10^{12} K/s) depending on the elements and the alloy composition. Even with the highest cooling rates as achieved by pulsed-laser quenching, crystallization during the cooling cannot be avoided for a large number of alloy melts.

The glass-forming ability, i.e., the ability of an alloy melt to be quenched into the glassy state with available cooling rates, is determined by energetic and kinetic factors, which are often interdependent [4.75]. Several different factors influencing the glass-forming ability in metallic melts have been discussed, among them, thermodynamics, kinetics, stoichiometry, concentration of valence electrons, atomic sizes of constituents and electronegativity. Also a *confusion principle* has been proposed, to the effect that complex mixtures of constituents have greater glass-forming tendency than binary mixtures. Frequently, the search for alloys that readily yield glassy metals is aided by the fact that eutectic compositions are favored.

Glass formation is generally thermodynamically favored in those systems and concentration ranges for which the free energy of the metastable amorphous state differs as little as possible from that of the corresponding metastable or stable crystalline state. Under these conditions, only small thermodynamic driving forces for crystallization processes exist for cooling below the liquidus temperature. In addition to the presence of small thermodynamic driving forces, glass formation is favored by kinetic factors which negatively influence nucleation and growth of crystalline phases during the quenching process. Generally, the nucleation of crystalline phases is impeded whenever the chemical short-range ordering or the composition of the associates does not coincide with that of one of the crystalline phases in the glass-formation region.

4.4.2 Chemical and Structural Homogeneity

Ideally, an amorphous metal alloy should be chemically and structurally isotropic. Isotropic surfaces would be ideal for catalytic applications, since any structural and chemical heterogeneity can impede the selectivity with which a particular chemical reaction is catalyzed. However, several studies [4.18, 50, 57–59] reveal that the surface of glassy metals prepared by melt-spinning may not be as homogeneous as expected. Structural as well as chemical inhomogeneities have been observed. Chemical and structural anisotropies in the bulk and surface region not only lower the catalytic selectivity of such materials but also lead to nonuniform propagation of the solid-state reactions occurring during the transformation of the amorphous precursor to the active catalyst. Inhomogeneities are frequently due to either too-slow cooling rate [4.18], contamination [4.18, 50] and thermally and chemically induced segregation of constituents [4.50]. All these phenomena largely impede the control of the structural and chemical properties of the alloy surface, an important requirement for catalytic applications.

4.4.3 Thermal Stability and Crystallization Behavior

The thermal stability is a severe limitation if the metallic glass is to be used in as-quenched state for catalysis; however, that is not necessarily the case if the glassy alloy is used as catalyst precursor. The thermal stability is mainly influenced by the chemical composition of the metallic glass and the medium to which it is exposed. It has been shown that the crystallization temperature can be significantly lowered in the presence of a hydrogen atmosphere [4.23, 24, 31, 50] or an adsorbed organic compound [4.76].

Glassy metals have been found to crystallize by nucleation and growth processes. The driving force is the difference in free energy between the glass and the appropriate crystalline phase(s). Depending on the composition, crystallization may occur by: (i) primary crystallization, where one crystal phase with a composition different from the amorphous matrix is produced; by (ii) polymorphic crystallization, where one phase with the same composition as the glass is crystallized (occurs only in concentration ranges near the pure elements or compounds); or by (iii) eutectic crystallization, where two crystalline phases grow concomitantly by a discontinuous reaction. Most glassy metals can crystallize by two or more different reactions. The route by which crystallization occurs depends not only on the thermodynamic driving force (difference in free energy), but also on the kinetics of the possible routes. In the case of pretreatment of the metallic glass precursor in reactive gas atmospheres, solid-gas phase reactions are likely to influence the expected crystallization behavior.

During rapid solidification as well as during annealing treatments, surfaces are expected to catalyze nucleation as the crystalline phase replaces a portion of the surface, thus reducing the total energy required for nucleation. An important factor for crystallization is the oxygen content near the surface. Oxygen may stabilize a number of crystalline phases, thus increasing the driving force for crystallization. Selective oxidation of one of the components, e.g., the metalloids at the surfaces of metal-metalloid glasses, is likely to result in excessive crystallization of the metal (e.g., copper [4.21, 31, 50], palladium [4.16, 24], nickel [4.13, 45] and iron [4.32, 44] in binary zirconium alloys). Selective oxidation is likely to exhibit the strongest influence on surface crystallization. Even at temperatures far below any crystallization event in the bulk glass, primary crystallization of the transition metal has been observed with metal-metalloid glasses [4.77].

It should also be noted that the crystallization behavior of melt-spun ribbons may be different on both ribbon sides [4.18]. Nucleation for primary crystallization of the transition metals is observed on both sides of the glassy ribbons, while other crystallization reactions have been observed to prefer usually either the free surface or the contact side of the ribbon [4.77]. This phenomenon may lead to different structural and chemical properties of the two ribbon sides, and consequently also to large anisotropy in the catalysts prepared from such ribbons [4.18, 19].

4.4.4 Oxidation Behavior

The oxidation behavior of glassy metals is of importance in many of their practical applications and particularly in catalysis. In view of this, surprisingly little is known about their oxidation behavior at the fundamental level. Most of the investigations reported in the literature deal with the oxidation of metal-metalloid alloys [4.18, 78–81] and group VIII transition metals–zirconium alloys [4.64, 65, 82–86]. When discussing the oxidation behavior and its influence on the catalytic applications of glassy metals, it is meaningful to distinguish between surface and bulk oxidation.

Surface oxidation behavior is particularly important with regard to the use of glassy metals in as-quenched state. A general observation made with binary alloys is that the more electropositive constituent of the alloy tends to segregate to the surface upon oxidation. This procedure can occur already at lower temperature, and consequently the surface of freshly prepared alloys is likely to be covered with a thin layer of oxides of this constituent. This phenomenon has certainly contributed to controversy with regard to the catalytic properties of glassy metal surfaces, since in many of the earlier investigations little care was taken of this behavior and authors tacitly assumed that the surface composition resembles the bulk composition of the quenched materials.

The effect of oxygen exposure during the quenching process has nicely been demonstrated by *Guczi* and coworkers [4.17, 18], who studied the structural and chemical properties of Fe–B alloys prepared by melt spinning under atmospheric conditions. They found drastic differences in the chemical and structural properties between the dull (in contact with copper wheel) and shiny side (exposed to air) of as-prepared ribbons. Such a behavior is likely to occur with alloys containing constituents with largely different heat of formation.

According to the theoretical models of *Mott-Cabrera* [4.87] and *Grimly* [4.88], it is expected that the constituent with the largest heat of formation for the corresponding oxide segregates to the surface. This behavior was evidenced experimentally by *Walz* [4.66], who studied the surface oxidation behavior of Ni–Zr, Pd–Zr and Fe–Zr glassy metal alloys using photoelectron spectroscopy (XPS and UPS). He found that the segregation of zirconium depends largely on the oxygen exposure and the chemical composition of the alloy, but only little on temperature. The zirconium oxides formed upon oxidation are nonstoichiometric ZrO_x ($1 < x < 2$), as was evidenced by the chemical shifts of the Zr 3d binding energies. Prevalent formation of oxygen-deficient zirconia in the surface and subsurface region of metal zirconium alloys has also been detected in other investigations [4.44, 45, 72] and seems to be characteristic for these alloys.

Under more drastic oxidation conditions (higher temperatures and oxygen partial pressures), bulk oxidation occurs. Bulk oxidation generally requires the transfer of oxygen from the surface to the bulk; consequently the compactness of the oxide surface layer plays an important role with regard to the bulk oxidation kinetics. Compact oxide surface layers are likely to slow down or even inhibit bulk oxidation due to slow diffusion of oxygen through the dense oxide layer to the unreacted core of the alloy. Mostly the bulk oxidation leads to partial crystallization and phase separation already at temperatures far below the crystallization temperature of the amorphous alloy due to the removal of the oxidized constituent from the alloy, a likely occurance if the concentration range of the constituents where glassy metals can be formed is narrow.

The bulk oxidation generally leads to large stresses in the materials due to the differences in the specific volumes of the unreacted part of the alloy and the oxidized part of the material. This stress eventually results in embrittlement of the material, a behavior which has to be accounted for if catalysts are prepared by oxidation of metallic glass precursors.

4.4.5 Dissolution of Gases in Glassy Metals

The dissolution of gases is frequently different in glassy metals than in their crystalline counterparts due to the marked differences in structural and electronic properties. As regards glassy metals, present knowledge concentrates mainly on the absorption of hydrogen. Since most glassy metals or catalysts derived from them were exposed to a hydrogen containing atmosphere either during reaction or activation, understanding the interaction and solid-state reactions induced by hydrogen seems to be crucial. *Maeland* et al. [4.89] have shown that the solubility (absorption capacity) of hydrogen in glassy metals of the general formulas $Ti_{1-x}Cu_x$ and $Zr_{1-x}Cu_x$ ($x = 0.3$–0.7) is larger than in corresponding crystalline alloys. Besides its possible direct influence on the catalytic properties of glassy metals as a hydrogen source, the absorption of hydrogen generally enhances the formation of metal hydrides, which have been shown to be crucial intermediates in the preparation of catalysts from amorphous [4.50] and crystalline [4.90, 91] metal alloys. Unfortunately, no similar studies are presently available for the solution of other gases in glassy metals.

4.4.6 Segregation Phenomena

The surface composition of metal alloys is often different from that of the bulk. Major driving forces for surface segregation revealed by model calculations [4.92] are: different surface free energies of the components and size mismatch in the case of clean surfaces, as well as different heats of chemisorption and reaction of components in the presence of adsorbates. Surface segregation induced by selective oxidation is well known for crystalline and amorphous alloys of the type A–B, where A is an early transition metal or rare earth metal (e.g., Zr, Ti, lanthanides or actinides), and B a group VIII (e.g., Ni, Fe, Pd) or group IB metal, (e.g., Cu, Au). Upon exposure of the alloy to oxygen, component A (the more electropositive element) is oxidized and enriched at the surface [4.44, 50, 66, 92–95]. As a result of this, phase separation may occur, and the remaining atoms of component B may cluster together and precipitate. The phase separation is crucial for the formation of oxide supported metal particles.

Similar segregation phenomena may also occur by adsorption or absorption of hydrogen. However, the enthalpies of hydride formations are much smaller when compared to those of oxide formation. The results on hydrogen-induced surface segregation are rather controversial. The exclusion of oxygen traces in the bulk and surface of the alloys is crucial for proper studies of surface segregation induced by hydrogen.

No significant segregation effect has been measured in crystalline $Cu_{30}Zr_{70}$ and $Cu_{70}Zr_{30}$ [4.93], $LaNi_5$ [4.96], Mn_2Zr, Cr_2Zr, V_2Zr [4.95] and amorphous $Ni_{26}Zr_{76}$, $Cu_{30}Zr_{70}$ [4.93] and $Fe_{24}Zr_{76}$ [4.97], whereas strong surface segregation was reported for amorphous $Cu_{70}Zr_{30}$ [4.50] and Pd–Zr alloys after exposure to a hydrogen atmosphere [4.98].

4.5 Conclusions and Outlook

About a decade has elapsed since glassy metals were introduced to catalysis research. The numerous studies reported in the literature have brought interesting information with regard to the potential and limitations of these materials for catalytic applications, and there are already some rather well-studied reactions and catalysts. In several investigations, it has been shown that glassy metals possess considerable potential as catalytically active materials. From the studies reported so far, two principal opportunities for the use of glassy metals in catalysis emerge, namely, in the as-quenched state or as catalyst precursor.

The motivation for using them in the as-quenched state is based on the ability to tailor the electronic properties and the special surface structure which ideally should be structurally and chemically isotropic and exhibit no long-range ordering of the constituents. The latter property seems, however, partly questionable, at least for surfaces of glassy metals prepared by melt-spinning, as investigations using STM indicated. Surfaces of glassy metals tend to undergo structural relaxation already at temperatures far below the crystallization temperature, and the presence of adsorbed reactants is likely to enhance this relaxation. Thus, in order to make use of the special surface structure of glassy metals, catalytic reactions have to be performed at low enough temperatures, which is a severe limitation, in particular in the light of the intrinsically small surface area of these materials. Hence, progress towards successful technical application of these materials in catalysis will depend on overcoming some of the structural and stability problems inherent in metallic glasses. Although the low-surface-area problem has been attacked by several chemical and physical roughening techniques, further progress towards the preparation of glassy metals with larger surface areas is necessary to make use of the catalytic potential of these materials in technical catalysis. This problem may be overcome by some amorphous alloys prepared by new methods. Promising routes include novel chemical techniques [4.99–103], atomization techniques [4.104] and comminution methods (milling and chopping), which produce amorphous powders with higher specific surfaces. The desired techniques should provide the possibility to deposit the glassy metals in high dispersion on a support material.

The stability problem seems to be receding as more glassy metals are examined under reaction conditions and some are found to be remarkably stable. However, too little information about the factors determining the stability of glassy materials under reaction conditions is still available. There are methods, however, suitable for improving the thermal stability of amorphous materials. Alloying of properly selected components can result in glassy alloys with improved thermal stability or increased activity, which permit low-temperature application.

The use of glassy metals as catalyst precursors has been shown to open new avenues for the preparation of supported metal catalysts with unusual chemical and structural properties. This potential resides mainly on the high reactivity and isotropic nature of these materials compared to their crystalline counterparts. Several efficient supported metal catalysts have been prepared by exposing the precursor alloy to an oxygen containing gas atmosphere at higher temperature and subsequent reduction. *In situ* activation, i.e., the transformation of the precursor alloy to the catalytically active material under reaction conditions, has also been applied successfully. The aim of these (pre)treatments is to transform the alloy into a supported metal catalyst by oxidizing the more electropositive component. Catalysts prepared as such have been shown to possess reasonably large BET surface areas (up to $100 \, m^2/g$) and metal surface areas for technical use. A major difference in their structure compared to conventionally prepared supported metal catalysts is that both the active metal as well as the oxidic support form intimately intergrown structures and consequently the interfacial area between metal particles and support is extremely large leading to very strong metal–support interactions. For most systems, such structures are difficult to prepare by conventional preparation methods.

The solid-state reactions occurring during pretreatment of the glassy metals are complex, and the structure of as-prepared catalysts is influenced by several factors determined by the intrinsic properties of the metallic glass and the conditions of pretreatment. Relevant intrinsic properties are: chemical composition, chemical and structural homogeneity, thermal stability, crystallization and oxidation behavior, dissolution of gases, and segregational phenomena. Better understanding of all these phenomena is a necessary prerequisite for progress in the use of glassy metals as catalyst precursors.

Although not covered in this review, it should be emphasized that glassy metals have also attracted much interest in electrochemistry in recent years, mainly on account of their resistance to corrosion (due to the absence of grain boundaries), but also because of the possibilities of specific electrocatalysis originating from the unusual elemental composition that can be generated in a single phase (see Table 4.1 and [4.9]). Large surface areas are not necessarily required in electrocatalysis, and consequently the materials can be used in as-quenched state, which enables to take advantage of the unusual surface structure of the glassy materials. Glassy metals have been used particularly as electrode materials in water electrolysis, electrolysis of sodium chloride, and in fuel cells for the oxidation of C_1 compounds. These research activities have been reviewed by *Molnar* et al. [4.7].

Studies performed on metallic glass surfaces are likely to aid the answering of some long-standing fundamental questions in catalysis. Glassy metals were suggested to be ideal model systems [4.5] for the study of several catalytic problems, among them, the role of bi-metal and multimetallics, the role of short-range ordering, the electronic and geometric structure of defects, the influence of promoters, and surface segregation and clustering.

Finally, it should be stressed that so far relatively little effort has been undertaken to develop glassy alloys with compositions specially designed for catalytic applications. Most glassy metals studied in catalysis have been those that were prepared for other technical applications. When properly designed with regard to structural and chemical properties, glassy metals could result in catalysts with outstanding activity and selectivity. The results gathered so far with glassy alloys designed for other purposes indicate this clearly. This strategy was nicely demonstrated by *Hara* et al. [4.105–107] for the development of the multicomponent Pd–Ir–Ti–Rh(Ru)–P alloys electrode materials for the electrolysis of sodium chloride. Investigations started with Pd–P alloys; then, with the addition of other components, gradual improvement of different properties was attained, eventually leading to alloys with compositions showing excellent corrosion resistance and higher activity than that of the traditional RuO_2/Ti electrode.

Acknowledgements. Thanks are due to H.-J. Güntherodt for initiating this work. The financial support of our work by Lonza AG and the Swiss National Science Foundation is acknowledged.

References

4.1   H. Komiyama, A. Yokoyama, H. Inoue, T. Masumoto, H. Kimura: Sci. Rep. Res. Inst. Tohoku Univ. Ser. A **28**, 217 (1980)

4.2   G.V. Smith, W.E. Brower, M.S. Matyaszczyk, T.L. Pettit: In Proc. 7th Intl Congr. Catal, ed. by T. Seiyama, K. Tanabe (Elsevier, New York 1981) Vol. A, p. 355

4.3   A. Yokoyama, H. Komiyama, H. Inoue, T. Masumoto, H. Kimura: J. Catal. **68**, 355 (1981)

4.4   R. Schlögl: In *Rapidly Quenched Metals*, ed. by S. Steeb, H. Warlimont (Elsevier, New York 1985) Vol. II, p. 1723

4.5   C. Yoon, D.L. Cocke, J. Non-Cryst. Solids **79**, 217 (1986)

4.6   M. Shibata, T. Masumoto: In Stud. Surf. Sci. Catal. (Prep. Catal. IV), ed. by B. Delmon, P. Grange, P.A. Jacobs, G. Poncelet (Elsevier, New York 1987) Vol. 31, p. 353

4.7   A. Molnar, G.V. Smith, M. Bartok: Adv. Catal. **36**, 329 (1989)

4.8   A. Baiker: Faraday Discuss. Chem. Soc. **87**, 239 (1989)

4.9   C. Suryanarayana: *Rapidly Quenched Metals: A Bibiliography, 1973–1979* (IFI/Plenum, New York 1980)

4.10   A. Yokoyama, H. Komiyama, H. Inoue, T. Masumoto, H. Kimura: Scr. Metall. **15**, 365 (1981)

4.11   A. Yokoyama, H. Komiyama, H. Inoue, T. Masumoto, H. Kimura: ACS Symp. Ser. **196**, 237 (1982)

4.12   M. Peuckert, A. Baiker: J. Chem. Soc. , Faraday Trans. 1 **81**, 2797 (1985)

4.13   Y. Shimogaki, H. Komiyama, H. Inoue, T. Masumoto, H. Kimura: Chem. Lett. (Jpn.) 661 (1985)

4.14   A. Yokoyama, H. Komiyama, H. Inoue, T. Masumoto, H. Kimura: Chem. Lett. (Jpn.) 195 (1983)

4.15   A. Yokoyama, H. Komiyama, H. Inoue, T. Masumoto and H. Kimura: J. Non-Cryst. Solids **61 & 62**, 619 (1984)

4.16   M. Shibata, N. Kawata, T. Masumoto, H. Kimura: Chem. Lett. (Jpn.), 1605 (1985)

4.17   G. Kisfaludi, K. Lazar, Z. Schay, L. Guczi, C. Fetzer, G. Konczos, A. Lovas: Appl. Surf. Sci. **24**, 225 (1985)

4.18   G. Kisfaludi, Z. Schay, L. Guczi, G. Konczos, L. Lovas, P. Kovacs: Appl. Surf. Sci. **28**, 111 (1987)

4.19   G. Kisfaludi, Z. Schay, L. Guczi: Appl. Surf. Sci. **29**, 367 (1987)

4.20   H. Yamashita, M. Yoshikawa, T. Funabiki, S. Yoshida: J. Catal. **99**, 375 (1986)

4.21   M. Shibata, Y. Ohbayashi, N. Kawata, T. Masumoto, K. Aoki: J. Catal. **96**, 296 (1985)

4.22   S.J. Bryan, J.R. Jennings, S.J. Kipling, G. Owen, R.M. Lambert, R.M. Nix: Appl. Catal. **40**, 173 (1988)

4.23   D. Gasser, A. Baiker: Appl. Catal. **48**, 279 (1989)

4.24   A. Baiker, D. Gasser: J. Chem. Soc., Faraday Trans. 1 **85**, 999 (1989)

4.25   G. Carturan, G. Cocco, E. Baratter, G. Navazio, C. Antonione: J. Catal. **90**, 178 (1984)

4.26   G.V. Smith, O. Zahraa, A. Molnar, M.M. Khan, B. Rihter, W.E. Brower, J. Catal. **83**, 238 (1983)

4.27   A. Molnar, G.V. Smith, M. Bartok: J. Catal. **101**, 540 (1986)

4.28   S. Yoshida, H. Yamashita, T. Funabiki, T. Yoezawa: J. Chem. Soc., Faraday Trans. 1 **80**, 1435 (1984)

4.29   H. Yamashita, M. Yoshikawa, T. Funabiki, S. Yoshida: J. Chem. Soc., Faraday Trans. 1 **81**, 2485 (1985)

4.30   J.C. Bertolini, J. Brissot, T. Le Mogne, H. Montes, Y. Calvayrac, J. Bigot: Appl. Surf. Sci. **29**, 29 (1987)

4.31   A. Baiker, H. Baris, H.J. Güntherodt: J. Chem. Soc., Chem. Commun., 930 (1986)

4.32   H. Yamashita, M. Yoshikawa, T. Funabiki, S. Yoshida: J. Chem. Soc., Faraday Trans. 1, **83**, 2895 (1987)

4.33   W.E. Brower, M.S. Matyjaszczyk, T.L. Pettit, G.V. Smith: Nature **301**, 497 (1983)

4.34   B.C. Giessen, S.S. Mohmoud, D.A. Forssyth, M. Hediger: In *Rapidly Solidified Amorphous and Crystalline Alloys*, ed. by B.H. Kear, B.C. Giessen, M. Cohen (Elsevier, New York 1982) p. 255

4.35   T. Takahashi, Y. Nishi, N. Otsuji, T. Kai, T. Masumoto, H. Kimura: Cdn. J. Chem. Eng. **65**, 274 (1987)

4.36   S. Yoshida, H. Yamashita, T. Funabiki, T. Yonezawa: J. Chem. Soc., Chem. Commun. 964 (1982)

4.37   H. Yamashita, M. Yoshikawa, T. Funabiki, S. Yoshida: J. Chem. Soc., Faraday Trans. 1, **82**, 1771 (1986)

4.38   H. Yamashita, T. Kaminade, T. Funabiki, S. Yoshida: J. Mater. Sci. Lett. **4**, 1241 (1985)

4.39   H. Yamashita, M. Yoshikawa, T. Funabiki, S. Yoshida: J. Chem. Soc., Faraday Trans. 1, **83**, 2883 (1987)

4.40   A. Baiker, H. Baris, H.J. Güntherodt: Appl. Catal. **22**, 389 (1986)

4.41   H. Yamashita, M. Yoshikawa, T. Kaminade, T. Funabiki, S. Yoshida: J. Chem. Soc., Faraday Trans. 1, **82**, 707 (1986)

4.42   A. Molnar, G.V. Smith, M. Bartok: J. Catal. **101**, 67 (1986)

4.43   S.S. Mahmoud, D.A. Forsyth, B.C. Giessen: Mater. Res. Soc. Symp. Proc. **58**, 131 (1986)

4.44   A. Baiker, R. Schlögl, E. Armbruster, H.J. Güntherodt: J. Catal. **107**, 221 (1987)

4.45   A. Armbruster, A. Baiker, H.J. Güntherodt, R. Schlögl, B. Walz: In Stud. Surf. Sci. Catal. (Prep. Catal. IV), ed. by B. Delmon, P. Grange, P.A. Jacobs, G. Poncelet (Elsevier, New York 1987) Vol. 31, p. 389

4.46   R. Lamprecht: Katalytische Reduktion von Stickstoffmonoxid mit den Reaktanden CO oder H_2 an amorphen und kristallinen Legierungen. University of Basel (1987)

4.47   A. Baiker, D. Gasser, J. Lenzner: J. Chem. Soc., Chem. Commun. 1750 (1987)
        A. Baiker, D. Gasser, J. Lenzner, A. Reller, R. Schlögl: J. Catal. **126**, 555 (1990)
        P. Barnickel, A. Wokaun, A. Baiker: J. Chem. Soc., Faraday Trans. **87**, 333 (1991)

4.48   H. Yamashita, T. Kaminade, M. Yoshikawa, T. Funabiki, S. Yoshida: C_1 Mol. Chem., **1**, 491 (1986)

4.49   G. Kisfaludi, K. Matusek, Z. Schay, L. Guczi: 6th Int'l Symp. on Heterogeneous Catalysis, Varna, Hungary (1987)

4.50   A. Baiker, H. Baris, F. Vanini, M. Erbudak: In Proc. 9th Int'l Congr. Catal. (Catalysis: Theory and Practice), ed. by M.J. Phillips, M. Ternan (Chem. Inst. Canada, Ottawa 1988) Vol. 4, p. 1928

4.51   H.J. Güntherodt: In *Rapidly Quenched Metals*, ed. by S. Steeeb, H. Warlimont (Elsevier, New York 1985) Vol. II, p. 1591

4.52   F.E. Luborsky: *Amorphous Metallic Alloys* (Butterworths, London 1983)

4.53   R.B. Diegle: J. Non-Cryst. Solids, **61 & 62**, 601 (1984)

4.54   G.A. Somorjai: Adv. Catal. **26**, 1 (1977)

4.55   P. Oelhafen: In *Glassy Metals II* ed. by H. Beck, H.J. Güntherodt, Topics Appl. Phys., Vol. 53 (Springer, Berlin, Heidelberg 1983) p. 283

4.56   K. Hashimoto, T, Masumoto: In *Ultrarapid Quenching of Liquid Alloys*, ed. by H. Herman (Academic, New York 1981) p. 291

4.57   R. Wiesendanger, M. Ringger, L. Rosenthaler, H.R. Hidber, P. Oelhafen, H. Rudin, H.J. Güntherodt: Surf. Sci. **181**, 46 (1987)

4.58   B. Walz, R. Wiesendanger, L. Rosenthaler, H.J. Güntherodt, M. Düggelin, R. Guggenheim: Mater. Sci. and Eng., **99**, 501 (1988)

4.59   R. Schlögl, R. Wiesendanger, A. Baiker: J. Catal. **108**, 452 (1987)

4.60   R. Hauert, P. Oelhafen, R. Schlögl, H.J. Güntherodt: Solid State Commun. **55**, 583 (1985)

4.61   L. Guczi, Z. Zoldos, G. Schay: J. Vac. Sci. Technol. A, **5** (1987) 1070

4.62   A. Baiker, H. Baris, R. Schlögl: J. Catal. **108**, 467 (1987)

4.63   A. Shamsi, W.E. Wallace: Ind. Eng. Chem. Prod. Res. Dev. **22**, 582 (1983)

4.64   M. Maciejewski, A. Baiker: J. Chem. Soc., Faraday Trans. **89**, 843 (1990)

4.65   A. Baiker, J. De Pietro, M. Maciejewski, B. Walz: Stud. Surf. Sci. Catal. **67**, 169 (1991)

4.66   B. Walz: Oxidation amorpher Uebergangsmetal-Zirkonium Legierungen. Dissertation, University of Basel (1989)

4.67   B. Walz, P. Oelhafen, H.J. Güntherodt, A. Baiker: Appl. Surf. Sci. **37**, 337 (1989)

4.68   Z. Altounian, Tu Guo-Hua, J.O. Strom Olsen: J. Appl. Phys. **54**, 311 (1983)
4.69   F. Vanini: Electron Spectroscopy Studies of Metallic and Alloid Zirconium. Dissertation No. 9006, ETH Zürich (1989)
4.70   B. Walz, H.J. Güntherodt: Catal. Lett. **3**, 191 (1989)
4.71   A. Baiker, D. Gasser, J. Lenzner, A. Reller, R. Schlögl: J. Catal. **126**, 555 (1990)
4.72   R. Schlögl, G. Lohse, M. Wesemann, A. Baiker: J. Catal. **137**, 139 (1992)
4.73   T. Rayment, R. Schlögl, J.M. Thomas, G. Ertl: Nature (London) **351**, 311 (1989)
4.74   R.B. Seymour, S.R. Montgomery: In *Heterogeneous Catalysis-Selected American Histories*, ed. by B.H, Davis, W.P. Hettinger, ACS Symp. Series, No. 222, (Am. Chem. Soc., Washington, D C 1983) p. 491
4.75   F. Sommer: In *Rapidly Quenched Metals*, ed. by S. Steeeb, H. Warlimont (Elsevier, New York 1985) Vol. I, p. 153
4.76   W. Kowbel, W.E. Brower: J. Catal. **101**, 262 (1986)
4.77   U. Köster: Z. Metallkde **75**, 691 (1984)
4.78   P.V. Nagarkar, S.K. Kulkarni, E. Umbach: Appl. Surf. Sci. **29**, 194 (1987)
4.79   J. Tamaki, H. Tagaki, I. Imanaka: J. Catal. **108**, 256 (1987)
4.80   J. Fusy, P. Pareja: J. Non-Cryst. Solids **89**, 131 (1987)
4.81   S. Myhra, J.C. Riviere, L.S. Welch: Appl. Surf. Sci. **32**, 156 (1988)
4.82   P. Sen, D.D. Sarma, R.C. Budhani, J.L. Chopra, C.N.R. Rao: J. Phys. F **14**, 565 (1984)
4.83   S. Sinha, S. Badrinarayanan, A.P.B. Sinha: J. Less Comm. Met. **125**, 85 (1986)
4.84   K. Aoki, T. Masumoto, C. Suryanarayana: J. Mater. Sci. **21** 793 (1986)
4.85   G. Wei, B. Cantor: Acta Metall. **36**, 167 (1988)
4.86   A. Garcia Escorial, A.L. Greer: J. Mater. Sci. **22**, 4388 (1987)
4.87   Cabrera, Mott, Repts. Prog. Physi. **12**, 163 (1949)
4.88   T.B. Grimley: In *Chemistry of the Solid State* (Butterworths, London 1955), p. 336
4.89   A.J. Mealand, L.E. Tanner, G.G. Libowitz: J. Less Comm. Met. **74**, 279 (1980)
4.90   R.M. Nix, T. Rayment, R.M. Lambert, J.R. Jennings, G. Owen: J. Catal. **106**, 216 (1987)
4.91   D.L. Cocke, M.S. Owens. R.B. Wright: Appl. Surf. Sci. **31**, 341 (1988)
4.92   A.R. Miedema: Z. Metall. **69**, 455 (1978)
4.93   F. Vanini, S. Büchler, Xin-nan Yu, M. Erbudak, L. Schlapbach, A. Baiker: Surf. Sci. **189/190**, 1117 (1987)
4.94   L. Schlapbach: Nato ASI B **136**, 397 (1986)
4.95   F. Spit, K. Blok, E. Hendriks, G. Winkels, W. Turkenburg, J. W. Drijver, S. Radelaar: In Prov. 4th Int'l Conf. Rapidly Quenched Met., ed. by T. Masumoto, K. Suzuki (Jpn. Inst. Met., Sendai 1982) p. 1635
4.96   L. Schlapbach, A. Seiler, F. Stucki, H.C. Siegmann: J. Less Comm. Met. **73**, 145 (1980)
4.97   S.M. Fries, H.G. Wagner, S.J. Campell, U. Gonser, N. Blaes, P. Steiner: J. Phys. F**15**, 1179 (1985)
4.98   P. Oelhafen, R. Lapka, U. Gubler, J. Krieg, A. DasGupta, H.J. Güntherodt, T. Mizoguchi, C. Hague, J. Kübler, S.R. Nagel: In Proc. 4th Int'l Conf. Rapidly Quenched Met., ed. by T. Masumoto, K. Suzuki (Jpn. Inst. Met., Sendai 1982) p. 1259
4.99   J. van Wonterghem, S. Morup, S.W. Charles, S. Wells, J. Villadsen: Phys. Rev. Lett. **55**, 410 (1985)
4.100 J. van Wonterghem, S. Morup, C.I.W. Koch, S.W. Charles, S. Wells: Nature **322**, 622 (1986)
4.101 Deng Jingfa, Zhang Xiping: Appl. Catal. **37**, 339 (1988)
4.102 S. Shina, S. Badrinarayanan, A.P. Shina: J. Less-Comm. Met. **125**, 1179 (1986)
4.103 H. Bönnemann, W. Brijoux, T. Joussen: Angew. Chem. **102**, 324 (1989)
4.104 S.A. Miller: In *Amorphous Metallic Alloys*, ed. by F.E. Luborsky (Butterworths, London 1983) p. 506
4.105 M. Hara, K. Hashimoto, T. Masumoto: J. Appl. Electrochem. **13**, 295 (1983)
4.106 M. Hara, K. Hashimoto, T. Masumoto: Electrochim. Acta **25**, 1091 (1980)
4.107 M. Hara, K. Hashimoto, T. Masumoto: J. Non-Cryst. Solids **54**, 85 (1983)

5. Interrelations Between Electronic and Ionic Structure in Metallic Glasses

P. Häussler

With 31 Figures

In this chapter we present a survey of our current understanding of interrelations between the electronic and ionic structure in late-transition–polyvalent-element metallic glasses. Evidence of a strong influence of conduction electrons on the ionic structure, and vice versa, of the ionic structure on the conduction electrons, is presented. We discuss as well the consequences to phase stability, the electronic density of states, dynamic properties, electronic transport, and magnetism. A scaling behaviour of many properties versus $\bar{Z}$, the mean electron number per atom, is the most characteristic feature of these alloys. Crystalline alloys which are also strongly dominated by the conduction electrons are often called electron phases or Hume–Rothery phases. The amorphous alloys under consideration are consequently described as an *Electron Phase* or *Hume–Rothery Phase with Amorphous Structure*. Similar theoretical concepts as applied to crystalline Hume–Rothery alloys are used for the present amorphous samples.

5.1 Introductory Remarks

Even at the very beginning, when *Buckel* and *Hilsch* [5.1] first observed the metallic amorphous state in pure Bi and Ga using the vapour-quenching technique, the behaviour of the structure was the focus of research. When *Duwez* and coworkers [5.2], several years later, first succeeded in preparing metallic glasses by a liquid-quenching technique and this new field came into the focus of modern research, structural questions kept their central interest; it has remained so till the present [5.3].[1] The state of the art of structure modeling of non-magnetic metallic glasses is now based on selfconsistent calculations which themselves are predominantly based on indirect interactions. Review articles are given by *Hafner* [5.4–6]. Indirect ion–ion interaction, for example, causes an oscillatory effective pairpotential $\phi_{\text{eff}}(r)$ versus $r = |r|$ due to the electric polarisability of the electron cloud around each ion screening its local charge. The oscillations are known as Friedel oscillations. A matching of ionic positions at short- and medium-range distances with the minima of $\phi_{\text{eff}}(r)$ can be seen as

---

[1] In this chapter we use *amorphous state* and *metallic glasses* synonymously.

Topics in Applied Physics, Vol. 72
Beck/Güntherodt (Eds.)
© Springer-Verlag Berlin Heidelberg 1994

the most obvious indication of the electronic influence on structure [5.7–9]. Correspondingly, in k-space an electronically induced peak occurs in the structure factor $S(K)$ at K_{pe}, very close to, or in agreement with $2k_F$, the diameter of the Fermi sphere [5.5, 8, 10, 11].

Since the glassy state is a metastable state, questions concerning its relative stability became important. Stability itself is closely related to structural features. An electronic influence was first proposed by *Nagel* and *Tauc* [5.12] in the frame of the nearly-free-electron model (NFE) as a Fermi sphere–Brillouin zone (FsBz) effect. Those effects are well known in crystalline Hume–Rothery phases [5.13–15]. A structure-induced minimum in the electronic density of states (MDOS) at E_F, postulated by *Nagel* and *Tauc* as a result of FsBz-effects, has not been observed for a long time. This failure can be attributed to the use of metallic glasses containing transition elements with large d-state contributions at E_F [5.16]. Indirect indications as, for example, the applicability of *Ziman's* approach to the resistivity [5.17] have also been used. A high resistivity ρ and a large negative value of $1/\rho \cdot \partial\rho/\partial T$ should coincide with a good glass-forming ability whenever $K_p = 2k_F$. But even this analysis was strongly hindered by the lack of a clear knowledge of $2k_F$ in these alloys and by d-state scattering.

Much effort therefore was directed onto simple alloys like $Mg_{70}Zn_{30}$ with no d-states at E_F. In r- as well as in k-space, the electronic influence on ionic structure has been observed [5.5], and quite recently an MDOS was found experimentally as well as theoretically [5.18, 19]. Unfortunately, there is no change of $\bar{Z}$ by changing the composition because both elements are divalent. In order to have this control, we focused our research on vapour-quenched alloys between noble metals with d-states well below E_F and polyvalent simple elements. Using Au and Sb, for example, $\bar{Z}$ changes from 1 e/a (Au) to 5 e/a (Sb). Different alloys were systematically investigated during the last decade by *Mizutani* et al. [5.20] and by the author [5.10] and may now serve as model systems in this field.

After presenting the sample preparation in Sect. 5.2, we give an introduction to the theoretical background in Sect. 5.3. In Sect. 5.4, we briefly review the electronic influence on structure and phase stability of crystalline Hume–Rothery phases. In Sect. 5.5, we discuss the properties of non-magnetic amorphous alloys of the type just mentioned. The electronic influence on structure (5.5.1) and consequences for the phase stability (5.5.2) are also discussed. Structural influences on the electronic density of states are shown in 5.5.3. Electronic transport properties versus composition indicate additionally the electron–structure interrelation (5.5.4), and those versus temperature, the influence of low-lying collective density excitations (5.5.5). An extension of the model of the electronic influence on structure and stability was proposed by *Häussler* and *Kay* [5.21, 22] whenever local moments are involved as, for example, in Fe-containing alloys. In Sect. 5.6, experimental indications for such an influence are presented, and additional consequences on phase stability and magnetic properties are briefly discussed.

The amorphous state has many similarities to the liquid state and can in fact be considered an undercooled liquid. Physical properties such as the electronic and ionic structure as well as electronic transport properties are temperature dependent and can be extrapolated from one state to the other. In this paper close relationships between both are shown.

5.2 Preparation

A vapour-quenching technique, in combination with substrates held at the temperature of liquid helium, is the most effective method to force metallic alloys over large concentration ranges into the amorphous state. A *sequential flash vapour-quenching technique* has mostly been used for the present alloys. This technique uses *one* vapour source, although the vapour pressure of the elements may differ by orders of magnitude. The source mainly consists of a hot filament (W, Ta, Mo), shaped as seen in Fig. 5.1a with two edges in order to get a homogeneous temperature distribution in the curved region, and a tube, tapped inside, which allows, due to rotation, the feeding of small pieces of the material onto the filament. Each piece has to have the overall nominal composition and gets flash evaporated. As each contributes one monolayer or less to the film thickness, even complete segregation of each piece during evaporation finally ends up with a homogeneous sample of the nominal composition. The current through the film, used as a thickness monitor, shows steps with progressing deposition (Fig. 5.1b). Figure 5.1c shows the film composition, measured with different methods on different systems, in good agreement with the nominal composition. Other preparation techniques such as sputtering and

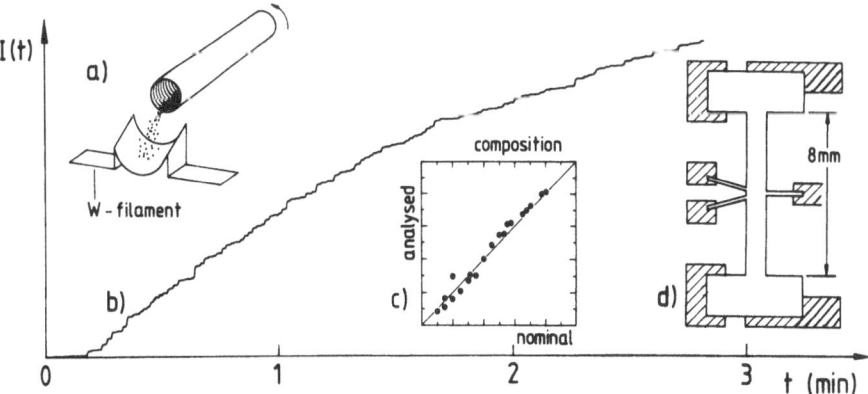

Fig. 5.1. a) Source for a sequential flash vapour-quenching; b) current $I(t)$ through the film during deposition at constant voltage; c) composition of some films as measured and nominal; d) film geometry for combined Hall coefficient and resistivity measurements

co-evaporation have also been used with no major differences of the film properties [5.22]. A mask at the substrate gives a well-defined geometry (see Fig. 5.1d as an example). *Au*-contacts are pre-evaporated before the film is deposited, with wires glued on it with silver paste.

5.3 Theoretical Background

In crystalline materials, long-range order gives, as a consequence, sharp Brillouin-zone boundaries in *k*-space. In glassy systems, long-range order is absent and sharp zone boundaries no longer exist. Short- and medium-range order, on the other hand, causes *pseudo* Brillouin zones with broad boundaries [5.23, 24]. Anisotropies disappear and the zones are sphere-like, drastically affecting many properties as electronic states and collective excitations as phonons and magnons. Electronic transport properties are strongly affected under conditions where the Fermi sphere is close to the zone boundary.

5.3.1 Structure and Stability

In polyvalent liquid elements as well as in many metallic glasses and liquid alloys, the first peak in the structure factor $S(K)$ with $K = |K| = |k' - k|$ is split into two (Figs. 5.2a, 7 and 29a). There is a peak at K_p and another one at K_{pe} close to $2k_F$. The latter is electronically induced [5.8] and essential to the understanding of disordered metals.

Pseudopotential theory relates rather directly physical properties to

$$v(K) = \frac{1}{\Omega} \cdot \int v(r) \cdot e^{-iK \cdot r} d^3r \quad \text{(Fig. 5.2c)}, \tag{5.1}$$

the Fourier transform of the pseudopotential [5.25]. Ω is the atomic volume and $v(r)$ the bare pseudopotential in *r*-space. For a discussion of structure and stability, the total energy

$$U_{tot} = U_v + (U_E + U_{bs})_{str} \tag{5.2}$$

must be considered. U_v is a volume-dependent term giving approximately 95% of the total energy. U_E as well as U_{bs} are structure dependent and responsible for eventual structure stability. In magnetic systems, other structure-dependent terms may have to be taken additionally into account (5.6). U_E, the so-called Ewald energy, contains part of the electrostatic energy and prefers simple crystalline phases. Under thermal equilibrium, this contribution dominates but necessitates the possible diffusion of the constituents. Since diffusion is suppressed, we assume the second structure-dependent term U_{bs} as dominant. U_{bs} is the band-structure energy which arises from gaps in the band structure, i.e., from deviations of the electron gas from the free-electron gas (5.3.2). U_{bs} depends

Fig. 5.2. a) Static structure factor of an amorphous or liquid metal; b) band-structure characteristic; c) pseudopotential; d) perturbation characteristic for two different Fermi-sphere diameters. a–c are qualitatively drawn in arbitrary units

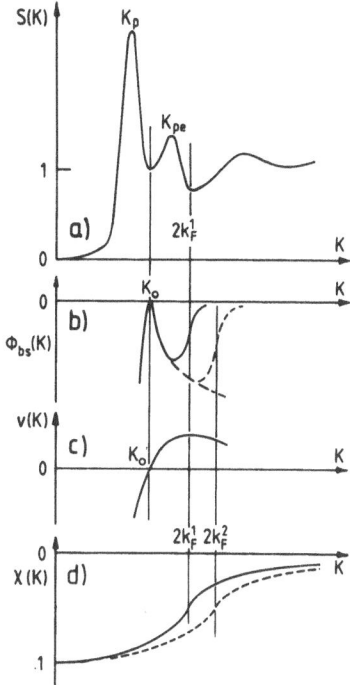

much on short- and medium-range order. This term might even be enhanced due to the isotropy of the disordered state if K_{pe} is close to $2k_F$, and therefore deviations of the electronic states from the free-electron behaviour occur uniformly in any direction. A virtual increase of U_E seems to be compensated by the increase of U_{bs}, which itself is negative and hence decreases U_{tot}. U_{bs} can be expressed in terms of $S(K)$, $v(K)$ and two functions $\varepsilon(K)$ and $\chi(K)$, which take into account the electronic response to the localized charge of the ions. $\varepsilon(K)$ is the screening factor or Lindhard's dielectric function with modifications due to exchange and correlation [5.25]:

$$\varepsilon(K) = 1 - \frac{8\pi e^2}{\Omega K^2}[1 - f(K)] \cdot \chi(K) , \tag{5.3}$$

with

$$\chi(K) \propto -\left(\frac{1}{2} + \frac{4k_F^2 - K^2}{8Kk_F} \cdot \ln\left|\frac{K + 2k_F}{K - 2k_F}\right|\right) . \tag{5.4}$$

$\chi(K)$, the perturbation characteristic, is shown in Fig. 5.2d. It is negative and approaches zero for $K > 2k_F$. At $2k_F$, there is a logarithmic singularity responsible for the Friedel oscillations of $\phi_{eff}(r)$. $f(K)$, the local-field correction, takes care of the modifications due to correlation and exchange. For crystalline systems, U_{bs} is a sum over all reciprocal lattice vectors g:

$$U_{bs} \propto \sum_g S(g) \cdot |v(g)|^2 \cdot \varepsilon(g) \cdot \chi(g) \ . \tag{5.5}$$

For disordered systems, g has to be replaced by $K = |K|$ and the summation by $\int K^2 \, dK$.

The band-structure characteristic or energy–wavenumber characteristic $\phi_{bs} = |v|^2 \cdot \varepsilon \cdot \chi$ is the Fourier transform of the indirect ion–ion interaction mediated by the conduction electrons [5.5]. $\phi_{bs}(K)$ is shown in Fig. 5.2b. The zero at K_0 is given by the zero of $v(K)$ and the increase for $K > 2k_F$ by $\chi(K)$, both causing a minimum for $K_0 < K < 2k_F$. U_{tot} is small and the stability high if a peak of $S(K)$ coincides with this minimum. Peaks close to $2k_F$ contribute most to U_{tot} (5.3.2). Changing $2k_F$ by alloying causes a change of the minimum and therefore a change of K_{pe}. The concept of *Jones zones* is helpful in this discussion [5.14, 15, 26]. The Jones zone is a zone built by those parts of different Brillouin zones which are in closest contact to the surface of the Fermi sphere. The pseudo Brillouin zone of disordered metals with the diameter of K_{pe}, for example, is a Jones zone.

In r-space, the electronic influence is seen even more directly [5.7]. A Fourier transform of $S(K)$ gives $g(r)$, the reduced pair-distribution function (5.5.1b).[2] The Fourier transform of $\phi_{bs}(K)$ can be written as an effective pair potential $\phi_{eff}(r)$ with Friedel oscillations, rising from the logarithmic singularity of $\chi(K)$, or more physically, from the screening behaviour of the electron cloud. $\phi_{eff}(r)$ is responsible for the indirect ion–ion interaction over large distances and can be expressed as

$$\phi_{eff}(r) \propto \frac{\cos(2k_F \cdot r + \Theta)}{r^3} \ , \tag{5.6}$$

with the Friedel wavelength $\lambda_F = 2\pi/2k_F$. At large distances, Θ approaches zero but may differ significantly from that for short and medium distances [5.27]. In r-space representation, U_{bs} is large if the maxima of $g(r)$ coincide with the minima of $\phi_{eff}(r)$ [5.5, 7, 9]. The change of λ_F with composition induces changes of atomic positions (5.5.1b).

5.3.2 Electronic States

In perfect crystals, Bloch's theorem applies for the calculation of electronic states as well as for elementary excitations such as phonons or magnons. Electronic states, for example, can be described as eigenstates of the system with a well-

---

[2] Unfortunately, the common nomenclature leads to three different meanings of g throughout this chapter which, hopefully, will not confuse the reader. $g, g = |g|$ is the reciprocal lattice vector and its absolute value; $g(r)$, the reduced pair-distribution function; and $g = N(E_F)/N_0(E_F)$, Mott's g-factor.

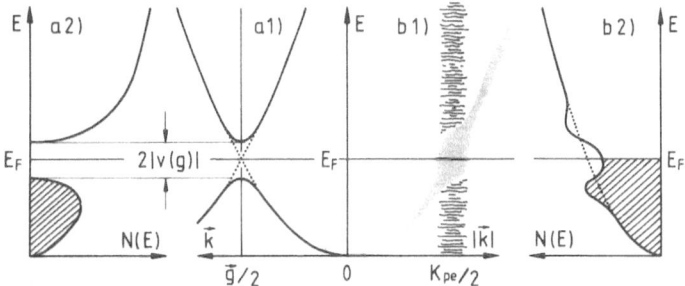

Fig. 5.3. Schematic dispersion relation of electrons and the corresponding DOS, for the crystalline state (a1, a2), and for the amorphous state (b1, b2). The dotted curves indicate the free-electron behaviour. The influence of the pseudo Brillouin zone is drawn by the shadowed area. K_{pe} indicates the position of a peak in $S(K)$ close to $2k_F$

defined dispersion relation (Fig. 5.3a1). Band gaps are produced by Bragg reflection of the electrons at certain directions (Fig. 5.3a2). In lowest order, the width of the gaps equals $2|v(g)|$, called the band-gap or pseudopotential coefficient [5.25]. Small band-gap coefficients imply weak scattering.

In disordered metals, the electrons couple with the static structure at any K-value and are heavily damped close to pseudo Brillouin-zone boundaries. A definite dispersion relation does not exist (Fig. 5.3b1) and Bloch's theorem is no longer valid. Electronic states cannot be described as eigenstates of the system and, strictly speaking, cannot be translated to the reduced-zone scheme.

There is long standing theoretical interest in the question of how this affects the electronic DOS [5.18, 28–33]. Due to the inapplicability of Bloch's theorem, calculations are extremely difficult. Using Green's function techniques, *Ballentine* [5.28] could show for liquid metals that distinct deviations from the free-electron-like behaviour may occur whenever $v(K)$ is significantly large at a peak of $S(K)$ (see Fig. 5.2a, c). The width of the so-called *pseudo gap* may then correspond to $2|v(K)|$ as in crystalline matter, and the depth to the intensity of $S(K)$, the structural weight. Theoretical considerations by *Nicholson* and *Schwartz* [5.30] as well as recent work by *Frésard* [5.32], *Beck* et al. [5.33], and *Hafner* et al. [5.18] could also show the structural effects on the DOS.

According to the structural results mentioned above, we expect two structure-induced pseudo gaps in the DOS of disordered metals. One should correspond to K_p and another one to K_{pe}. Experimentally, strong indications for both have been observed (5.5.3). *Beck* et al. showed cluster calculations with different minima due to different structure peaks [5.33].

The electronic influence on structure and stability is easily explained by Fig. 5.3. The electronic states get raised as well as depressed at any zone boundary. If all the states up to infinite energy are occupied, the presence of the gaps has no effect on the total energy. This is different if the states are occupied up to $g/2$ or $K_{pe}/2$ (g or $K_{pe} = 2k_F$). The decrease of the energy of the occupied states is not counterbalanced because the states above E_F are empty; a net lowering of U_{tot} occurs. Therefore, peak positions in $S(K)$ close to $2k_F$ have the largest influence on stability [5.34] and consequently are most favored. The structure might even adjust to a changing k_F in order to stay optimally maximized.

Due to the pseudo gap, the center of gravity of the DOS shifts to lower energies compared to the free-electron behaviour (Fig. 5.3b2), the original interpretation of the electronic influence on stability according to *Nagel* and *Tauc* [5.12]. The pseudo gap is a measure of the lowering of U_{tot}.

Electronic transport properties are strongly influenced by a touch of the Fermi sphere with the zone boundary, in the crystalline as well as in the disordered state. Exhaustive reviews on this subject have been given by *Massalski* and *Mizutani* [5.35] and *Mizutani* [5.20]. In the same way as sharp zone boundaries in crystalline materials are responsible for *umklapp* processes, in amorphous systems we can talk in terms of *diffuse umklapp* processes caused by the pseudo Brillouin-zone boundary. This description was first introduced by *Hafner* [5.36].

5.3.3 Collective Excitations

Neutron-inelastic-scattering experiments on $Mg_{70}Zn_{30}$ metallic glasses, performed by *Suck* et al. [5.37] have revealed energetically low-lying short-wavelength collective-density excitations at wavenumbers Q_{pe}, where the main peak in the structure function lies (Fig. 5.4). To make the nomenclature easier to handle, following *Handrich* and *Resch* [5.38], we refer to these states as *phonon-rotons*. Their energy $\hbar\omega_0$ or characteristic temperature T_0 goes down to a few meV or some 10 K, respectively. With increasing intensity of the peak at K_{pe} the minimum shifts to smaller energies (temperatures) [5.38]. In crystals,

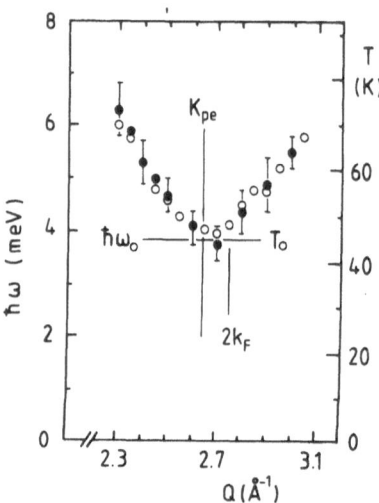

Fig. 5.4. Collective density excitations (phonon-roton type) of $Mg_{70}Zn_{30}$ metallic glasses [5.37]. K_{pe} indicates the position of the experimentally determined peak in $S(K)$ close to $2k_F$

those states go down to $\hbar\omega = 0$ and are identical with those at long wavelength by a translation into the first Brillouin zone. In Fig. 5.5, a schematic drawing is shown for the crystalline (a1) as well as for the disordered case (b1). Since translations into the first pseudo Brillouin zone are not strictly valid, the conceptional difficulty arises that the wavelength of these excitations is smaller than the mean interatomic distance. We therefore are reminded of the *diffuse umklapp* scattering mentioned above. In analogy to the crystalline state, short wavelength excitations get transferred into inelastic long-wavelength excitations where part of the momentum gets elastically transferred to the whole *lattice*.[3] There might even be a hierarchy of inelastic phonon-rotons with different energies $\hbar\omega_i$ when multiples of K_{pe} are transfered. Higher phonon-rotons indeed are predicted by the theory [5.39]. Phonon-rotons have become the subject of many theoretical publications in the last few years [5.31, 39–43] and have been observed in amorphous [5.37] as well as liquid systems [5.44]. They are now believed to be a general feature of disordered systems. The corresponding density of states is different from the Debye model and enhanced at low ω. A proportionality with ω instead of ω^2 was often observed (Fig. 5.5b2).

In the present alloys Q_{pe} equals $2k_F$. Under this condition, the phonon-rotons can easily interact with electrons for $T > T_0$ causing *inelastic* umklapp scattering of the electrons. Below T_0, only elastic umklapp scattering and inelastic scattering with normal phonons occur. Above T_0, phonon-rotons can be excited thermally as well as by electron scattering. Electronic transport properties versus temperature may therefore be strongly affected (5.5.4).

Concluding this section we would like to emphasize that the three character-istic wavenumbers agree in the metallic glasses under consideration: 1) $2k_F$, the Fermi-sphere diameter, 2) K_{pe}, the diameter of the pseudo Brillouin zone, 3) Q_{pe}, the wavenumber of phonon-rotons.

This intimate relationship between electronic, structural and dynamical properties makes these alloys very special.

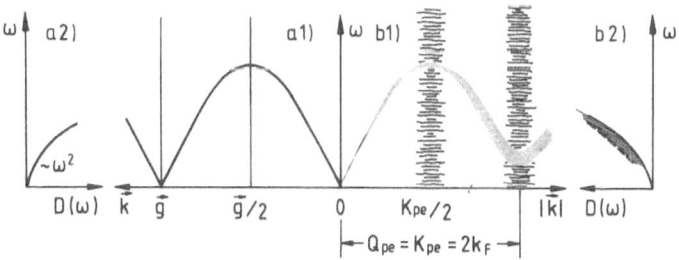

Fig. 5.5. Schematic dispersion relation of collective density excitations in the crystalline case (a1) as well as in the disordered case (b1). Phonon-roton states are shown in b1) at K_{pe}. Cörresponding dynamic density of states are included (a2, b2)

---

[3] To call those states *short wavelength* as above and in many papers is therefore somehow misleading.

5.4 Crystalline Electron Phases

The phase stability of crystalline electron phases or Hume–Rothery (HR) phases has been explained by the afore-mentioned band-structure effects. For this purpose, the k- as well as r-space representation have been used successfully [5.45, 46]. Crystalline HR-phases are well documented and excellent text books or reviews exist in this field [5.13, 14, 35]. In the present section, only a few facts are mentioned in order to show how glassy metals belong to this class of phases.

Under equilibrium conditions, the alloys considered in this paper show several crystalline HR-phases between $\bar{Z} = 1\,e/a$ and approximately $\bar{Z} = 1.8$–$2.0\,e/a$ (Table 1: $\alpha - \varepsilon$).

Although the composition, which depends on the valence of the polyvalent element, might be quite different, structurally similar phases exist for different alloys in similar $\bar{Z}$-regions. In the lower part of Fig. 5.6, this is shown for some $Cu_{100-x}M_x$ alloys with M = Zn, Be, Al, Sn chosen for the different valencies. Such a scaling behaviour of physical properties with $\bar{Z}$ is the most characteristic feature of HR-phases. As an additional indication, effects on the DOS have often been observed experimentally and minima are well established [5.35, 47].

In the upper part of Fig. 5.6, some Jones zones are shown. For fcc and bcc, they are identical with the first Brillouin zone. For both, most of the Fermi surface is well inside the corresponding zone boundary. A FsBz-effect occurs in a few k-directions [5.45]. In r-space, ionic positions match with the Friedel minima in the corresponding r-directions [5.46]. Due to the increase of $2k_F$ with the content of the polyvalent element, several effects may occur in order to minimize the total energy of the system [5.14, 25]. U_{bs} might increase due to a more intimate contact of E_F with the zone boundary. Lattice distortions may occur in order to keep the Jones zone boundary in an optimal position relative to $2k_F$, or lattice defects may occur in order to slow down the increase of $2k_F$ with composition. Finally the phase becomes unfavourable and a structural change may occur, for example, from fcc to bcc or bcc to γ-brass and the game starts again in the new phase. Hence, the increase of $2k_F$ with $\bar{Z}$ causes the scaling behaviour of the phase stability shown in Fig. 5.6.

Whereas fcc contains 4 and bcc 2 atoms per unit cell, γ-brass is rather complex with 52 atoms per cell. $\bar{Z}$ became close to 1.8 e/a, and the Jones zone is

Table 1. Hume–Rothery phases

Phase type	Structure	Approx. $\bar{Z}$	Ref.
α	fcc	1.00–1.42	[5.14]
β	bcc	1.36–1.59	[5.14]
ζ	hcp	1.32–1.83	[5.14]
γ-brass	Complex cubic	1.54–1.7	[5.14]
δ	Complex cubic	1.55–2.0	[5.14]
ε	hcp	1.65–1.89	[5.14]
Amorphous	Disordered	1.80–	[5.10]

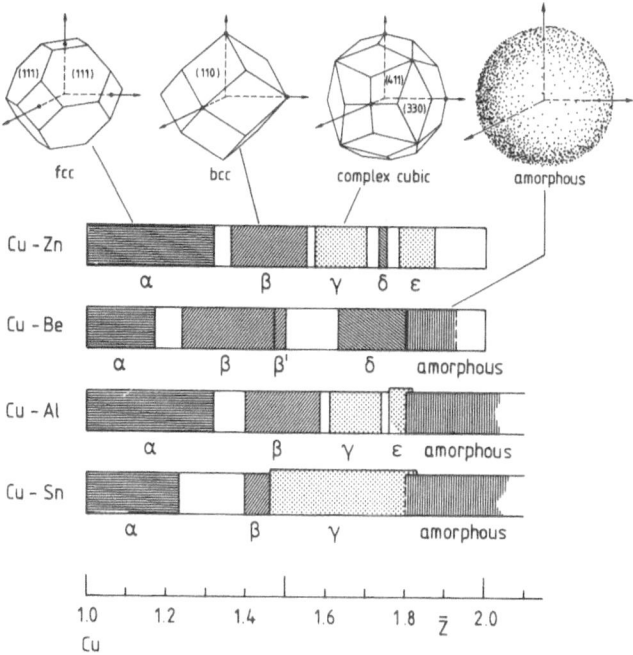

Fig. 5.6. Phase regions of some Hume–Rothery alloys vs. $\bar{Z}$ and the corresponding Brillouin or Jones zones

already nearly a sphere with many faces in contact with E_F. Following this trend, we may expect an even more spherical Jones zone for $\bar{Z}$ above 1.8 e/a. Contrarily, a further increase causes phase segregation. Homogeneous alloys, in general, no longer exist for this $\bar{Z}$-region under thermal equilibrium [5.48]. If segregation, on the other hand, is not allowed by a quench down to low temperatures, the amorphous phase will occur. The metallic glasses under consideration can therefore be seen as the limiting case of the crystalline HR-phases [5.10, 49]. The Jones zone became a sphere with soft boundaries (Fig. 5.6). $2k_F$ touches this zone boundary in every direction. In r-space, the ionic positions are optimally matching with the Friedel minima of $\phi_{eff}(|r|)$ in every direction (5.5.1b). Hence the amorphous phase is a model system with respect to the electronic influence on structure and phase stability.

5.5 Nonmagnetic Amorphous Alloys

5.5.1 Structure

Ion positions in disordered systems are not as well defined as in crystalline systems, and so the Jones zone can adapt to the electronic constraints over large concentration ranges. In this respect, all structure data of amorphous and liquid noble-metal–polyvalent-element alloys, known so far in the literature, have been re-analysed and summarized by *Häussler* in previous works [5.9, 10]. In principle, for binary alloys, three partial structure factors $S_{ij}(K)$ have to be taken into account for a complete description. As they are not known, only total-structure functions have been analysed and already show interesting features:

a) *k*-space

As an example, $S(K)$ of Au–Sb reported by *Leitz* and *Buckel* [5.50] is shown in Fig. 5.7. In this system, $\bar{Z}$ changes from 1 e/a to 5 e/a with the amorphous state above 1.8 e/a ($x \geq 20$). At $\bar{Z} = 1.8$ e/a, a non-split first peak exists, with the position in exact agreement with $2k_F$. With increasing $\bar{Z}$, an additional peak at

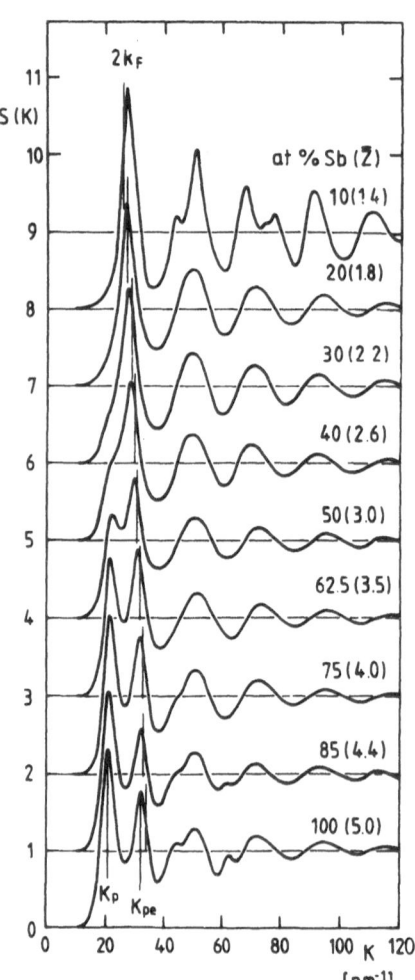

Fig. 5.7. Total structure factor $S(K)$ of quench-condensed Au–Sb alloys [5.50]. The parameters indicate the Sb-content and $\bar{Z}$, respectively. $2k_F$ is given by solid vertical lines

K_p is present well below $2k_F$. The k_F of an alloy $A_{100-x}B_x$ is estimated with the FEM[4]

$$k_F = \sqrt[3]{3\pi^2 \bar{Z} \cdot \bar{n}_0} \quad \text{with} \quad \bar{Z}(x) = \left(1 - \frac{x}{100}\right) \cdot Z_A + \frac{x}{100} \cdot Z_B \ . \tag{5.7}$$

Z_A and Z_B are the valences of the pure elements counting s and p-states only. The mean number density $\bar{n}_0 = 1/\bar{\Omega}_0$ of the *liquid* state has been used as an approximation[5] assuming the ionic volume Ω_i of the individual atoms as concentration independent:

$$\bar{\Omega}_0(x) = \left(1 - \frac{x}{100}\right) \cdot \Omega_A + \frac{x}{100} \cdot \Omega_B \quad \text{with} \quad \Omega_i = \frac{A_i}{L \cdot \rho_i} \ . \tag{5.8}$$

A_i and ρ_i are the atomic weights and the mass densities of the pure elements. L is Avogadro's number.

Convincing evidence of the electronic influence on structure is the shift of K_{pe} nearly parallel to the changing $2k_F$ over the whole amorphous region. The pseudo Brillouin zone adjusts to the Fermi sphere as mentioned above. In order to distinguish between the peak at K_p and the electronically induced peak at K_{pe}, the small e was added to the latter. The shift of K_{pe} parallel to $2k_F$ and the characteristic scaling with $\bar{Z}$ is better seen in Fig. 5.8 (Au–Sb is at the far right). The vertical solid lines in these and some of the following figures are a mark for $\bar{Z} = 1.8$ e/a. Whenever $\bar{Z}$ equals 1.8 e/a *Nagel-Tauc's* condition is fulfilled, which has now to be written as $K_{pe} = 2k_F$. This is true for the amorphous [5.9, 10, 52] as well as for the liquid state [5.53]. For $\bar{Z} > 1.8$ e/a, K_{pe} in the amorphous state is closer to $2k_F$ than in the liquid state. This and the fact that this peak is more intense in the amorphous state [5.51] can be interpreted as a stronger electronic influence within the former. In Fig. 5.9, the differences $K_{pe} - 2k_F$ of all liquid and amorphous alloys known so far are plotted versus $\bar{Z}$. There is a uniform behaviour at $\bar{Z} = 1.8$ e/a, in the sense that the pseudo Brillouin zone is in excellent agreement with the Fermi sphere. Above this value, K_{pe} stays slightly below $2k_F$ over large $\bar{Z}$-values.

b) r-space

In polyvalent liquid elements, the existence of an electronically induced peak or shoulder at the high K-value side of the first peak close to $2k_F$ is quite common [5.54] and has successfully been ascribed to preferred positions of the ions at medium-range distances [5.8, 11]. The matching of the ionic positions with

---

[4] Hall-effect data can, in general, not be used for a deduction of k_F because of a strong FsBz-effect on R_H (Sect. 5.5.4a).

[5] For some alloys such as Au–Sb [5.50], Fe–(Sb, Ge) [5.22] at $\bar{Z} = 1.8$ e/a and Cu–Sn [5.51] over the whole amorphous region, this assumption has been confirmed by analysing the reduced pair-distribution function at small r-values. There are more structural similarities of the amorphous to the liquid than to the crystalline state.

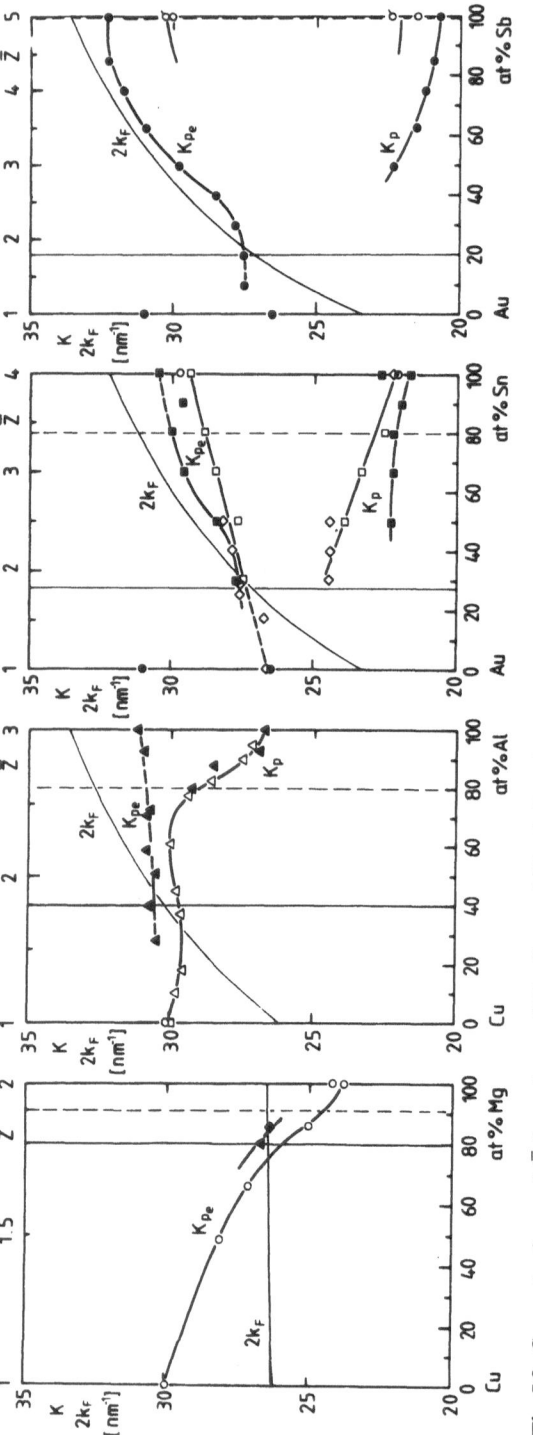

Fig. 5.8. Concentration and $\bar{Z}$ dependence of K_p, K_{pe}, and $2k_F$. Full and open symbols correspond to metallic glasses and liquid alloys, respectively. The vertical solid and dashed lines include the regions of the homogeneous amorphous phase. References were given in [5.9, 10]

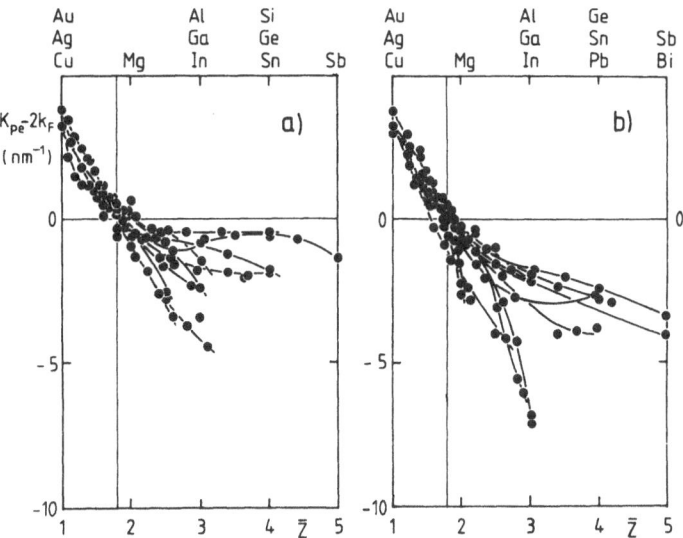

Fig. 5.9. Distance between K_{pe} and $2k_F$ vs. $\bar{Z}$ for a) metallic glasses, and b) the liquid state. References were given in [5.9, 10]

the minima of $\phi_{eff}(r)$ is equivalent to a large band-structure energy

$$U_{bs} \propto \int r^2 \cdot [g(r) - 1] \cdot \phi_{eff}(r) \, dr \ . \tag{5.9}$$

$g(r)$ is the reduced pair-distribution function, deduced from the Fourier transform of $S(K)$. The same description is used for the present alloys [5.9, 10, 50]. Those ionic arrangements are most favourable with $g(r)$ having the same periodicity as $\phi_{eff}(r)$; in other words, if $2\pi/K_{pe} = 2\pi/2k_F$, Nagel-Tauc's criterion in r-space [5.4, 7, 8]. Figure 5.10 shows $g(r)$ of Au–Sb. A phase shift $\Theta = \pi/2$ has been found empirically [5.9] with (5.6) becoming

$$\phi_{eff}(r) \propto -\frac{\sin(2k_F \cdot r)}{r^3}. \tag{5.10}$$

The position of the n'th minimum is hence at $r_F^n = (n + 1/4) \cdot \lambda_F$. According to Fig. 5.11, the optimal matching of the ionic position with the $(-)$ sin-like behaviour of $\phi_{eff}(r)$ is obvious. At $\bar{Z} = 1.8\,e/a$, all neighbours up to large distances are optimally positioned and we call these alloys the *ideal* amorphous state. U_{bs} is large and the phase should be relatively stable. At this special $\bar{Z}$-value, r_1 equals $r_F^1 = 5/4 \cdot \lambda_F$. At medium-range distances, the adjustment to the constraints given by the electronic screening can well be fulfilled and ionic positions change drastically with composition. The nearest-neighbour distance on the other hand seems to be hindered by the hard core of the ions.

In order to show the agreement of r_1 with $5/4 \cdot \lambda_F$ at $\bar{Z} = 1.8\,e/a$ as a characteristic feature of the alloys considered here, we show in Fig. 5.12 r_1 versus $\bar{Z}$ for all amorphous and liquid noble-metal–polyvalent-element alloys known to us.

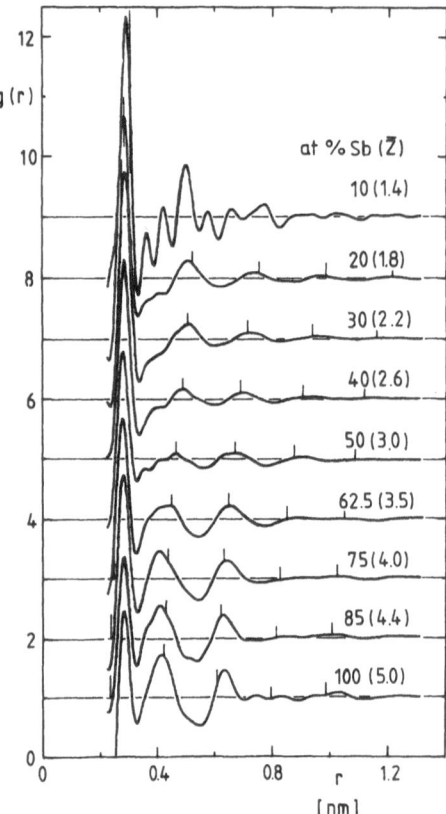

Fig. 5.10. Total reduced pair-distribution function $g(r)$ of quench-condensed Au–Sb [5.50]. Vertical lines indicate the position of the Friedel minima in $\phi_{\text{eff}}(r)$ according to (5.10)

The scaling behaviour is obvious, although there is no theoretical explanation for the systematic phase shift of $\Theta = \pi/2$ yet. An electronic influence on the hard-core diameter due to screening was demonstrated by *Hafner* and *Heine* [5.27], although the behaviour shown above is not reported.

5.5.2 Stability

Questions related to stability have two aspects: one is the glass-forming ability, i.e., the concentration range where the metallic glass can be prepared, and the other the thermal stability, the stability against crystallisation during annealing. The stability of metallic glasses is influenced by both the amorphous and the crystalline state.

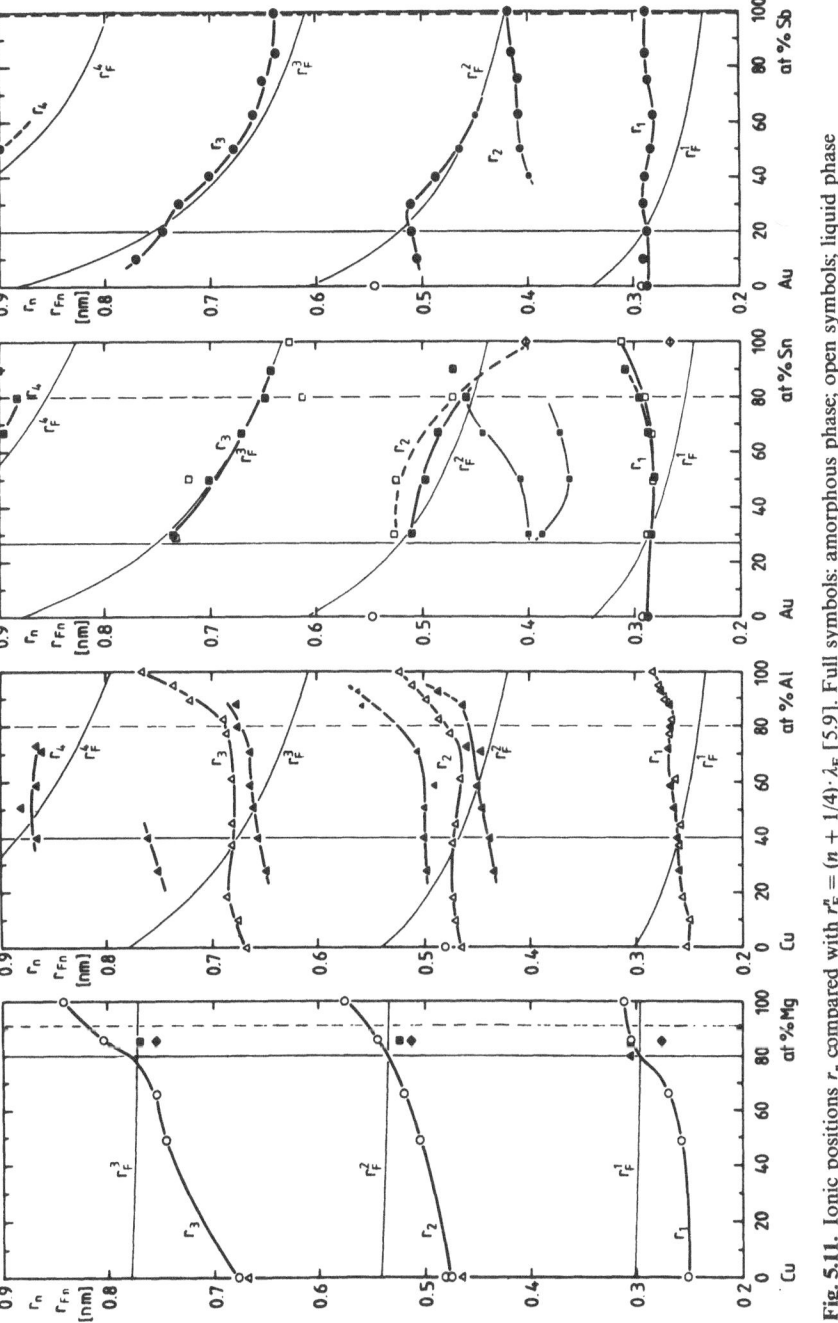

Fig. 5.11. Ionic positions r_n compared with $r_F^n = (n + 1/4) \cdot \lambda_F$ [5.9]. Full symbols: amorphous phase; open symbols; liquid phase

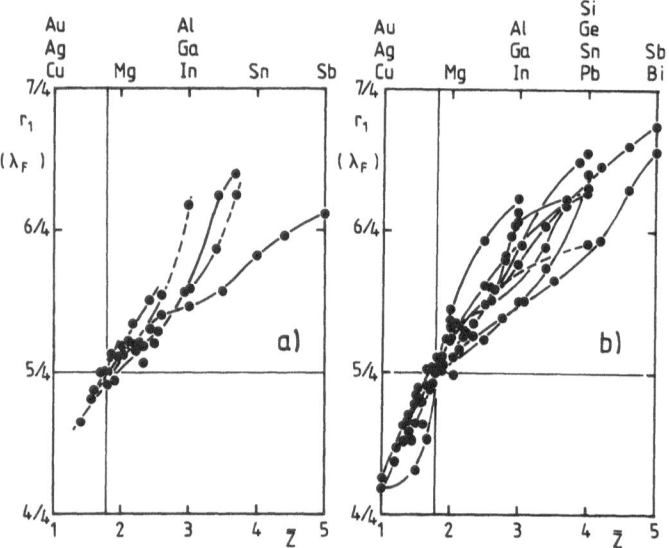

Fig. 5.12. r_1 in units of λ_F vs. $\bar{Z}$ for a) metallic glasses, and b) the liquid state. References are given in [5.9, 10]

a) Glass-Forming Ability

For $\bar{Z} < 1.8$ e/a simple crystalline phases exist. The Ewald energy as well as the band-structure energy, considerable in a few directions, contribute to the phase stability. For $\bar{Z} \approx 1.8$ e/a, where simple crystalline structures no longer exist, U_E is less favorable and the system starts to segregate. If diffusion is suppressed due to quenching, U_{bs} finally dominates, forcing the system into the amorphous state. The band structure contribution occurs now in any direction because of the isotropy of the disordered state. For $\bar{Z} > 1.8$ e/a, a further increase of $2k_F$ forces the amorphous structure to adjust to the changing electronic constraints. The role of the noble metals for the glass-forming ability is not yet clear. However, a minimal content of 20 at. % Au, Ag, Cu was found to be necessary in order to make metallic polyvalent elements amorphous [5.10, 55]. This minimal content is not necessary for elements with covalent bonds [5.56].

b) Thermal Stability

In general, a direct comparison of different alloys cannot be made and might, if at all, only be allowed under restricting conditions. One of these restrictions is that stable crystalline compounds be absent. Consequently, the crystalline state should be phase segregated and the crystallization process itself should be diffusion controlled. As already mentioned, this is fulfilled for the alloys considered here at $\bar{Z} \approx 1.8$ e/a. In Fig. 5.13, the crystallisation temperatures T_K of (Au, Ag, Cu)–Sn alloys, already published elsewhere [5.55, 57], are redrawn.

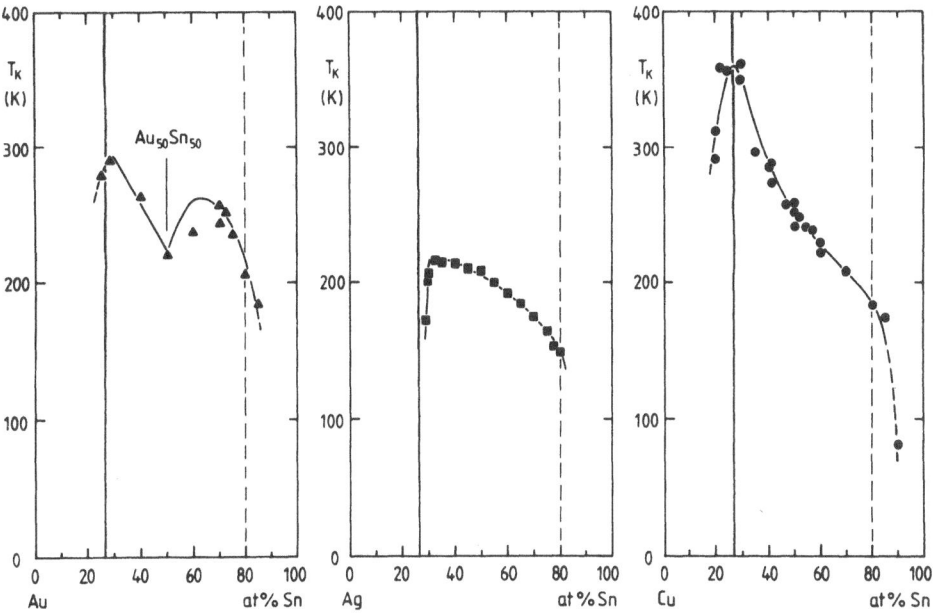

Fig. 5.13. Crystallisation temperatures of quench-condensed (Au, Ag, Cu)–Sn alloys. References to the individual systems are given in [5.55, 57]

These alloys are further on representative for many other systems. The most stable state exists at 25–30 at. % Sn content or at $\bar{Z} = 1.8$ e/a, stated above as the *ideal* amorphous phase. Below this composition, crystalline HR-phases are more favorable and above the ideal positions of the ions within the Friedel minima are lost more and more with consequences on stability. The noble metals have different influences which might be ascribed to the band-gap coefficients $2|v(K)|$ and the structural weights $S(K)$ close to $2k_F$. Au–Sn has a compound at $x = 50$ and therefore a less stable amorphous state in this region [5.58].

The uniform behaviour of the alloys under consideration is even reflected in their thermal stability. Figure 5.14 shows T_K versus $\bar{Z}$ of different Au-alloys. The alloys at $\bar{Z} = 1.8$ e/a crystallise all at $T_K = 300k \pm 10$ K. With increasing $\bar{Z}$, T_K becomes different with the different polyvalent elements. Influences of crystalline compounds are seen in this region. Similar behaviour exists for alloys with Ag and Cu, although T_K at $\bar{Z} = 1.8$ e/a itself is different [5.10].

5.5.3 Electronic Density of States

Structure-induced effects on the DOS may exist for *s*- and *p*-states with large dispersion; flat *d*-bands, for example, should be less influenced. The metallic glasses considered here are best suited for those measurements. *d*-states are well below E_F and the structure of these glasses is well known (5.5.1). First evidence

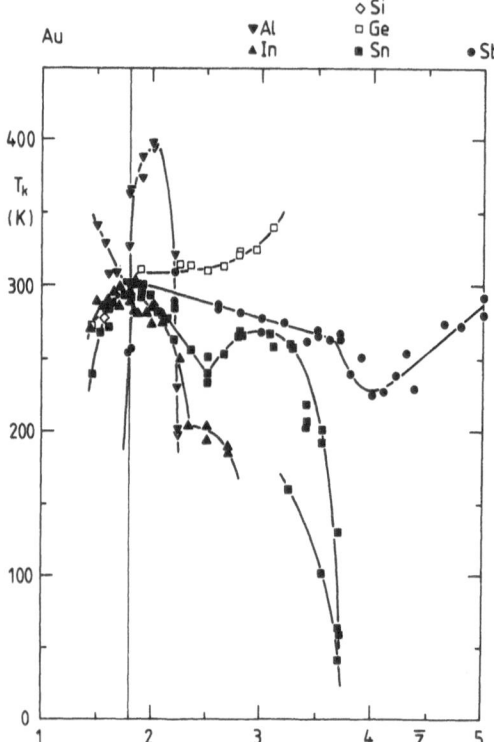

Fig. 5.14. Crystallisation temperature vs. $\bar{Z}$ for different metallic glasses. The explanation of the symbols is given on top of the figure. As an example, ▲ represents values of amorphous Au–In alloys with $1 \leq \bar{Z} \leq 3$. References are given in [5.10]

for an MDOS at E_F corresponding to the peak at K_{pe} was indeed observed for Au–Sn [5.59]. Minima in the DOS well below E_F were found by *Indlekofer* quite recently [5.60] and can be interpreted as induced by K_p, the first peak in $S(K)$.

Photoelectron spectroscopy and measurements of the specific heat and magnetic susceptibility are methods available to gain information about the DOS. In disordered systems, the former gives electron-distribution curves (EDC) reflecting the band density of states [5.61]. Whereas the UPS-method reveals the shape of the DOS and hardly gives the absolute value of the density of states, specific heat as well as susceptibility measurements, on the other hand, give only $N(E_F)$.

a) Photoelectron Spectroscopy

UPS spectra of three systems are plotted in Fig. 5.15. The increase in the intensity at large binding energies E_B is caused by Au 5d-states. Attention is focused on relative changes close to E_F, where s- and p-states show a decrease towards E_F.

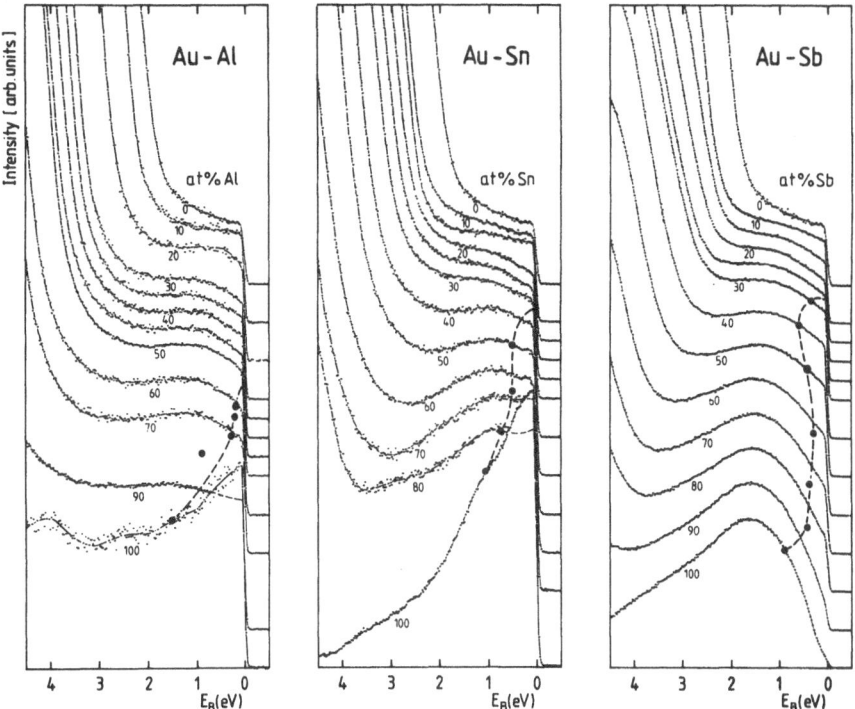

Fig. 5.15. UPS He I ($hv = 21.2$ eV) valence-band spectra of Au–Al, Au–Sn, Au Sb alloys taken at 77 K after deposition [5.57, 59, 62]. The large dots indicate $E_B(K_{pe}/2)$, the position of structure-influenced electronic states

s, p–d-hybridization effects have been proposed by *Mizutani* et al. [5.63] for the decreasing DOS towards E_F. In a recent publication [5.64], those effects were excluded because the d-band binding energies vary with composition by several eV and their intensity decreases with decreasing Au-content, whereas the MDOS keeps its position at E_F and for Au–Al and Au–Sb gets even more pronounced.

Accordingly, we discuss below the MDOS as structure induced as proposed by several authors [5.18, 28, 30–33]. The large dots in Fig. 5.15 indicate binding energies of electronic states at the pseudo Brillouin-zone boundary, roughly estimated with the E versus k relation of free electrons

$$E_B\left(\frac{K_{pe}}{2}\right) = \frac{\hbar^2}{2m} \frac{(2k_F - K_{pe})^2}{4} . \tag{5.11}$$

m is the mass of free electrons. K_{pe}, which changes parallel to $2k_F$ (Fig. 5.8), obviously induces the MDOS at E_F over large composition ranges. Differences between the systems correlate with structural differences. For example, in Au–Al the structural weight $S(2k_F)$ was found as nearly concentration independent [5.65] whereas in Au–Sn it decreases from $x = 30$ to $x = 80$ [5.51], both in

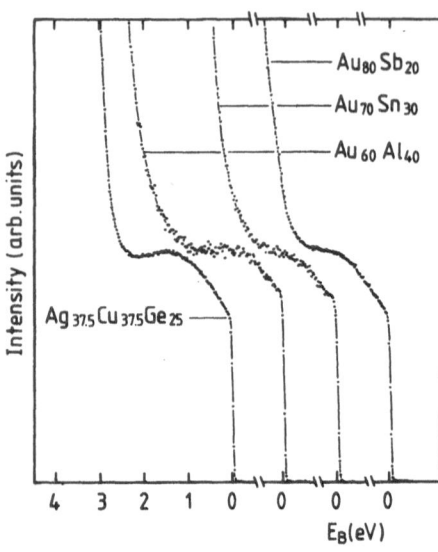

Fig. 5.16. UPS-spectra close to E_F for different alloys with $\bar{Z} \approx 1.8$ e/a [5.62, 64]. The spectra are normalized to the same intensity at $E_B = 1.5$ eV

agreement with the variation of the depth of the MDOS at E_F with composition. In Au–Sb $S(2k_F)$ is quite large even at pure Sb (Fig. 5.7) and in addition, a transition from the metallic to the semiconducting behaviour occurs causing a deep gap at E_F.

In Fig. 5.16, the uniform behaviour of the DOS at E_F of different alloys with $\bar{Z} \approx 1.8$ e/a is clearly seen. The $S(2k_F)$ of the Au-alloys were found to be nearly identical [5.57] obviously causing identical influences on the DOS. The ternary $Ag_{37.5}Cu_{37.5}Ge_{25}$ metallic glass behaves quite similarly. Structure measurements exist for this system [5.66], but $S(2k_F)$ itself is unfortunately not known to us.

For (Au, Ag, Cu)–Sn, a normalization procedure has been applied to the EDCs [5.57], giving $g_{UPS} = N(E_F)/N_0(E_F)$, Mott's g-value, as a function of composition (Fig. 5.17). $N_0(E_F)$ is the free-electron value. For alloys with the highest stability (Fig. 5.13) the MDOS is most pronounced. The increase of g with increasing Sn-content is in agreement with the decrease of $S(2k_F)$. This direct indication of the structural influence on the DOS is shown in Fig. 5.18. There is a proportionality between $S(2k_F)$ and $(1 - g)$, the measure of the depth of the MDOS at E_F:

$$(1 - g) = 1 - \frac{N(E_F)}{N_0(E_F)} \propto S(2k_F) \ . \tag{5.12}$$

The structural weight at $2k_F$ obviously corresponds to depth as expected in (5.3.2). We believe this to hold valid whenever *metallic* polyvalent elements are involved. In the following sections we may use $(1 - g)$ instead of $S(2k_F)$ because structure data of Ag–Sn metallic glasses are not available.

UPS-spectra of corresponding liquid alloys have been performed by *Indlekofer* et al. [5.70]. A fairly close general similarity between the valence

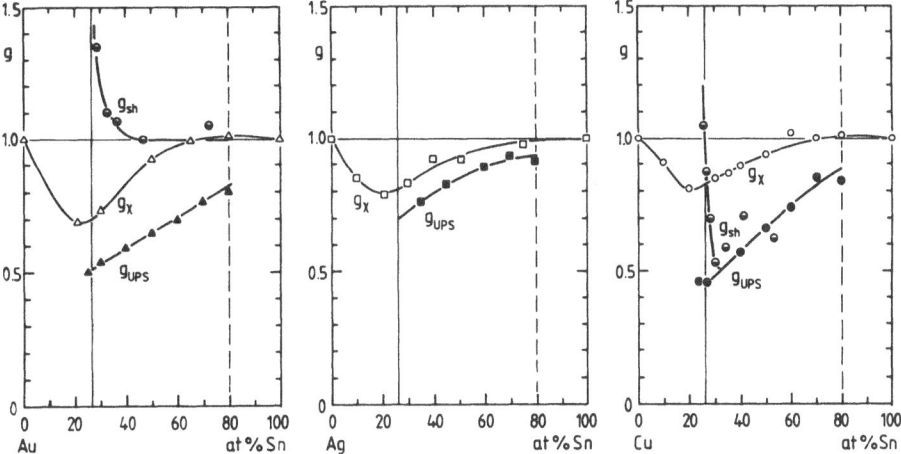

Fig. 5.17. Electronic density of states at E_F normalized to the FEM-value. g_{UPS}: metallic glasses, deduced from UPS-measurements [5.57]; g_{sh}: metallic glasses, deduced from specific-heat data (Sect. 5.5.3b) [5.67, 68]; g_x: liquid state, deduced from susceptibility data (Sect. 5.5.3c) [5.69]

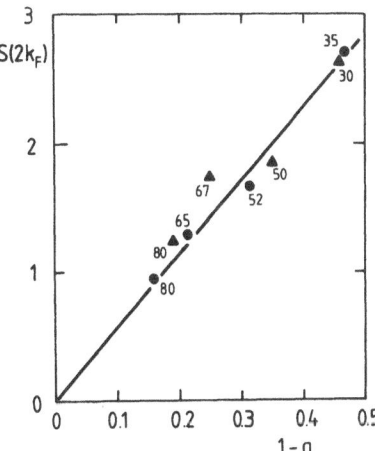

Fig. 5.18. Relations between $S(2k_F)$ and $(1 - g)$, the depth of the MDOS at E_F, for (Au, Cu)–Sn metallic glasses [5.57]. The numbers indicate the Sn-content in at. %. ▲: Au–Sn, ●: Cu–Sn

bands in the two states is observed (Fig. 5.19). Temperature dependences including the amorphous state are most obviously seen as a continuous shift of the Au $5d$-states (Fig. 5.19b). This shift is attributed to changes of the mass density and of short-range order with temperature [5.70]. The missing MDOS at E_F in the liquid state can be understood as the weaker electron–structure relation as stated above (5.5.1a). However, even noble-metal-rich alloys seem to show no depression at E_F [5.70, 71], although magnetic susceptibility (5.5.3c) and electronic transport properties (5.5.4) indicate a small MDOS.

Whereas the peak in $S(K)$ close to $2k_F$ is much smaller in the liquid compared to the amorphous state, the peak at K_p becomes even larger [5.51]. In

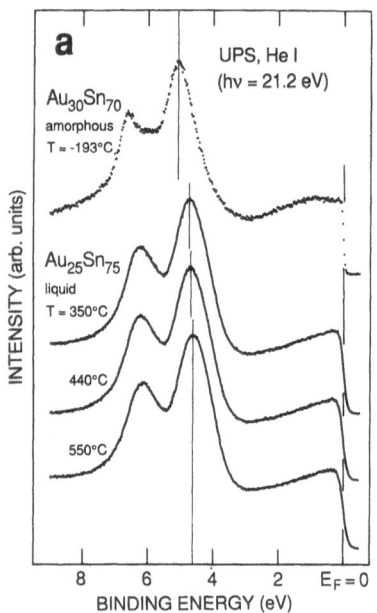

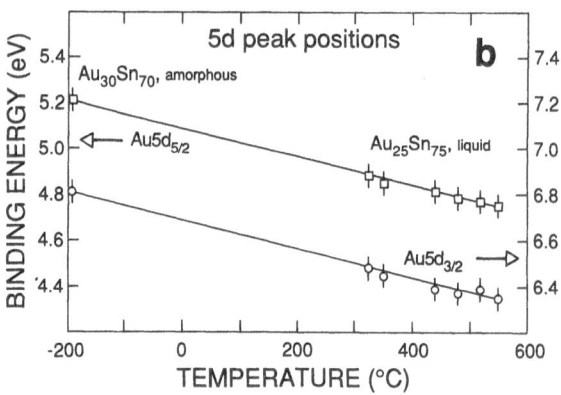

Fig. 5.19. Temperature dependence of the Au $5d$-state binding energies of Au–Sn: a) UPS HeI ($h\nu = 21.2$ eV) valence band spectra of Sn-rich liquid and glassy alloys; b) Au $5d$ peak positions as a function of temperature [5.70]

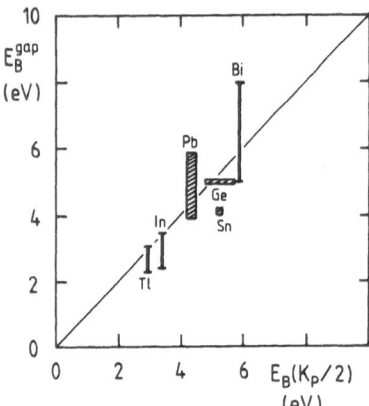

Fig. 5.20. Comparison of the position of structure-induced minima in the UPS spectra well below E_F with the estimation according to (5.11)

the present alloys the corresponding energy region of the DOS is dominated by noble-metal d-states and a minimum is not observable. Liquid polyvalent elements, on the other hand, show similar structural features; K_p as well as K_{pe} exist. *Indlekofer* could show that striking minima in the DOS of some liquid elements exist far below E_F [5.60]. A detailed discussion of systematic trends along the periodic table is given in his work. *Beck* et al. [5.33] could ascribe these minima by cluster calculations to structure effects. Selfconsistent *ab initio* calculations of the atomic *and* the electronic structure by *Jank* and *Hafner* [5.72] could explain the trends found by *Indlekofer*, but a direct relation between

K_p and the position of the minima according to (5.11) has not been reported by these authors. Such a simple relation indeed exists and is shown in Fig. 5.20. The horizontal width of the symbols corresponds to structure uncertainties if data of several sources are taken into account. The vertical length represents the width of the MDOS as reported by *Indlekofer*. This width should correspond to the band-gap coefficient $2|v(K_p)|$. Large coefficients indeed exist for Tl, In, Ge, Sn, Pb, and in particular for Bi, whereas Al, Ga, and Si show smaller values by factors 2–4 [5.73]. For the latter an MDOS could not be resolved [5.60, 74].

b) Electronic Specific Heat

In this section we will show that some specific–heat measurements support and others contradict the existence of an MDOS at E_F. After the subtraction of the lattice contribution, a term $\gamma_{exp} \cdot T$ is left at low temperatures. The experimentally derived γ-value consists of simple metallic glasses of several contributions [5.20, 74]:

$$\gamma_{exp} = \gamma_0 \cdot (1 + \lambda) + \gamma_{dis} \quad \text{with} \quad \gamma_0 = \frac{\pi^2 k_B^2}{3} \cdot N(E_F) \ . \tag{5.13}$$

γ_0 is the electronic contribution. The electron–phonon coupling constant λ takes care of corrections due to electron–phonon interaction [5.75]. γ_{dis} is a non-electronic contribution associated with the disordered state due to energetically low-lying excitations (see also Section 5.5.5). In the past, γ_{dis} for simple metallic glasses was found to be orders of magnitude smaller than γ_0 [5.74] and λ as 0.3 in $Mg_{70}Zn_{30}$, for example [5.76]. Ignoring both, *Mizutani* found γ_{exp}/γ_0-ratios for alloys of the present type between 1.01 and 1.6 and so claimed that the $N(E_F)$ of metallic glasses is in good agreement with $N_0(E_F)$ [5.20]. (Au, Cu)–Sn metallic glasses show λ-values between 0.4–1.1 [5.67, 68], which by no means can be neglected.

Measurements performed without magnetic field need a proper extrapolation from above T_c^{sc}, the transition to the superconducting state, down to $T = 0\,K$. These data, corrected for λ, are included in Fig. 5.17 for Cu–Sn. Above 30 at. % Sn, there is an excellent agreement with our estimation from UPS. Pseudo binary $(Ag_{0.5}Cu_{0.5})$–Ge metallic glasses indicate a DOS which is also at least 20–30% lower than the free-electron value [5.77].

In order to make the extrapolation to $T = 0\,K$ more reliable, quite recently measurements of Au–Sn metallic glasses were performed down to $300\,mK$ in an applied magnetic field [5.67]. After taking into account the λ-correction, a surprisingly good agreement with $N_0(E_F)$ was obtained (Fig. 5.17). Specific-heat measurements therefore give contradictory results. For Au–Sn, the tremendous change near $\bar{Z} = 1.8\,e/a$ (Fig. 5.17) could be attributed to γ_{dis} by measuring the same samples within the superconducting state well below T_c^{sc} [5.67]. After the subtraction of γ_{dis}, the agreement with $g = 1$ becomes excellent over the complete amorphous region.

c) Magnetic Susceptibility

Making the situation even more confusing, susceptibility measurements taken in the liquid state just above the melting point are also included in Fig. 5.17. Again deviations occur especially at $\bar{Z} = 1.8$ e/a [5.69], but are less pronounced than in the amorphous state in accordance with theory [5.30, 31, 78]. Experimentally, similar results were obtained for $(Ag_{0.5}Cu_{0.5})_{100-x}Ge_x$ with $20 \leq x \leq 30$ ($\bar{Z} \approx 1.8$ e/a). In the amorphous state of this system, the depression of the DOS at E_F is quite deep, whereas in the liquid state the minimum gets weaker and tends to be smeared out as temperature increases [5.79]. This behaviour led to the conclusion that the free electron value of the DOS at E_F is valid in the liquid state at high temperatures well above the melting point, and that a depression becomes evident with lowering temperature. In contrast to the amorphous state (Fig. 5.16), UPS again could not resolve an MDOS at E_F in the corresponding liquid state [5.71].

5.5.4 Electronic Transport

Electronic transport properties depending on electron–structure relations and vice versa, are useful for the revelation of their role in metallic glasses. Extensional work on the present alloys has been reported and discussed elsewhere [5.10, 55, 57]. Figure 5.21 shows, exemplary, the resistivity $\rho(T)$, the thermopower $S(T)$, and the Hall coefficient $R_H(T)$ of Sn-rich metallic glasses. During annealing there is hardly any change in $\rho(T)$ and $R_H(T)$ up to T_K, where pronounced steps occur (but see in Fig. 5.27 such data in an enlarged version). R_H for Sn-rich alloys is in good agreement with the free-electron value

$$R_H^0 = -\frac{1}{|e|n} \quad \text{with } n = \bar{Z} \cdot \bar{n}_0 . \tag{5.14}$$

$S(T)$ is linear with T for $T_c^{sc} < T < T_0$. At T_0 there is a bending over, again giving $S(T)$ linear with T for another 120 K.[6] We distinguish between $S^{\ell}(T)$, the low-temperature thermopower below T_0, and $S^h(T)$, the high-temperature value, For Sn-rich alloys, $S^{\ell}(T)$ is close to the free-electron value

$$S_0(T) = -\frac{\pi^2 k_B^2}{|e|E_F} \cdot T. \tag{5.15}$$

First, we will discuss concentration dependences and the scaling behaviour of these properties with $\bar{Z}$. Later, we will discuss temperature dependencies taking into account effects due to the inelastic excitation of phonon-rotons.

---

[6]The tiny hump near T_0 is artificial due to the reference material (Pb) [5.80, 82].

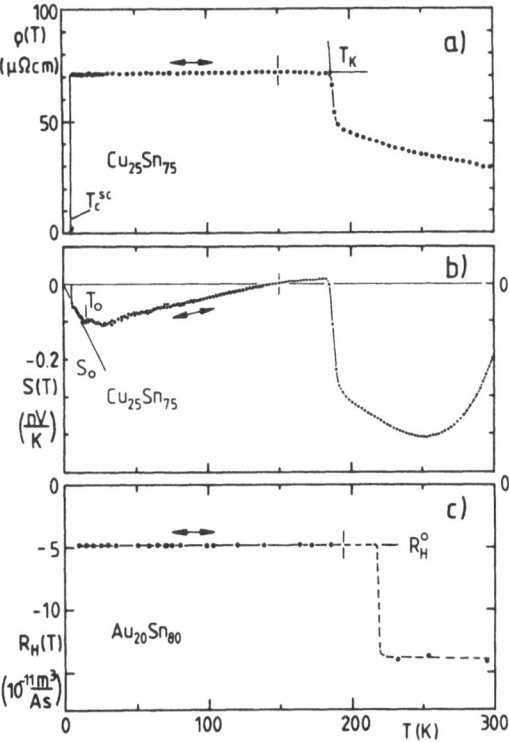

Fig. 5.21. Electronic transport properties of Sn-rich metallic glasses vs. temperature. a) Resistivity and thermal stability [5.80]; b) thermopower [5.80]; and c) Hall coefficient [5.81]. In all figures, only the region below the vertical bar is in an annealed state and is hence reversible

a) Versus Composition

Figure 5.22 shows ρ, R_H, and $S^\ell(T)/T$ for disordered (Au, Ag, Cu)–Sn systems versus x. T_K and the MDOS have already been presented above (Fig. 5.13 and Fig. 5.17). A close overall similarity exists among the different alloys as well as to the corresponding liquid state. For alloys with 27–30 at. % Sn, where $K_{pe} = 2k_F$ is fulfilled, ρ is maximal while R_H and $S^\ell(T)/T$ deviate from the corresponding free-electron value. These deviations are obviously related to FsBz-effects showing that E_F or k_F cannot be deduced from Hall-effect data in alloys with a high peak at K_{pc} as mentioned above. Equations (5.7) and (5.8) are always found to be a better approach.

Concentration as well as temperature dependences of ρ in both the amorphous [5.86–89] and the liquid state [5.90–92] are understood in the framework of the diffraction or extended Faber–Ziman model [5.90, 92, 93]. Experimentally determined total as well as model partial-structure functions have been successfully applied. Disregarding inelastic-scattering effects and extensions of

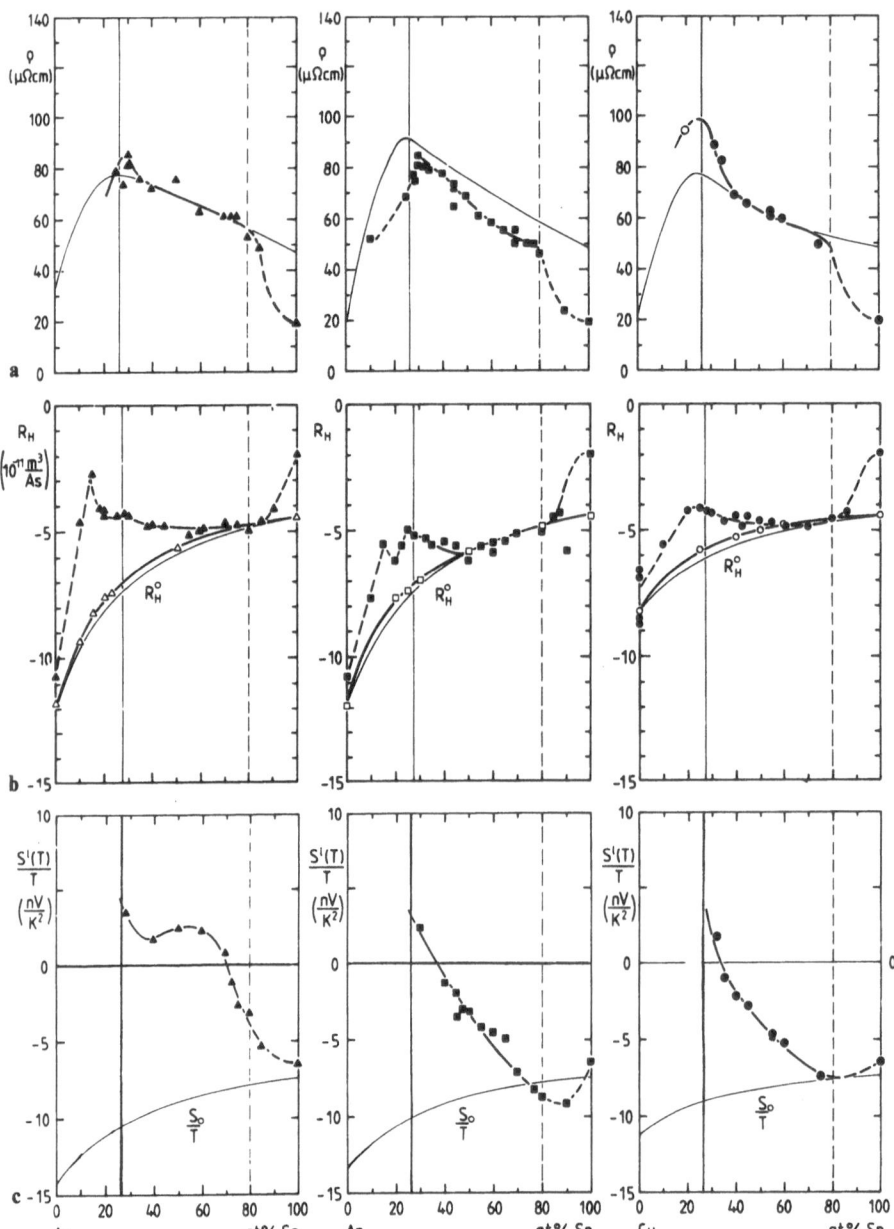

Fig. 5.22. Electronic transport properties of (Au, Ag, Cu)–Sn metallic glasses and the corresponding liquid state vs. composition. a) Resistivity: metallic glasses (full symbols) [5.80, 83, 84], liquid state (thin curves) [5.53, 85]; b) Hall coefficient: metallic glasses (full symbols) [5.55], liquid state (open symbols) [5.53]; c) Low-temperature thermopower: [5.80, 83, 84]. In Figs. b, c the thin curves represent the free-electron values (5.14, 15)

the formalism to alloys, the diffraction model gives approximately [5.17]

$$\rho_z = \frac{3\pi\Omega_0 m^2}{e^2\hbar^3 k_F^2} \cdot \langle S(K) \cdot |v(K)|^2 \rangle, \tag{5.16}$$

$$\langle S(K) \cdot |v(K)|^2 \rangle = \int_0^1 S(K) \cdot |v(K)|^2 \cdot 4 \cdot \left(\frac{K}{2k_F}\right)^3 \cdot d\left(\frac{K}{2k_F}\right). \tag{5.17}$$

The decrease of $S(2k_F)$ with increasing Sn-content (Fig. 5.18) explains the decrease of ρ with x. We will not go into further details but refer instead to the literature mentioned above.

The Hall effect of disordered systems still remains a challenging problem which is, by no means, solved theoretically yet. If deviations to R_H^0 are observed, such deviations cannot be explained in terms of two-band mechanisms because of the Fermi-surface blurring. Hole-like states require a clearly defined dispersion relation at the upper band edge. The relaxation-time solution of the Boltzmann equation itself gives R_H^0 even for the NFE-model [5.33, 94–97]. Using the generalized transport equation and assuming a finite spectral width of the electronic states, *Beck* et al. [5.33] could explain deviations of R_H from the free-electron value. The effect is relatively small yielding deviations of the order of a few percent for reasonable pseudopotentials. In liquid (Au, Ag, Cu)–Sn alloys, small deviations (Fig. 5.22b) exist and hence may be explained by this theory. The small deviations of the liquid state were earlier explained by density anomalies versus composition [5.53]. In the amorphous state, deviations become so large that both these explanations seem to be unlikely. In concentration ranges where large deviations exist, temperature dependences occur, and vice versa [5.55]. Although the temperature dependence was found a factor of four too small to bring R_H of the amorphous state in agreement with the liquid state by extrapolation, this trend might relate to changes of the MDOS at E_F with T. In the glassy state such correlations between R_H and $N(E_F)$ have empirically been revealed versus composition [5.57] (Fig. 5.23). Obviously $\Delta R_H/R_H^0$ obeys

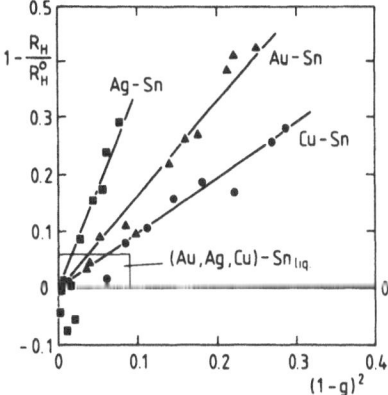

Fig. 5.23. Deviations of the Hall coefficient to R_H^0 vs. $(1-g)^2 \propto S^2(2k_F)$ [5.57]. Deviations of the corresponding liquid samples (Fig. 5.7) are confined within the box close to zero. g_x-data were used for this purpose

a parabolic relation with $(1 - g)$ or $S(2k_F)$:

$$1 - \frac{R_H}{R_H^0} \propto (1 - g)^2 \propto S^2(2k_F) .\qquad(5.18)$$

ΔR_H seems to depend only on the electronic states at E_F, which have raised a controversy for a long time [5.95, 96, 98, 99]. Additional support for this finding comes from the liquid elements where R_H is very close to R_H^0 in accordance with a diminishing MDOS at E_F [5.53], although minima in the DOS exist well below E_F (5.5.3a). The deviations of some heavy liquid elements as Hg, Tl, Pb, Bi and the existence of these minima well below E_F (Fig. 5.20) are not in contradiction because these elements also show structures in the DOS close to E_F [5.60]. Theoretical work on the Hall coefficient of liquid elements is reviewed by *ten Bosch* [5.95] and *Ballentine* [5.96] and new theories are discussed by *Itoh* [5.99].

In contrast to ρ and R_H, the thermopower of disordered systems has scarcely been studied. Most measurements have been done on metallic glasses with d-states at E_F. Besides some simple alloys such as (Ca, La)–Al [5.100] and Mg–Zn [5.101], systematic studies have been done on (Au, Ag, Cu)–Sn metallic glasses [5.80, 83, 84, 102].

The basic expression is given by the Mott formula [5.103]

$$S(T) = -\frac{\pi^2 k_B^2}{3|e|E_F} \cdot \xi \cdot T \quad \text{with} \quad \xi = E_F \cdot \frac{\partial \ln \rho(E)}{\partial E}\bigg|_{E_F} .\qquad(5.19)$$

If the scattering mechanism of the electrons is T-independent, ξ is constant and $S(T)$ becomes proportional to T [5.103]. Many-body effects, on the other hand, may cause non-linearities at low temperatures due to electron–phonon mass-enhancement effects, giving $S(T) = [1 + \lambda(T)] \cdot S_b(T)$ [5.104]. S_b is now the bare thermopower of (5.19) without mass-enhancement. Higher correction terms have been proposed by *Kaiser* [5.105]. The resistivity should not be influenced by mass-enhancement effects [5.106].

Ag–Sn [5.84] as well as Au–Sn [5.83] were originally described along these lines and the bend at T_0 was ascribed to $\lambda(T)$. However, serious discrepancies with the theory appeared. The high-temperature value, for example, is not really proportional to T (Fig. 5.21). $S_b(T)$ itself was found to be positive for any composition, although $(1 - g)$ (Fig. 5.17) as well as ΔR_H (Fig. 5.21) can be small for certain compositions, indicating an agreement with the FEM where the sign of $S(T)$ should be negative in accordance with the sign of R_H. Consequently, we assume for the present alloys this model as not applicable over the whole temperature range.

Hence we recently proposed a new description [5.57] taking into account the phonon-rotons as described above (5.3.3). The scattering mechanism changes with T and hence a proportionality with T over the whole temperature range can no longer be expected. According to Fig. 5.21b and Fig. 5.22c, below T_0 the thermopower is linear with T and approaches the free-electron value as required whenever $S(2k_F)$ becomes small. The bend at T_0 is interpreted as the onset of an inelastic umklapp scattering due to the excitation of phonon-rotons.

Starting with the diffraction model, we first describe $S^{\ell}(T)$. Phonon-roton influences do not exist below T_0. With ρ_z Mott's formula gives [5.103]

$$\zeta = 3 - 2q - \tfrac{1}{2}r ,\tag{5.20}$$

with

$$q = \frac{S(2k_F)\cdot|v(2k_F)|^2}{\langle S(K)\cdot|v(K)|^2\rangle}\tag{5.21}$$

and

$$r = \frac{\displaystyle\int_0^1 S\left(\frac{K}{2k_F}\right)\cdot k_F\cdot\left.\frac{\partial|v(K)|^2}{\partial k}\right|_{k_F}\cdot 4\cdot\left(\frac{K}{2k_F}\right)^3\cdot d\left(\frac{K}{2k_F}\right)}{\langle S(K)\cdot|v(K)|^2\rangle} .\tag{5.22}$$

According to (5.19) and (5.20), the free-electron value (5.15) is valid under conditions $q = 0$ and $r = 0$. Deviations of $S^{\ell}(T)$ to $S_0(T)$ can hence be expressed as

$$1 - \frac{S^{\ell}(T)}{S_0(T)} = \frac{2q + \tfrac{1}{2}r}{3} \propto S(2k_F) \propto (1 - g)\tag{5.23}$$

due to elastic umklapp scattering. The latter two proportionalities are valid if $q \neq 0$ and $r = 0$ are assumed. The deviation of $S^{\ell}(T)$ to $S_0(T)$ is indeed linearly varying with $(1 - g)$ or $S(2k_F)$ as seen in Fig. 5.24 for small $(1 - g)$- or $S(2k_F)$-values. Although the proportionality is not exactly fulfilled, we take this feature as a strong indication of the applicability of 5.23 below T_0. Taking into account mass-enhancement effects (replacement of $S^{\ell}(T)$ in (5.23) by $S^{\ell}(T)/(1 + \lambda)$), a strict proportionality to $S(2k_F)$ is observed for Cu–Sn [5.80, 107]. A detailed discussion will be published elsewhere [5.108]. Both the r-term as well as

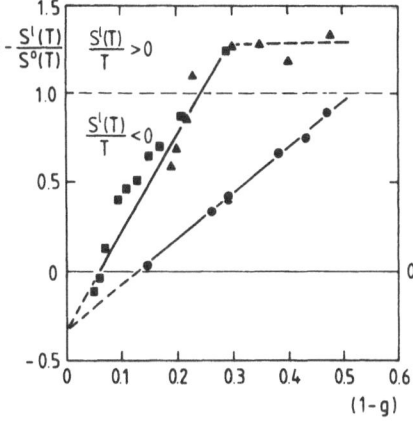

Fig. 5.24. Deviations of $S^{\ell}(T)$ to the FEM-value plotted vs. $(1 - g)$ [5.57]. ▲: Au–Sn [5.83], ■: Ag–Sn [5.84], ●: Cu–Sn [5.80]. The dashed horizontal line separates the region of positive from the region of negative thermopowers at low temperatures

influences of the phonon-rotons (5.5.4b–5.5.5) might be responsible for the non-linearities at large deviations.

In conclusion, below T_0 $S(T)$ is understood by the FEM-behaviour (5.15) and deviations to this model by *elastic* umklapp-scattering processes (5.23). Small-angle inelastic-scattering effects due to common long-wavelength phonons (normal processes) seem to be negligible in agreement with the theory [5.109]. Electron–phonon mass-enhancement effects might need to be accounted for $\lambda(T)$. If deviations to the FEM are not too large, the old standing problem that R_H and $S(T)$ show different signs seems to be solved if both are considered below T_0.

b) Scaling Behaviour Versus $\bar{Z}$

The characteristic scaling behaviour of physical properties of the Hume–Rothery alloys with $\bar{Z}$ is also valid for ρ, R_H, and $S^\ell(T)/T$ as seen in Fig. 5.25. Noting how different the systems are away from $\bar{Z} = 1.8$ e/a, it is surprising how remarkably close they get at this special $\bar{Z}$-value. ρ of pure Si and Ge, for example, differs by many orders of magnitude from pure Sn, whereas at $\bar{Z} = 1.8$ e/a the differences are less than 10%. With the exception of Au–In, at this $\bar{Z}$-value, R_H is also identical within the experimental resolution [5.10]. The values of ρ and R_H themselves are slightly different at $\bar{Z} = 1.8$ e/a if Au is replaced by Ag or Cu [5.10]. As only a few thermopower data are known, all alloys are compared with each other, independent of the type of the noble metal. The low-temperature value $S^1(T)/T$ seems to show a singularity at $\bar{Z} = 1.8$ e/a. Even $Mg_{70}Zn_{30}$ ($\bar{Z} = 2$ e/a) fits well into this general behaviour.

c) Versus Temperature

There is a different situation above T_0 due to inelastic scattering of the electrons with the low-lying phonon-rotons. In glassy $Mg_{70}Zn_{30}$, phonon-roton states have been measured by inelastic-neutron diffraction (Fig. 5.4). A bend in $S(T)$ at roughly the correct temperature is indeed observed [5.80, 101]. With the generalized *Faber–Ziman* theory (*Baym's* formula [5.93]), *Hafner* [5.24] could reveal for this system the influence of the phonon-rotons on the resistivity by taking into account the measured dynamic structure factor. The characteristic behaviour of $\rho(T)$ is shown in Fig. 5.26 [5.24, 88, 110]. The Debye temperature Θ_D, also influenced by the phonon-rotons, is strongly temperature dependent and separates the theoretical curves.

The resistivity of the alloys considered here show similar T-dependences [5.80, 111, 112]. $\rho(T)$ of $Cu_{20}Al_{80}$ at low temperatures is given as an example strongly enlarged in Fig. 5.27a. The corresponding thermopower is shown in Fig. 5.27c and for completeness the Hall coefficient in Fig. 5.27d. Below T_0 the absolute values of ρ as well as $S^\ell(T)$ are dominated by elastic umklapp scattering as discussed above. The increase of $\rho(T)$ with decreasing T has been ascribed to weak localization [5.113] or e–e interaction [5.114] as well as to the phonon-

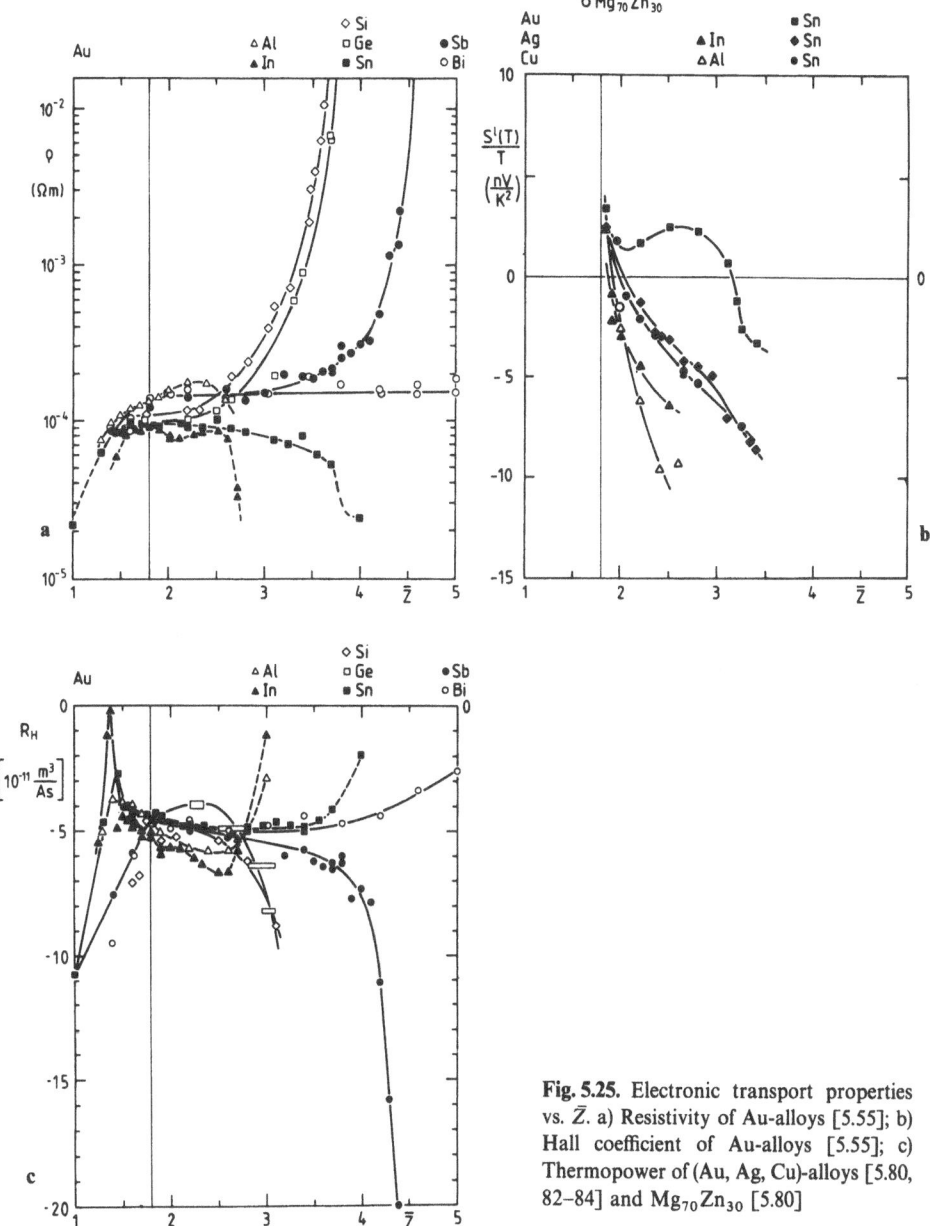

Fig. 5.25. Electronic transport properties vs. $\bar{Z}$. a) Resistivity of Au-alloys [5.55]; b) Hall coefficient of Au-alloys [5.55]; c) Thermopower of (Au, Ag, Cu)-alloys [5.80, 82–84] and $Mg_{70}Zn_{30}$ [5.80]

dispersion curve of the type shown in Figs. 5.4, 5 [5.24]. The increase with increasing T is caused by the inelastic scattering of the electrons with normal-long wavelength phonons. Theoretically, a T^2-dependence is expected in this region [5.86, 87] but not exactly fulfilled for the present sample. The T^2-dependence disappears with rising temperature when the electron mean free

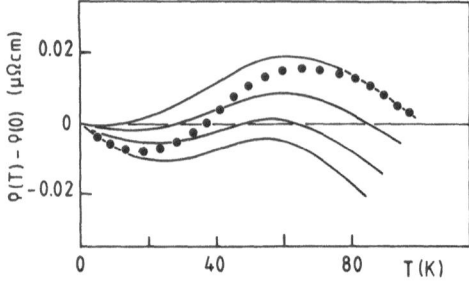

Fig. 5.26. Low-temperature resistivity of $Mg_{70}Zn_{30}$ [5.24]. Full dots: experimental data; thin curves: calculations with different Θ_D (lowest curve: small Θ_D, highest curve: large Θ_D)

path becomes short so that phonons with wavelength longer than this value become ineffective electron scatterers (saturation effect) [5.115, 110]. The second derivative of $\rho(T)$ (Fig. 5.27b), indicating the deviation from the increasing resistivity, shows the inflection point at $T = T_0$, the same temperature where the bend in $S(T)$ occurs. $R_H(T)$ seems to be uninfluenced, although we would like to emphasize that the resolution is more than a factor 10 worse than for the resistivity. A gradual change of approximately 0.3% from $T = 0\,K$ to $T = 120\,K$ seems to occur. At much higher temperatures than T_0, $\partial\rho/\partial T$ becomes linearly decreasing with T. The second derivative of $\rho(T)$ approaches zero. The linear variation of the resistivity is ascribed to the decrease of $S(K_{pe})$ according to the Ziman model. The Debye–Waller factor describes the decrease of $S(K_{pe})$ with T due to inelastic excitations in a solid. For the present alloys, the decrease of $\rho(T)$ with increasing T seems to be caused by the *inelastic* umklapp scattering of the electrons with the phonon-rotons starting at $T = T_0$. In-elastic umklapp scattering caused by these states is obviously less effective on ρ than the elastic umklapp scattering.

Approaching $\bar{Z} = 1.8\,e/a$ ($x = 60$), the characteristic temperature T_0 shifts to lower temperatures and at $\bar{Z} = 1.8\,e/a$, finally, a negative temperature coefficient of $\rho(T)$ exists down to the lowest temperatures and $S(T)$ is positive over the whole temperature region. A detailed discussion about concentration dependences is reported elsewhere [5.80, 112].

5.5.5 Dynamic Properties, Low-Lying Excitations

According to the decrease of T_0, with composition we expect low-lying non-electronic excitations in the specific heat especially close to $\bar{Z} = 1.8\,e/a$ due to the thermal excitation of phonon-rotons. In Sect. 5.5.3.b, we already mentioned that those excitations indeed have been observed for Au–Sn. γ_{dis}-values extracted from specific-heat measurements well below T_c^{sc} are shown in Fig. 5.28 for (Au, Cu)–Sn, clearly showing the increasing contribution of phonon-rotons at this value. The strong increase of the linear specific heat for Cu–Sn close to $\bar{Z}$

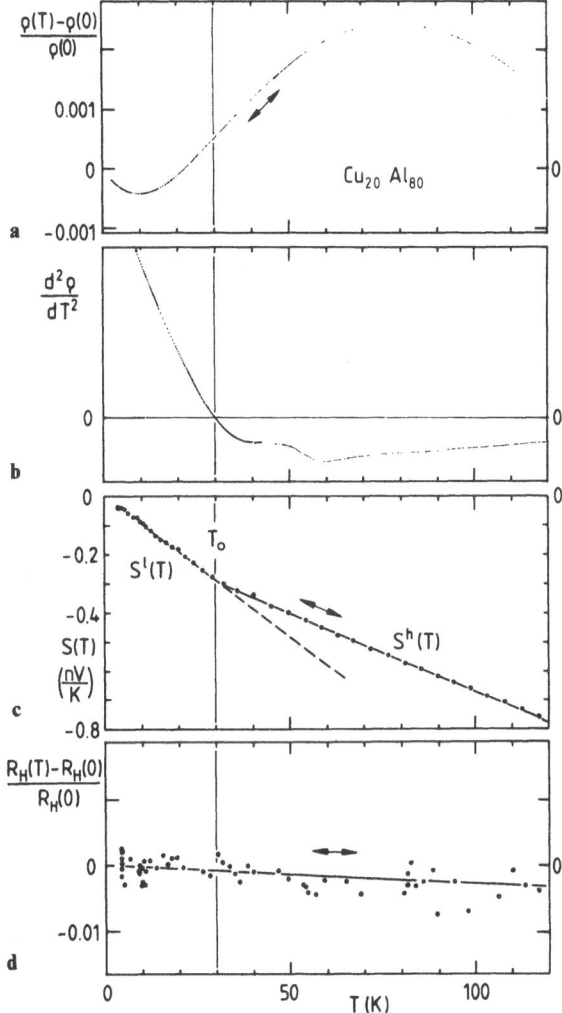

Fig. 5.27. a–d Enlarged plot of transport properties of $Cu_{20}Al_{80}$ metallic glass. a) Resistivity [5.112]. Below $T = 15$ K a magnetic field has been applied in order to suppress superconductivity; b) The second derivative of the resistivity data in a); c) Thermopower [5.80]; d) Hall coefficient [5.112]. All measurements have been performed with the samples well annealed

$= 1.8$ e/a (Fig. 5.17) is also believed to have the same non-electronic origin, although this has not yet been proved and therefore these data are not included in Fig. 5.28.

The stability of the metallic glasses under consideration with decreasing Sn-content seems to be influenced by the decrease of T_0 or ω_0, respectively. If ω_0 becomes very low for alloys with $\bar{Z}$ approaching 1.8 e/a, indicated by the increase of the γ_{dis}-contribution, the system finally crystallizes. Phonon-roton

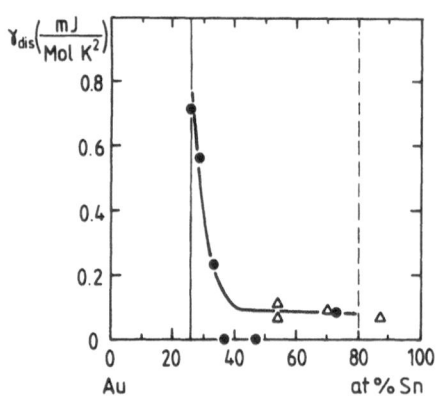

Fig. 5.28. Non-electronic low-lying contributions γ_{dis} to the T-linear term of the specific heat (5.13). ● : Au–Sn [5.67] and; △ : Cu–Sn metallic glasses [5.116]

effects on stability have been discussed by *Beck* and *Oberle* [5.7] as well as by *Egorushkin* et al. [5.117].

5.6 Magnetic Amorphous Alloys

After having discussed the non-magnetic alloys, we finally go to the more complex systems with Fe, Co and Ni or Pd instead of the noble metals. Extensive research on these metallic glasses has been done for many years by *Piecuch* et al. and *Geny* et. al. [5.118]. Ni–P and Pd–Si and many other alloys have successfully been prepared by different methods. For the samples reported here, a co-deposition with two magnetron sources as well as the sequential flash-evaporation technique has been used with no major differences in the physical properties [5.22].

The stabilizing role of conduction electrons may become twofold according to their charge and spin. Replacing, for example, Au, Ag, or Cu by Fe, close to local moments the electron cloud gets spin polarized giving rise to spin-density oscillations and therefore to an oscillatory indirect *magnetic* interaction, the so-called RKKY-interaction [5.119]. U_m, a magnetic energy, structure dependent as well, may result. The common RKKY-interaction, based on long mean-free-path conditions, changes drastically in systems with short electron mean free path [5.120]. The effective indirect exchange-coupling coefficient $J_{eff}(r)$ is no longer *cosine*-like and has to be replaced by

$$J_{eff}(r) \propto \frac{\sin(\beta \cdot r)}{r^3} \cdot e^{-\alpha r} \,, \tag{5.24}$$

with both α and β dependent on $2k_F$. To first approximation, $\beta = 2k_F$ has been assumed [5.21, 22], causing the same wavelength λ_F as for Friedel oscillations. $J_{eff}(r)$ causes ferromagnetic order if the moments are at distances r_F^n (5.5.1.b), or antiferromagnetic order if the moments are at $r_F^n + \lambda_F/2$. If maxima of $g(r)$

coincide with maxima of $J_{\mathrm{eff}}(r)$, an additional depression of the total energy will result. The magnetic influence due to ferromagnetic interaction will depress the total energy and will enhance the stability of the disordered state even more, compared to the case where only Friedel minima in $\phi_{\mathrm{eff}}(r)$ are effective [5.21, 22].

The Curie temperature T_c obeys [5.120]

$$T_c \propto \int_0^\infty g(r) \cdot J_{\mathrm{eff}}(r) \cdot r^2 \, dr \, , \tag{5.25}$$

indicating additionally such correlations between $g(r)$ and $J_{\mathrm{eff}}(r)$. Below we show enhanced crystallization temperatures T_K whenever a high T_c occurs.

5.6.1 Structure

The main features of magnetic metallic glasses are shown, in an exemplary manner, on Fe–Sb (Fig. 5.29). The amorphous phase exists for $x > 23$–27 [5.22]. The electronic-induced peak in $S(K)$ close to $2k_F$ is clearly shown (Fig. 5.29a). $2k_F$, calculated with the assumption that $Z_{\mathrm{Fe}} = 1.0$ e/a, is indicated by thin vertical lines. The finding of an effective valence for pure Fe is one of the main problems in this respect. Calculations using $Z_{\mathrm{Fe}} = 1.1$ e/a are included in Figs. 5.29c, 30. Over the whole region of the amorphous state, K_{pe} shifts close to $2k_F$ (Fig. 5.30), indicating again the electronic influence on structure. *Nagel-Tauc's* criterion is best fulfilled near 25 at. % Sb. At this composition, $S(K_{\mathrm{pe}})$ itself is largest, indicating already the optimal matching of the position of the ions (local moments) with $\phi_{\mathrm{eff}}(r)$ (and/or $J_{\mathrm{eff}}(r)$), respectively.

In Figs. 5.29b, c peaks in $g(r)$ are compared with r_F^n. With changing r_F^n due to a change of $\bar{Z}$, the peak positions also change. Between r_1 and r_2, there is additional intensity in the antiferromagnetic position, not observed for example in Au–Sb or pure Sb (Fig. 5.7). This peak is also small below 36 at. % Sb content showing preferred ferromagnetic positions. The correlation between r_n and r_F^n versus composition is most obvious in Fig. 5.29c. Between both vertical lines (1.96 e/a $< \bar{Z} < 2.1$ e/a) their matching seems to be reasonably fulfilled. As in Au–Sb (Fig. 5.11), r_n can shift with r_F^n at medium range. Following the characteristic behaviour of these alloys (Fig. 5.12) r_1 itself coincides with $r_F^1 = 5/4 \cdot \lambda_F$ at 25 at. % Sb. Near this composition the ferromagnetic positions dominate and a high Curie temperature – and correspondingly a high crystallization temperature – can be expected.

5.6.2 Magnetism and Stability

This indeed can be seen in Fig. 5.31. In the region where T_c increases strongly, T_K shows an additional increase and is approximately 160 K higher than in Au–Sb. Fe–Sn as well as Fe–Ge show similar behaviour from the structural

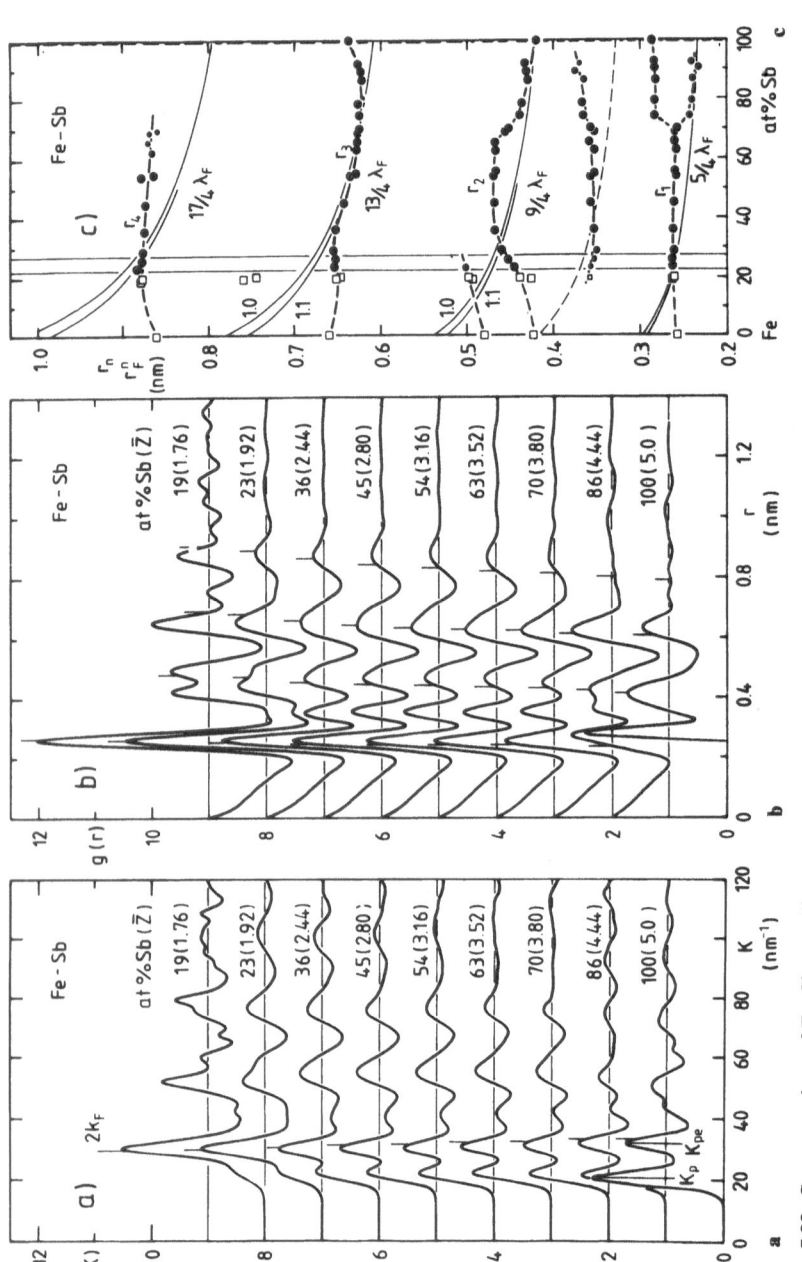

Fig. 5.29. Structure data of Fe–Sb metallic glasses [5.22]. Pure Sb is taken from [5.50]. a) Structure factor; b) pair-distribution function. Vertical lines in both figures indicate $2k_F$ or r_F^n, respectively. $Z_{Fe} = 1$ e/a was used for their calculation; c) ion (local moment) positions vs. composition, compared with the position of minima in $\phi_{eff}(r)$ and maxima of $J_{eff}(r)$ calculated with $Z_{Fe} = 1.0$–1.1 e/a (thin solid curves). The thin dashed curve indicates a favourable antiferromagnetic position of local moments. The thin vertical lines in c) indicate approximately the composition where the *Nagel-Tauc* criterion is best fulfiled

Fig. 5.30. Concentration dependence of K_p, K_{pe}, and $2k_F$. The vertical lines indicate the regions where $K_{pe} \approx 2k_F$, $r_1 \approx 5/4\lambda_F$, $T_K = T_K^{max}$, $T_c = T_c^{max}$ is fulfiled

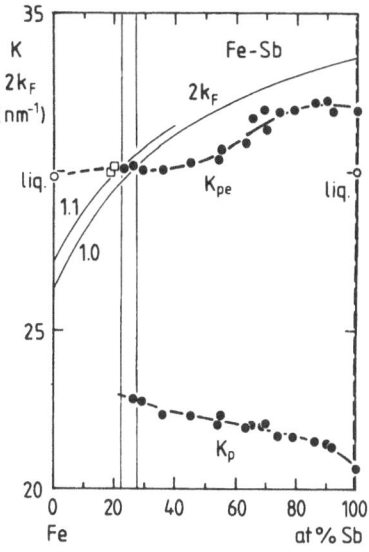

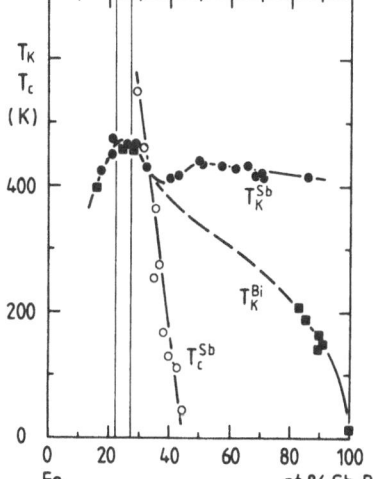

Fig. 5.31. Crystallisation temperature and Curie temperature of quench-condensed Fe–Sb and Fe–Bi metallic glasses [5.22]

point of view, although the antiferromagnetic position is not as occupied as in Fe–Sb [5.22]. Fe-alloys with C, Si, Sn and Ge show the enhanced stability in the region of highest T_c. T_K and T_c intersect with each other at compositions with $K_{pe} = 2k_F$, strongly indicating the magnetic influence [5.22]. In this region, the local moments obviously are ferromagnetically situated (Fig. 5.29). For larger Sb content the system is non-magnetic, but shows spin-glass behaviour at low temperatures due to frustrated spin position. The mean value $\bar{Z}$ at the optimal composition seems to be larger ($\bar{Z} \approx 1.9$–2.1 e/a) than in the case of noble-metal alloys ($\bar{Z} = 1.8$ e/a), but too few measurements are available to generalize this statement.

Transport properties of the magnetic alloys considered here are much more complex than discussed in 5.5.4. The maximal resistivity and the largest negative temperature coefficient of the resistivity neither occurs at the same composition nor at compositions where the thermal stability is maximal and $K_{pe} = 2k_F$ fulfilled [5.118]. Those differences are attributed to strong d-state scattering and magnetic phase transitions. UPS measurements show large d-state intensities at E_F and an MDOS has not been observed [5.121]. R_H becomes positive for amorphous alloys [5.122] as in the liquid state [5.123]. In Pd–Sn, $R_H \approx 0$ [5.122], clearly indicating again that Hall-effect data cannot be used for the deduction of E_F or k_F. The scaling behaviour is not yet established for these alloys due to the lack of a reasonable number of systematic measurements. Additional low-lying excitations of magnetic origin (magnon-rotons [5.124]) were observed by inelastic-neutron scattering as well [5.125], but will not be discussed here.

5.7 Concluding Remarks

It was our aim to give a comprehensive overview of a large class of binary disordered metals. (Au, Ag, Cu)–Sn and (Au, Fe)–Sb metallic glasses have served as model systems in this respect. This qualitative discussion was based only on pseudopotential theory and total-structure functions.

We have clearly demonstrated electronic influences on structure due to indirect interactions. In non-magnetic alloys, electric polarization of the electron cloud around localized charges (ions) causes Friedel oscillations and hence a preferred position of the ions. In magnetic alloys, evidence of another indirect interaction due to spin polarization around localized local moments causing RKKY oscillations has been shown.

The electronic model for phase stability has been extrapolated from the region of the crystalline HR-phases to the amorphous state considered here, indicating the latter as a limiting case of the crystalline Hume–Rothery phases for $\bar{Z} \geq 1.8$ e/a. The scaling behaviour with $\bar{Z}$ of all properties is explained along these lines.

The electronic influence on the stability of metallic glasses, first proposed by *Nagel* and *Tauc* for non-magnetic alloys, has been extended to magnetic alloys. Additional stabilizing effects and effects on magnetic properties have been discussed.

Structural effects on the DOS have been shown. There is an MDOS close to E_F due to the electronic induced peak at K_{pe} as well as an MDOS well below E_F due to the peak at K_p. The latter could only be shown for liquid elements. The depth of the MDOS at E_F is proportional to the structural weight $S(2k_F)$.

Deviations of electronic transport properties as R_H and $S'(T)$ from the FEM are proportional to $S^2(2k_F)$ and $S(2k_F)$, respectively. Mass-enhancement effects on the thermopower at low temperatures might also have to be taken into account. Low-lying phonon-roton states are found to be responsible for the T-dependence of $S(T)$ and $\rho(T)$ as well as for the non-electronic excitations seen with the specific heat. Detailed thermopower calculations, especially those taking into account the full dynamic information where the phonon-roton states are included, still await implementation.

The interrelation between electronic and ionic structure contains in our opinion the key to the understanding of the metallic glasses discussed in the present work.

Acknowledgements. I would like to express my gratitude to F. Baumann, H. Beck, W. Buckel, H.-J. Güntherodt, J. Hafner, M. Itoh, E. Kay, P. Oelhafen as well as G. Indlekofer and H.-G. Boyen for many stimulating discussions. Particular acknowledgement is also given to M. Burkhardt, C. Lauinger, T. Moser, and P. Rieger for making their unpublished results available. I would also like to thank Ms. M. Bielfeld for making the drawings and the Deutsche Forschungsgemeinschaft as well as IBM for financial support.

Addendum

The phase shift of the Friedel oscillations by $\pi/2$ from a *cosine*-function to a minus *sine*-function was recently reported as a common effect of the short mean-free path of the electrons [5.126], an interpretation quite similar to that argued by *Kaneyoshi* for the effective exchange interaction described in the review. Structure-induced pseudo gaps at E_F in amorphous Mg–Zn alloys mentioned above, were confirmed and were also found in the mean time for Ca–Zn [5.127].

A detailed comparison of the amorphous with the liquid state was presented by us for Pb–Bi and Tl–Bi [5.128]. These alloys show no d-state contributions within the valence band and hence are predestinated for UPS-measurements of structural effects on the DOS. In accordance with the existence of two peaks in $S(K)$ at $0 < K \leq 2k_F$, clear indications for the existence of two MDOS for $0 < E \leq E_F$ were observed. One is far below E_F and clearly related to the peak in $S(K)$ at K_p, quite similar to the case of the liquid elements reported above, and another one at E_F caused by the peak at $2k_F$. Temperature dependences of both were observed, which are in qualitative agreement with the change of the structural features with T. The MDOS far below E_F shifts closer to E_F from the amorphous to the liquid state as the peak at K_p shifts closer to $2k_F$ with rising temperatures. The MDOS at E_F gets weaker and finally disappears well above the melting point in accordance with the disappearance of the corresponding peak in $S(K)$ close to $2k_F$. The latter variation with T, observed to our knowledge for the first time, might have strong impact on the theory of electronic transport versus temperature.

In a sequence of three papers, *Jank* and *Hafner* [5.129] reported the theoretical basis for the electronic structure of polyvalent liquid elements. Pseudo gaps far below E_F are attributed to strong non-localities of the pseudo-potential resulting from relativistic effects.

References

5.1   W. Buckel, R. Hilsch: Z. Phys. **138**, 109 (1954)
5.2   P. Duwez, R.H. Williams, W. Klement: J. Appl. Phys. **31**, 1136 (1960)
5.3   A review was given by P.H. Gaskell: In *Glassy Metals II* ed by H. Beck, H.-J. Güntherodt, Topics Appl. Phys., Vol. 53 (Springer, Berlin, Heidelberg 1983) p. 5
5.4   J. Hafner: In *Glassy Metals I*, ed by H. Beck, H.-J. Güntherodt, Topics Appl. Phys., Vol. 46 (Springer, Berlin, Heidelberg 1983) p. 93
5.5   J. Hafner: In *Amorphous Solids and the Liquid State*, ed by N.H. March, R.A. Street, M. Tosi (Plenum, New York 1985) p. 91
5.6   J. Hafner: *From Hamiltonian to Phase Diagrams – The Electronic and Statistical-Mechanical Theory of sp-bonded Metals and Alloys* Springer Ser. (Solid-State Sci. Vol. 70) (Springer, Berlin, Heidelberg 1987)
5.7   H. Beck, R. Oberle: *Proc. 3rd Int'l Conf. on Rapidly Quenched Metals*, ed. by B. Cantor (The Metal Society, London 1978) Vol. 1, p. 416
5.8   R. Oberle, H. Beck: Solid State Commun. **32**, 959 (1979) H. Beck, R. Oberle: J. Phys. **41**, C8-289 (1980)
5.9   P. Häussler: J. Phys. **46**, C8-361 (1985)
5.10   P. Häussler: Z. Phys. B **53**, 15 (1983); P. Häussler: Der Einfluss der mittleren Electronenkonzentration auf Transporteigenschaften und Stabilität amorpher binärer Legierungen des Typs $Sn_{100-x}Au_x$, Dissertation, Univ. Karlsruhe, Germany (1981)
        P. Häussler, F. Baumann: Physica **108B**, 909 (1981)
5.11   G. Kahl, J. Hafner: Solid State Commun. **49**, 1125 (1984)
5.12   S.R. Nagel, J. Tauc: Phys. Rev. Lett. **35**, 380 (1975); Solid State Commun. **21**, 129 (1977)
5.13   W. Hume-Rothery: *The Structure of Metals and Alloys* (Inst. Metals, Monogr. Rep. Ser. No. 1, London, 1956)
        H. Jones: Proc. Phys. Soc. A **47**, 250 (1937)
5.14   T.B. Massalski, H.W. King: Progr. Mat. Sci. **10**, 1 (1961);
        C.S. Barrett, T.B. Massalski: *Structure of Metals* (McGraw-Hill, New York 1966)
5.15   N.F. Mott, H. Jones: *The Theories of the Properties of Metals and Alloys* (Clarendon, Oxford 1936) p. 310
5.16   A review was given by P. Oelhafen: In *Glassy Metals II*, ed by H. Beck, H.-J. Güntherodt, Topics Appl. Phys., Vol. 53 (Springer, Berlin, Heidelberg 1983) p. 283
5.17   J.M. Ziman: Adv. Phys. **16**, 551 (1967)
5.18   J. Hafner, S.S. Jaswal, M. Tegze, J. Krieg, A. Pflugi, P. Oelhafen, H.-J. Güntherodt: J. Phys. F **18**, 2583 (1988)
5.19   J. Hafner, S.S. Jaswal: Phil. Mag. A **58**, 61 (1988); J. Phys. F **18**, L1 (1988)
5.20   U. Mizutani: Prog. Mat. Sci. **28**, 97 (1983)
        U. Mizutani: Mat. Sci. Eng. **99**, 165 (1988)
        U. Mizutani, T.B. Massalski: *Noble Metal Alloys*, ed. by T.B. Massalski, W.B. Pearson, L.H. Bennett, Y.A. Chang (AIME, The Metallurgical Society 1986) p. 127:
        see many references in these papers
5.21   P. Häussler, E. Kay: Z. Phys. Chem. Neue Folge **157**, 377 (1988)
5.22   P. Häussler, D. Moorhead, R. Hauert, H. Poppa, E. Kay: J. Non-Cryst. Solids **117/118**, 293 (1990)
5.23   G.S. Grest, S.R. Nagel, A. Rahman: Phys. Rev. Lett. **49**, 1271 (1982)
5.24   J. Hafner: J. Non-Cryst, Solids **75**, 253 (1985)
5.25   V. Heine, D. Weaire: *Solid State Phys 24*, 249 (Academic, New York 1970)
5.26   J.R. Reitz; Solid State Phys. **1**, 1 (Academic, New York 1955)
5.27   J. Hafner, V. Heine: J. Phys. F **13**, 2479 (1983)
5.28   L.E. Ballentine: Can. J. Phys. **44**, 2533 (1966); Adv. Chem. Phys. **31**, 263 (1975)
5.29   N.C. Halder, K.C. Phillips: phys. stat. sol. (b) **115**, 9 (1983)

5.30   D.M. Nicholson, L. Schwartz: Phys. Rev. Lett. **49**, 1050 (1982)
        D.M. Nicholson, A. Chowdhary, L. Schwartz: Phys. Rev. B **29**, 1633 (1984)
5.31   V.E. Egorushkin, N.V. Melnikova: J. Phys. F **17**, 1379 (1987)
5.32   R. Frésard: Etude de la diffusion multiple de électrons dans les métaux désordonnés. Dissertation, University of Neuchâtel, Switzerland (1989)
5.33   H. Beck, R. Frésard, Y. Cuche: Z. Phys. B **68**, 237 (1987)
5.34   S.M.M. Rahman: J. Phys. F **11**, 1191 (1981); J. Phys. F **11**, 2301 (1981); phys. stat. sol. (b) **105**, K115 (1981); Z. Phys. B **45**, 307 (1982)
5.35   T.B. Massalski, U. Mizutani: Prog. Mater. Sci. **22**, 151 (1978)
5.36   J. Hafner: J. Phys. C **14**, L287 (1981); Helv. Phys. Acta **56**, 257 (1983)
5.37   J.-B. Suck, H. Rudin: In *Glassy Metals II*, ed. by H. Beck, H.-J. Güntherodt. Topics Appl. Phys., Vol. 53 (Springer, Berlin, Heidelberg, 1983) p. 217
        J.-B. Suck, H. Rudin, H.-J. Güntherodt, H. Beck: J. Phys. C **14**, 2305 (1981); Phys. Rev. Lett. **50**, 49 (1983)
5.38   K. Handrich, J. Resch: phys. stat. sol. (b) **108**, K57 (1981); phys. stat. sol. (b) **152**, 377 (1988)
5.39   N.S. Saxena, M. Rani, A. Pratap, P. Ram, M.P. Saksena: Phys. Rev. B **38**, 8093 (1988)
        M. Rani, A. Pratap, N.S. Saxena: phys. stat. sol. (b) **154**, K23 (1989);
        A.B. Bhatia, R.N. Singh: Phys. Rev. B **31**, 4751 (1985)
5.40   Y.R. Wang, A.W. Overhauser: Phys. Rev. B **38**, 9601 (1988)
        M. Rani, A. Pratap, N.S. Saxena: Ind. J. Pure Appl. Phys. **26**, 452 (1988)
        A.M. Belyayev, V.B. Bobrov, S.A. Trigger: J. Phys., Cond. Mat. **1**, 9665 (1989)
5.41   J.J. Rehr, R. Alben: Phys. Rev. B **16**, 2400 (1977)
5.42   J. Hafner: Phys. Rev. B **27**, 678 (1983)
        L. von Heimendahl: J. Phys. F **9**, 161 (1979)
5.43   L. von Heimendahl, M.F. Thorpe: J. Phys. F **5**, L87 (1975)
5.44   K. Tankeshwar, G.S. Dubey, K.N. Pathak: J. Phys. C **21**, L811 (1988)
5.45   R. Evans, P. Lloyd, S.M.M. Rahman: J. Phys. F **9**, 1939 (1979)
        D. Stroud, N.W. Ashcroft: J. Phys. F **1**, 113 (1971)
        S.M.M. Rahman: J. Phys. F **13**, 303 (1983)
5.46   A.P. Blandin: *Phase Stability in Metals and Alloys*, ed. by P.S. Rudman, J. Stringer, R.I. Jaffee (McGraw-Hill, New York 1967) p. 115
5.47   B.W. Veal, J.A. Rayne: Phys. Rev. **132**, 1617 (1963)
        K.R. Mountfield, J.A. Rayne: Solid State Commun. **49**, 1055 (1984)
        T.B. Massalski, U. Mizutani, S. Noguchi: Proc. Roy. Soc. London A **343**, 363 (1975)
        U. Mizutani, T.B. Massalski: Proc. Roy. Soc. London A. **343**, 375 (1975)
5.48   M. Hansen: *Constitution of Binary Alloys* (McGraw-Hill, New York 1958)
5.49   U. Mizutani, Y. Yazawa: Scr. Metall. **14**, 637 (1980)
5.50   H. Leitz, W. Buckel: Z. Phys. B **35**, 73 (1979)
5.51   H. Leitz: Z. Phys. B **40**, 65 (1980)
5.52   P. Häussler: *Proc. 5th Int'l Conf. on Rapidly Quenched Metals* (Würzburg, Germany 1984), ed. by S. Steeb, H. Warlimont (Elsevier, Amsterdam 1985) p. 797
5.53   G. Busch, H.-J. Güntherodt: Phys. Kondens. Mater. **6**, 325 (1967)
        G. Busch, H.-J. Güntherodt: Solid State Phys. **29**, 235 (1974)
5.54   C.N.J. Wagner: *Liquid Metals Physics and Chemistry*, ed. by Z. Beer (Dekker, New York 1972) p. 257
        E. Canessa, D.F. Mariani, J. Vignolo: phys. stat. sol. (b) **124**, 465 (1984)
5.55   P. Häussler, F. Baumann: Z. Phys. B **49**, 303 (1983)
5.56   P. Häussler, W.H.-G. Müller, F. Baumann: Z. Phys. B **35**, 67 (1979)
        B. Stritzker, H. Wühl: Z. Phys. **243**, 361 (1971)
5.57   P. Häussler: Phys. Rep. **222**(?), 65 (1992)
5.58   E. Blasberg, D. Korn, H. Pfeifle: J. Phys. F **9**, 1821 (1979)
5.59   P. Häussler, F. Baumann, J. Krieg, G. Indlekofer, P. Oelhafen, H.-J. Güntherodt: Phys. Rev. Lett. **51**, 714 (1983); J. Non-Cryst. Solids **61&62**, 1249 (1984)

5.60   G. Indlekofer: Systematics in the electronic structure of polyvalent liquid metals determined by electron spectroscopy. Dissertation, University of Basel (1987)
       G. Indlekofer, A. Pflugi, P. Oelhafen, D. Chauveau, C. Guillot, J. Lecante: J. Non-cryst. Solids **117/118**, 351 (1990)

5.61   D.E. Eastmann: Phys. Rev. Lett. **26**, 1108 (1971)

5.62   P. Häussler, F. Baumann, U. Gubler, P. Oelhafen, H.-J. Güntherodt: *Proc. 5th Int'l Conf. on Rapidly Quenched Metals* (Würzburg, Germany 1984), ed. by S. Steeb, H. Warlimont, (Elsevier, Amsterdam, 1985) p. 1007; Z. Phys. Chem. Neue Folge, **157**, 471 (1988)

5.63   U. Mizutani, R. Zehringer, P. Oelhafen, V.L. Moruzzi, H.-J. Güntherodt: J. Phys.: Condens. Matter. **1**, 1365 (1989)

5.64   H.G. Boyen, P. Häussler, F.Baumann, U. Mizutani, R. Zehringer, P. Oelhafen, H.-J. Güntherodt, V.L. Moruzzi: J. Phys.: Condens. Matter. **2**, 7699 (1990)

5.65   W. Folberth, H. Leitz, J. Hasse: Z. Phys. B **43**, 235 (1981)

5.66   U. Mizutani, T. Yoshida: J. Phys. F **12**, 2331 (1982)

5.67   P. Rieger, F. Baumann: J. Phys.: Condens. Matter **3**, 2309 (1991)

5.68   J. Dutzi, W. Buckel: Z. Phys. B **55**, 99 (1984)

5.69   R. Dupree, C.A. Sholl: Z. Phys. B **20**, 275 (1975)

5.70   G. Indlekofer, A. Pflugi, P. Oelhafen, H.-J. Güntherodt, P. Häussler, H.-G. Boyen, F. Baumann: Mater. Sci. Eng. **99**, 257 (1988)

5.71   P. Oelhafen, A. Pflugi, G. Indlekofer: J. Non-Cryst. Solids **117/118**, 267 (1990)
       P. Oelhafen, G. Indlekofer, H.-J. Güntherodt: Z. Phys. Chem. Neue Folge **157**, 483 (1988)

5.72   W. Jank, J. Hafner: J. Non-Cryst. Solids **117/118**, 304 (1990)

5.73   M.L. Cohen, V. Heine: *Solid State Physics* **24**, 37 (Academic, New York 1970)

5.74   H. von Löhneysen: Phys. Rep. **79**, 161 (1981); *Rapidly Quenched Alloys*, ed. by H.H. Liebermann (Dekker, New York 1990)

5.75   W.L. McMillan: Phys. Rev. **167**, 331 (1968)

5.76   R. van den Berg, S. Grondy, J. Kästner, H. von Löhneysen: Solid State Commun. **47**, 137 (1983)

5.77   U. Mizutani, K. Yoshino: J. Phys. F **14**, 1179 (1984)

5.78   M. Kuroha, Y. Waseda, K. Suzuki: J. Phys. Soc. Jpn. **42**, 107 (1977)

5.79   U. Mizutani, I. Nakamura: J. Phys. F **13**, 2685 (1983)

5.80   C. Lauinger: Thermokraft amorpher Cu_xSn_{100-x}-und Mg_xZn_{100-x}-Legierungen. Diploma Thesis, University of Karlsruhe, Germany (1990)

5.81   P. Häussler: unpublished data

5.82   T. Moser: Untersuchung der Thermokraft abschreckend kondensierter Ag_xIn_{100-x}-Schichten. Diploma Thesis, University of Karlsruhe, Germany (1989)

5.83   M. Burkhardt: Untersuchung der Thermokraft abschreckend kondensierter Au_xSn_{100-x}-Schichten. Diploma Thesis, University of Karlsruhe, FRG (1988)

5.84   E. Compans: Untersuchung der Thermokraft abschreckend kondensierter Ag_xSn_{100-x}-Schichten. Dissertation, Univ. Karlsruhe, Germany (1987)
       E. Compans, F. Baumann: Jpn. J. Appl. Phys. (Suppl.) **26**, 805 (1987)

5.85   A. Roll, H. Motz: Z. Metallkde. **48**, 435 (1957)

5.86   K. Fröböse, J. Jäckle: J. Phys. F **7**, 2331 (1977)

5.87   P.J. Cote, L.V. Meisel: Phys. Rev. Lett. **39**, 102 (1977) In *Glassy Metals I*, ed. by H. Beck, H.-J. Güntherodt, Topics Appl. Phys., Vol. 46 (Springer, Berlin, Heidelberg 1983) p. 141

5.88   L.V. Meisel, P.J. Cote: Phys. Rev. B **27**, 4617 (1983)

5.89   P. Wochner, J. Jäckle: Z. Phys. B **44**, 293 (1981)

5.90   T.E. Faber, J.M. Ziman: Phil. Mag. **11**, 153 (1965)

5.91   O. Dreirach, R. Evans, H.-J. Güntherodt, H.-U. Künzi: J. Phys. F **2**, 709 (1972)

5.92   H. Beck: In *Amorphous Solids and the Liquid State*, ed. by N.H. March, R.A. Street, M. Tosi (Plenum Press, New York 1985) p. 281

5.93   G. Baym: Phys. Rev. A **135**, 1691 (1964)

5.94   J.-P. Jan: Am. J. Phys. **30**, 497 (1962)

5.95   A. ten Bosch: Phys. Kondens. Mater. **16**, 289 (1973)

5.96   L.E. Ballentine: Inst. Phys. Conf. Ser. **30**, p. 188 (1977)

5.97   L.E. Ballentine: *The Hall Effect and Its Applications*, ed. by C.L. Chien, C.R. Westgate (Plenum, New York 1980) p. 201

5.98   T. Matsubara, T. Kaneyoshi: Progr. Theor. Phys. **40**, 1257 (1968)
       H. Fukuyama, H. Ebiswa, Y. Wada: Progr. Theor. Phys. **42**, 494 (1969)

5.99   M. Itoh: J. Phys. F **14**, L179 (1984)

5.100 D.G. Naugle, R. Delgado, H. Armbrüster, C.L. Tsai, T.O. Callaway, D. Reynolds, V.L. Moruzzi: Phys. Rev. B **34**, 8279 (1986)
       R. Delgado, H. Armbrüster, D.G. Naugle, C.L. Tsai, W.L. Johnson, A. Williams: Phys. Rev. B **34**, 8288 (1986)

5.101 M.N. Baibich, W.B. Muir, Z. Altounian, T. Guo-Hua: Phys. Rev. B **26**, 2963 (1982)
       T. Matsuda, U. Mizutani, W.B. Muir, M. From: J. Phys. F **14**, L21 (1984)

5.102 D. Korn, W. Mürer: Z. Phys. B **27**, 309 (1977)

5.103 R.D. Barnard: *Thermoelectricity in Metals and Alloys* (Taylor & Francis, London 1972)

5.104 J. Opsal, B. Thaler, J. Bass: Phys. Rev. Lett. **36**, 1211 (1976)
       J. Jäckle: J. Phys. F **10**, L43 (1980)

5.105 A.B. Kaiser: J. Phys. F **12**, L223 (1982); Phys. Rev. B **29**, 7088 (1984)
       A.B. Kaiser, G.E. Stedman: Solid State Commun. **54**, 91 (1985)

5.106 P.E. Nielsen, P.L. Taylor: Phys. Rev. B **10**, 4061 (1974)

5.107 C. Lauinger, J. Feld, E. Compans, P. Häussler, F. Baumann: Physica B **165 &166**, 289 (1990)

5.108 C. Lauinger, J. Feld, E. Compans, P. Häussler, F. Baumann: in preparation

5.109 Y.A. Ono, P.L. Taylor: Phys. Rev. B **22**, 1109 (1980)

5.110 T. Matsuda, U. Mizutani: J. Phys. F **12**, 1877 (1987)

5.111 U. Mizutani, K. Sato, I. Sakamoto, K. Yonemitsu: J. Phys. F **18**, 1995 (1988)

5.112 J. Feld: Widerstand und Halleffekt amorpher CuAl- und MgZn-Legierungen. Diploma Thesis, University of Karlsruhe, Germany (1990)

5.113 U. Mizutani, J. Hashizume, T. Matsuda: J. Phys. Soc. Jpn. **55**, 3188 (1986)

5.114 O. Rapp, S.M. Bhagatt, H. Gudmundsen: Solid State Commun. **42**, 741 (1982);
       An excellent review was given by M.A. Howson, B.L. Gallagher: Phys. Rep. **170**, 265 (1988)

5.115 P.J. Cote, L.V. Meisel: Phys. Rev. Lett. **40**, 1586 (1978)

5.116 M. Hofacker, W. Sander, C. Sürgers, H. von Löhneysen: Jpn. J. Appl. Phys. **26**, 737 (1987)

5.117 V.E. Egorushkin, A.I. Murzashev, V.E. Panin: Sov. Phys. J. **27**, 34 (1984)

5.118 M. Piecuch, J.-F. Geny, G. Marchal: In *Amorphous Metals and Non-Equilibrium Processing*, ed. by M.V. Allmen, (1984) p. 79
       J.-F. Geny, G. Marchal, Ph. Mangin, Chr. Janot, M. Piecuch: Phys. Rev. B **25**, 7449 (1982)
       J.-F. Geny, D. Malterrre, M. Vergnat, M. Picuch, G. Marchal: J. Non-Cryst. Solids **61 & 62**, 1243 (1984)

5.119 M.A. Ruderman, C. Kittel: Phys. Rev. **96**, 99 (1954)

5.120 T. Kaneyoshi: J. Phys. Jpn. **45**, 94 (1978)

5.121 H.G. Boyen: Photoelectronenspektroskopie an amorphen Übergangsmetall/Zinn-Schichten. Dissertation, University of Karlsruhe, Germany (1990)

5.122 G. Gantner: Elektronische Transporteigenschaften abschreckend kondensierter Pd_xSn_{100-x}- und Fe_xSn_{100-x}-Schichten. Diploma Thesis, University of Karlsruhe, Germany (1989)

5.123 G. Busch, H.-J. Güntherodt, H.-U. Künzi: Phys. Lett. **34**, 309 (1971)
       H.-J. Güntherodt, H.-U. Künzi: Phys. Kondens. Mater. **16**, 117 (1973)

5.124 L.A. Vakarchuk, I.F. Margolych: phys. stat. sol. (b) **149**, 301 (1988)

5.125 H.A. Mook, N. Wakabayashi, D. Pan: Phys. Rev. Lett. **34**, 1029 (1975)

5.126 P.D. Nguyen: Theorie der Friedel-Oszillationen der Ladungsverteilung in ungeordneten Systemen, Diploma Thesis, University of Karlsruhe, Germany (1990)

5.127 P. Häussler, R. Zehringer, P. Oelhafen, H.-J. Güntherodt: Mater. Sci. Eng. **133**, 115 (1991)

5.128 P. Häussler, G. Indlekofer, H.-G. Boyen, P. Oelhafen, H.-J. Güntherodt: Mater. Sci. Eng. **133**, 120 (1991); Europhys. Lett. **15** (7), 759 (1991)

5.129 W. Jank, J. Hafner: Phys. Rev. B**41**, 1496 (1990); Phys. Rev. B**42**, 6926, 11530 (1990)

6. The Polycluster Concept of Amorphous Solids

A.S. Bakai

With 21 Figures

This chapter is the introduction to the physics of polycluster amorphous solids. At first the polycluster model was developed as a constructive foundation to describe some properties of metallic glasses. The assumption about the presence of a comparatively perfect local order (LO) (not necessarily of one type) leads naturally to the definition of the locally regular cluster (LRC), after which one must make only one step (not quite ambiguous) to introduce the definition of the polycluster structure. From this definition, there evolves the description of structure defects (Sect. 6.4).

Studying the kinetics of the liquid–solid transformation shows that the polycluster-structure formation competes with crystallization and, under certain conditions, it becomes dominant and leads the liquid to transfer to the glassy state (Sect. 6.2).

In the phenomena of reversible and irreversible relaxation in metallic glasses, the low-energy excitations connected with the rearrangement of atomic configurations including the tunneling states play an essential role. The description of low-energy excitations and diffusion mechanisms in polyclusters is contained in Sects. 6.6 and 6.7.

In the physics of glasses, the clarification of the nature and the description of the glass–liquid transition are of particular interest. In polyclusters, the restoration of ergodicity begins with the *melting* of boundaries. The thermodynamics of this transition is considered in Sect. 6.8.

The mechanisms of plastic deformation and mechanical states of polyclusters are described in Sect. 6.9. The comparatively high density of cluster boundaries (experimental data point to the fact that cluster sizes are about $10^2 a$, a being the average interatomic distance), peculiarities of structure and displacement of dislocations under the action of stress determine the dominant deformation mechanisms in some region of temperature T and stress σ.

It needs to be established whether the concept and model developed are adequate to real objects. This question is answered to some extent by the experimental data given in Sects. 6.3 and 6.5. We also discuss briefly some other structure models of metallic glasses to establish the connection between them and the polycluster model (6.4.5).

Topics in Applied Physics, Vol 72
Beck/Güntherodt (Eds.)
© Springer-Verlag Berlin Heidelberg 1994

6.1 From the History of the Subject

One may say that the polycluster model has deep historical roots despite the fact that its formulation was not directly connected with the development and formalization of the ideas existing earlier about the structure of nonmetallic and metallic glasses. The common features of different approaches in the description of any object are usually prompted by the nature of the object itself and by the requirement that the assumptions suggested do not contradict the firmly established experimental facts.

Two papers not directly related to metallic glasses but which are the earliest and which influenced the development of ideas on the glass and liquid structure deeply deserve to be noted separately.

In his early paper, *Lebedev* [6.1], having revealed by the temperature dependences of the refractive index of light, the birefringence and the thermal expansion coefficient that sharp changes of properties in glasses occur almost in the same temperature region as for quartz (during the $\alpha \rightleftharpoons \beta$ transition), put forward the hypothesis of the crystallite structure of glass. As the information about the structure and properties of glasses grew, this hypothesis evolved as well. Later, *Lebedev* suggested the crystallites were not small crystals but formations containing parts with a structure close to a crystalline one. He argues in one of his later papers [6.2] that glass-like substances contain regions "with the relatively large degree of order in the location of atoms: these regions may be most suitably called the crystallites... The crystallites are formed from the hardened cybotactic groups being the intermediate chain between them and small crystals of colloidal sizes... All crystallites must be connected between themselves by their external, mostly distorted parts into one continuous but not quite disordered network." *Lebedev* himself did not oppose with his "crystallite hypothesis" the *Zahariasen* model of the continuous random network [6.3], as is seen from the citation given above as well as from his other papers [6.4].

The cybotaxis mentioned above forms the subject of another basic paper belonging to *Stewart* and *Morrow* [6.5] dealing with studies of the primary normal alcohol structure. This paper established the existence in a liquid of noncrystalline formations called cybotactic groups as well as that a liquid containing a noticeable quantity of cybotaxises was called cybotactic. Cybotaxis is an aggregate or cluster consisting of the molecules of a liquid. They appear and are destroyed. The conclusion about their noncrystalline structures is important. Note that the assumptions about the presence of noncrystalline clusters in metallic melts and the essential role of the latter in forming the glasses were put forward in a number of later papers [6.6].

In studying the structure of liquids, the model of dense random packing (DRP-model) suggested by *Bernal* [6.7] played an important role. In this model determined by the algorithm of the structure formation from spheres interacting according to some law), the correlations in the mutual location of atoms rapidly decrease with distance and the cybotaxis or crystallite orderings are absent.

DRP-models played an important role also in modeling and studying the properties of metallic glasses. Besides the DRP-models, so-called stereochemically determined models were considered, in which the dominant types of LO were assumed along with cluster models in which the structure of glass is constructed from the definite type of noncrystalline polyatomic clusters. The polycluster model is close to those stereochemically determined and to cluster models that go along with ideas about the crystallite structure of glasses belonging to *Lebedev* and about the cybotactic structures of liquids belonging to *Stewart* and *Morrow*. Some reviews and the analysis of the models mentioned above can be found in [6.8–10].

6.2 Glass Formation: General Consideration

The main purpose of this section is to give an account of general peculiarities of the microscopic scenarios of the supercooled liquid solidification and to show that formation of the polycluster glasses is a commonplace case. A more detailed consideration of the supercooled liquid structure, its thermodynamics and solidification kinetics is given in Sect. 6.10.

The choice of the structure model is suggested by the nature of the glassy state. A glass is a disordered solid in which the processes of structure relaxation leading to crystal formation occur very slowly. In contrast to a liquid, a glass is a system with broken ergodicity[1]. The liquid–glass transition separates the liquid and glass-like states and, along with this, some peculiarities of this transition (transition–temperature dependence on the past history and temperature variation rate, the presence of hysteresis, etc.) point to the nonequilibrium nature of the glassy state, to the fact that kinetic processes (but not the condition of the thermodynamic equilibrium) play the decisive role in glass formation. The process of glass formation competes with the crystallization process. Crystallization is one of the possible ways of forming solids from the melt leading to a stable equilibrium state. Besides there exist other ways of forming solids leading to metastable or nonequilibrium solid states. The very existence of glasses shows that melts exist for which one or another way of forming a solid becomes more probable depending on various conditions. Something like that occurs during the amorphization processes, when the transformation of the nonequilibrium crystalline structure into the amorphous one occurs. Let us try to clarify when glass formation becomes more probable than crystallization.

Crystalline as well as noncrystalline solid nuclei may form in the melt. The term *noncrystalline* means that the mutual location of atoms in the nucleus possesses no symmetry compatible with the translational invariance (e.g., it has

---

[1] The systems with broken ergodicity and the methods of their description are considered in the review article [6.11].

local five-fold symmetry axes). We will call the noncrystalline nuclei the clusters (in contrast to crystallites). Denote by $\Delta G^c(N, T, P)$ and $\Delta G^{cl}(N, T, P)$ the excesses of the free energy of the crystallite and the cluster, respectively:

$$\Delta G^c = G^c - G^l, \qquad \Delta G^{cl} = G^{cl} - G^l , \tag{6.1}$$

where G^c, G^{cl} and G^l are free energies of the crystallite, cluster, and the liquid, respectively. In ΔG^c one may separate the contributions from the volume and the surface of the nucleus:

$$\Delta G^c(N, T, P) = \Delta G_V^c(N, T, P) + G_S^c(N, T, P) \tag{6.2}$$

ΔG^{cl} may be presented similarly.

At temperatures below the crystallization point, ΔG_V^c is negative and it decreases $\sim N$ with the nucleus growth and G_S^c is positive and it grows $\sim N^{2/3}$ with N. ΔG^{cl} possesses similar properties. The value of the volume term depends essentially on the average bond energy of atoms in the nucleus, which, in turn, depends on LO (structure of coordination polyhedrons) and the interatomic forces. The noncrystalline structure of clusters may provide higher values than the average bond energies and lower surface energies than for crystallites. For small dimensions of nuclei, when the relative number of surface atoms is comparatively largely, the decrease in surface energy may be achieved by deformation, LO changing, or the appearance of noncrystalline structure elements. Then the total-energy gain may appear to be essential so that for small N such a cluster may be formed for which $\Delta G^{cl}(N, T, P) < \Delta G^c(N, T, P)$. (For large N, evidently, $\Delta G^c < \Delta G^{cl}$). Note that there can be many types of clusters differing in structure, the number of atoms N being fixed, and their number grows rapidly with the growth N. All of these structure states contribute to the partition function and lower the free energy of the cluster ensemble.

Because of the peculiarities of the nucleus free-energy dependence versus N mentioned above, it is clear that for crystallites as well as for clusters it is positive and growing at small N and it becomes negative for $N \to \infty$ in the temperature domain of glass stability ($T < T_g$, where T_g is the vitrification temperature). For certain values of N_c^c and N_c^{cl}, the functions $\Delta G^c(N)$ and $\Delta G^{cl}(N)$ are at their maximum (as usual, we assume that each of these functions has only one maximum). These values determine the critical sizes of crystalline and noncrystalline nuclei. The kinetics of the nucleus formation and liquid structure are characterized by two important quantities: the nucleation rate (the rate of nuclei formation with overcritical size) and the distribution function of subcritical nuclei. It is known from nucleation theory [6.12] that in the quasi-steady regime of the nucleation, when the influence of overcritical nuclei on the subcritical nuclei kinetics is negligible, the nucleation rate appears to be proportional to $\exp[-\Delta G^c(N_c^c)/k_B T]$ for the crystalline nuclei and to $\exp[-\Delta G^{cl}(N_c^{cl})/k_B T]$ for the noncrystalline ones (k_B is the Boltzmann constant).

In the formulae for the nucleation rates, the frequencies of atom attachment to nuclei come into play as well although the exponents pointed out above are dominant when comparing the kinetics of formation of crystallites and clusters because the frequencies of atom attachments to crystalline and noncrystalline nuclei are not likely to differ. As regards the distribution functions of subcritical nuclei, they appear to be proportional to $\exp[-\Delta G^c(N)/k_B T]$ and to $\exp[-\Delta G^{cl}(N)/k_B T]$.

If with $T \approx T_g$ the $\Delta G^c(N)$ and $\Delta G^{cl}(N)$ versus N dependences are such as shown in Fig. 6.1a, then the probability of formation of comparatively large noncrystalline clusters exceeds considerably the probability of formation of the critical crystalline nuclei. In this case, at rapid quenching when the formation of overcritical crystalline and noncrystalline nuclei is excluded, one may expect that the noncrystalline solid (glass) will be formed consisting of clusters with comparatively large sizes with small dissemination of the crystalline phase. In this case, the glass is a frozen polycluster liquid. Perhaps this is a case of glass-forming alloys. The glass structure then appears to be a polycluster one and like the liquid structure before freezing.

If the behavior of the ΔG^c and ΔG^{cl} with $T \to T_g$ becomes as shown in Fig. 6.1b, then the crystallization of the supercooled melt is generally hampered and has small probability because $\Delta G^c(N_c^c) > \Delta G^{cl}(N_c^{cl})$; thus, at first, a non-crystalline solid forms by nucleation and growing of noncrystalline precipitates.

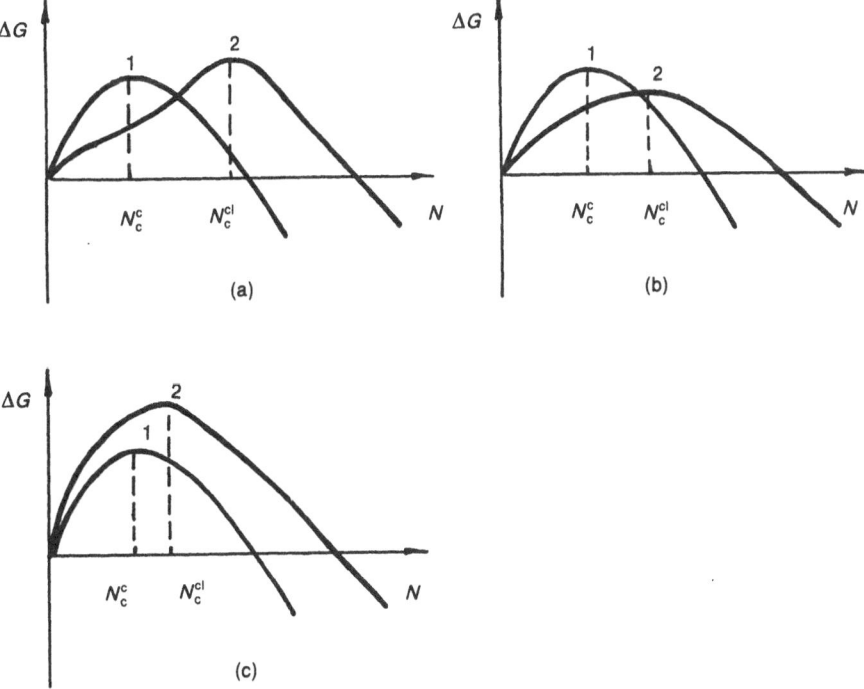

Fig. 6.1. Free energies of crystalline (curves 1) and non-crystalline (curves 2) nuclei vs the number of atoms in the nucleus for glass-forming (a, b) and nonglass-forming melts (c). N_c^{cr} and N_c^{cl} are the critical sizes of embryos

Of course, it may crystallize in the course of time because at large $N \Delta G^c(N) < \Delta G^{cl}(N)$. This case definitely is typical for the melted high-molecular compositions, the crystallization of which is possible with changing of the compositional and/or topological order in molecules and aggregates. Believably a lot of a good glass-forming metallic alloys possess this property. Glass in this case obviously has polycluster structure.

If ΔG^c and ΔG^{cl} depend at low temperatures on N as is shown in Fig. 6.1c, then the polycluster glass formation when cooling the melt does not occur because the polycrystal formation occurs with overwhelming probability.

Note that other scenarios of glass formation are not excluded. For example, as computer simulations show [6.8, 10], the liquid with Lennard-Jones potential of interatomic interactions can be vitrified under very high cooling rates ($> 10^{12}$ K/s) and rather low temperatures (about 50 K). This computer glass possesses a DRP-structure. Believably amorphous metallic films which can be obtained only by depositing the atom flux on a cryogenic substrate own the DRP-structure.

We see that to ascertain the scenario of supercooled liquid solidification and glass formation, it is necessary to calculate $\Delta G^c(N)$ and $\Delta G^{cl}(N)$ with comparatively small number of atoms in nuclei N when the relative fraction of boundary atoms and their contribution to the free energy are essential.[2]

Considerations similar to the ones given above may be used also to discuss the kinetics of some processes of amorphization of solid metallic systems that are perfectly reviewed by *Samwer* [8.13]. (Chap. 2).

Some suggestions and empirical rules characterized systems in which the formation of noncrystalline structures may occur early is outlined in [6.14, 15].

6.3 Some Experimental Data on the Structure of Real Metallic Glasses

We reproduce or mention here only some experimental results of metallic glass structures. Our aim is to discuss further the structure of real objects of interest.

Figure 6.2 presents the dark-field pattern of $Fe_{71}Si_7B_{12}$ metallic glass obtained by *Noskova* et al. [6.16]. The specimen was first deformed (the strain was 16%) and then annealed for 30 minutes at 450°C. We can observe the nonuniform structure, the fringes which the authors call the rough polycluster, with the characteristic nonuniformity scale of ~ 10 nm. The strongly spread-out halos on the diffractogram point to the amorphous structure.

Figure 6.3 presents the amorphous alloy obtained by *Hirotsu* et al. [6.17] with the help of high resolution TEM. Circles separate the regions of medium-range ordering (MRO) with the sizes ~ 2 nm in the amorphous alloy Pd–Cu–Si. When annealing is performed below the crystallization temperature, the dimen-

---

[2] See Sect. 6.10

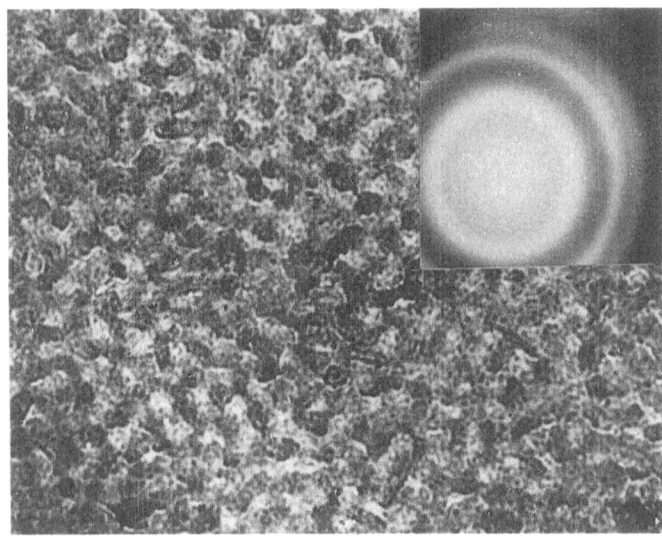

Fig. 6.2. Structure inhomogeneities of the alloy $Fe_{81}Si_7B_{12}$ (rough polycluster structure [6.16])

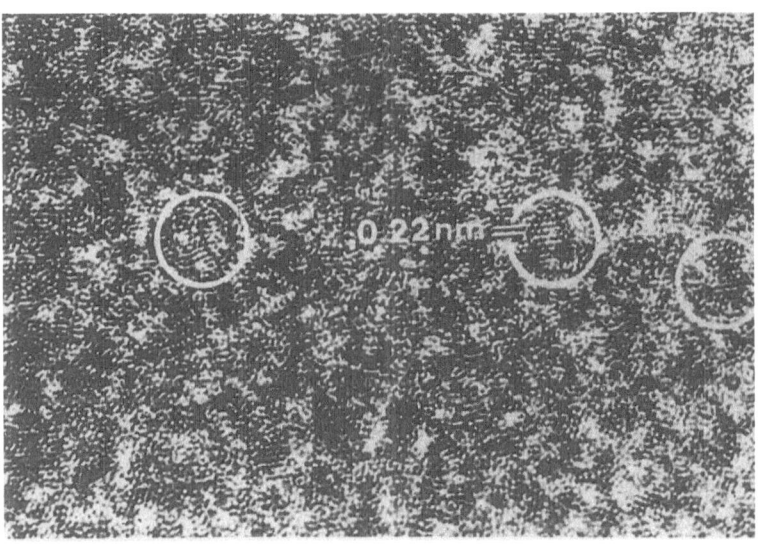

Fig. 6.3. Regions of medium-range order in the amorphous alloy Pd–Cu–Si [6.17]

sions of the MRO regions grow. A sharp boundary between the MRO regions and the surrounding amorphous matrix is not found.

Figure 6.4 presents the alloy $Fe_{40}Ni_{40}B_{20}$ obtained by *Mikhailovsky* et al. [6.18] with the help of the field-ion microscope (FIM). The light points are attributed to the images of boron atoms; the lines decorated by boron atoms are

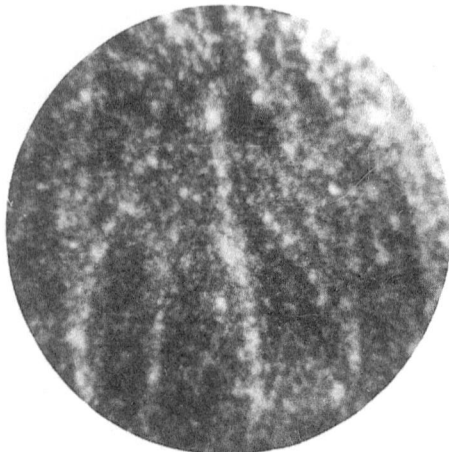

Fig. 6.4. FIM-image of the alloy $Fe_{48}Ni_{40}B_{20}$. Dense chains of light points indicate the outlets on the surface of cluster boundaries decorated with boron atoms [6.18]

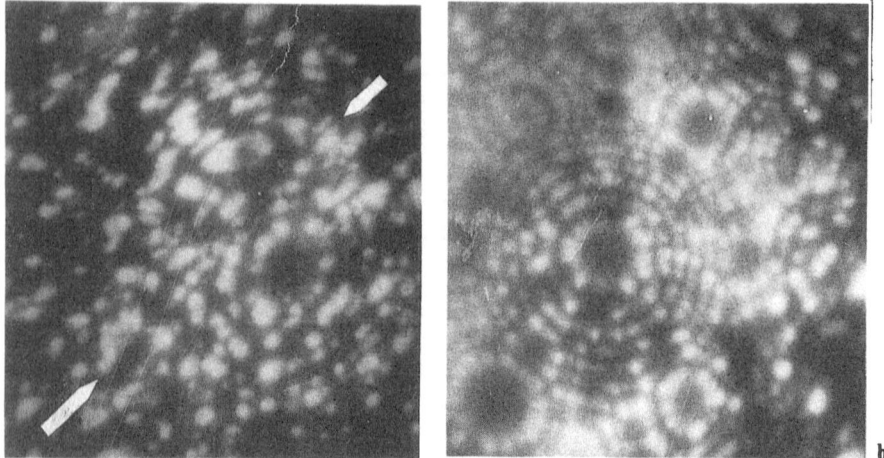

a   b

Fig. 6.5. FIM-image of the alloy $(SiTi)_{10}Nb_{20}$ in the glassy (a) and crystallized in the microscope column states (b). Arrows indicate the boundary

seen. Special investigations show that these lines are outlets on the surface of the defects which are the intercluster boundaries. The cluster appears to be elongated in the direction that is normal to the ribbon plane (the clusters are column-like). The longitudinal size of clusters range from 6 to 14 nm, the average transverse dimension is about 5 nm.

The same group of authors obtained with FIM snapshots of the alloy $(CuTi)_{40}Nb_{20}$ (Fig. 6.5). In the amorphous state, we scarcely see MRO regions, i.e., the portions of deformed atomic planes with rather dense packing (Fig. 6.5a). The sizes of the MRO regions are about 2 nm. Arrows point to the boundary which is easily found because of the neighborhood of two noncoherently conjugated MRO regions. Figure 6.5b shows the same area after crystallization in the microscope column. The system of concentric circular atomic steps is seen looking at the crystalline structure.

In [6.19], *Piller* and *Haasen* revealed with FIM in the glass $Fe_{40}Ni_{40}B_{20}$ the presence of the MRO regions with sizes $\sim 1-2$ nm, the region sizes having grown on annealing below the crystallization temperature.

Melmed and *Klein* [6.20] found two-dimensional defects (probably cluster boundaries) in the quasi-crystal Al–Mn. The sizes of clusters (quasi-crystallites) comprise $\sim 10^2 a$, and the boundary thickness equals approximately a.

6.4 Polycluster Structures

A structure model must be based on a noncontradictory, closed and complete definition. A definition is closed if it does not contain indefinite elements and notions, and it is complete if it includes the description of all structure elements. Thus, for instance, the model in which the amorphous structure is considered as a dislocationally disordered crystal [6.21, 22] becomes not closed if the dislocation structure (in particular, the one of dislocation core) is not defined. At high density of dislocations when their cores may overlap and their structure becomes very indefinite, the model is not closed. The free-volume model [6.23–25], in which the question about geometry and topology of atomic configurations is put aside, is not complete.

At the basis of the polycluster model, there lies the assumption of the presence of one or several types of atom LO in a solid. At this point, the polycluster model is close to the so-called stereochemical models [6.8]. Besides, it includes the definition of a cluster as a set of locally ordered atoms and the definition of the boundaries as the closing of this set. Finally, the assumption that clusters conjoin along common boundaries completes the definition of a polycluster structure. The polycluster model includes a rather wide set of structures. This model was suggested in [6.26, 27] and developed and applied while describing various properties of metallic glasses [6.28–33].

6.4.1 Local Order

The formulation of the polycluster model requires a sufficiently complete definition of LO. For simplification, we will only discuss briefly the presence of atoms of different kinds and the compositional (chemical) order, concentrating instead on the topology and geometry of the structure, though the consideration scheme may be easily generalized.

LO is the characteristic for the energy of an atom and it is determined by the configuration of its coordination polyhedron (CP). The CP structure is determined by interatomic interactions. The quantitative characteristics of the i-th atom LO may be the volume per atom (the volume of the Voronoi polyhedron [6.34]), the binding energy $\mathscr{E}_i$, and also the geometric properties of the coordination polyhedron (the number of vertexes, edges, faces, the angle values, etc.). Between CP's, one may separate sets of topologically equivalent ones.

Let $\{P_\mu\}$ be one of such sets, μ the index numbering the sets. The polyhedra are topologically equivalent if they have the same number of vertex, edges and faces with identical number of edges meeting in vertexes. In other words, the topologically equivalent polyhedra differ in edge length, angles at vortexes and also in angles between faces and between edges. Obviously, every polyhedron belonging to a given set $\{P_\mu\}$ may be transferred to another one of the same set by continuous deformations.

The main contribution to the binding energy is made by atom interactions with the nearest surroundings; therefore the binding energy is uniquely connected with the CP geometry. Let $\{\mathscr{E}_\mu\}$ be the set of binding energy values of atoms with topologically equivalent coordination of the μ type and $\mathscr{E}_\mu^0$ the value of the absolute maximum for this set. Denote by P_μ^0 the polyhedron on which this maximum is achieved and call it basic. The basic CP is described by a set of vectors (basis) $\{b_\mu^0\}$ consisting of vectors

$$r_{ij}^0 = X_j - X_i, \quad j = 1, 2, \ldots, z_\mu \tag{6.3}$$

where X_j are the coordinates of vertexes surrounding the i-th atom contained inside of P_μ^0.

The CP orientation is determined by three angles, the values of which determine the orientation of the coordinate axes system rigidly connected with the polyhedron. Assume that a method is given to choose the coordinate system connected with the polyhedron [6.29] and with this the orientation of each of them is determined. Denote by Ω_μ the set of angles giving the orientation of the polyhedron P_μ.

In the description of LO, it is important to account for elastic deformations. It is natural to assume that if in a solid LO is realized with CP from $\{P_\mu\}$, then they differ from the basic polyhedron P_μ^0 only through elastic deformations. The deformation of the polyhedron P_μ is considered to be elastic if the atom binding energy $\mathscr{E}_\mu$ changes monotonously in the process of deformation transfering P_μ^0 to P_μ. The deformation magnitude is described by a set of deformation vectors. Let $\{b_\mu^i\}$ be the polyhedron basis of the i-th atom $P_\mu(\Omega_i)$ with the Ω_i orientation and $P_\mu^0(\Omega_i)$ the basic polyhedron of the same orientation. Then the set of vectors

$$u_{ij} = r_{ij} - r_{ij}^0, \quad r_{ij} \in \{b_\mu^i\}, \quad r_{ij} \in \{b_\mu^{0i}\} \tag{6.4}$$

describes the local deformation, allowing to introduce the deformation tensor and separate the uniform and nonuniform parts of the deformation [6.29].

From the definitions of elastic deformation and the topological equivalence, there follows the definition of the admissible deformation region of coordination polyhedrons that we denote by $\mathscr{D}_\mu$. The boundary of the set of elastic deformations $\mathscr{D}_\mu$ possesses the obvious property that on it the atom binding energy achieves the minimum.

6.4.2 LRC's and Polyclusters

Let us have s types of CP-($\{P_\mu\}$, $\mathscr{D}_\mu$, $\mu = 1, 2, \ldots, s$). Assume the site (the equilibrium position of an atom) and the atom occupying it (that is coordinated according to one of the types noted) to be regular. If the site coordination does not coincide with any of the ones given, then it is irregular. The connected set of regular atoms together with their CP is called LRC. The types of LO and their number s are the characteristics of the given cluster. Note that besides the regular sites and atoms, the LRC contains also irregular ones but such that belong to the CP of regular sites. The set of irregular atoms forms the LRC boundaries. As is seen, the LRC lattice is the locally ordered random network with the intermittent LO and the boundary that is multi-connected in general.

The bounded cluster may be extended by placing atoms on the boundary surface. The completion of boundary site CP's such that these sites become regular is called the regular LRC continuation. The regular continuation of the network is generally ambiguous. The uncertainty of the continuation is connected with the admissible variations of elastic deformations of the extended network and also with the possible arbitrariness in the choice of the local ordering of boundary sites, which gives rise to structure frustrations. By the regular continuation we mean that with which the sum of binding energies of boundary-layer atoms achieves maximum.

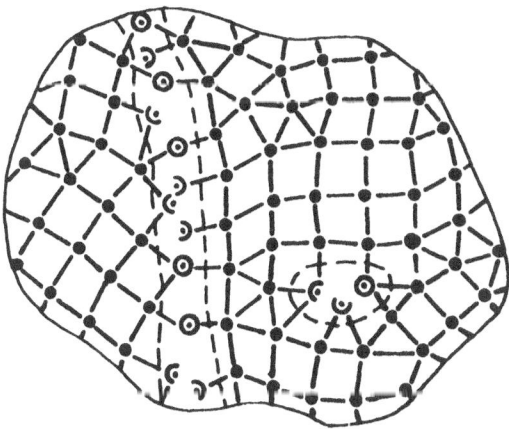

Fig. 6.6. A fragment of the 2-D polycluster Intercluster and inner boundaries are shown. ●-regular sites, ⊙-coincident sites, ⌣-noncoincident sites

Two LRC's or two sections of the same LRC are called conjoined if they possess a common boundary. Consider the structure being a set of conjoining LRC's such that all irregular sites lie on the surface, on the boundaries of cluster conjoining, or on inner boundaries (the boundaries are called inner ones if the sections of one and the same cluster conjoin along them). *The structure consisting of the set of conjoining LRC's is called a polycluster, and the solid possessing such a structure is a polycluster.* Figure 6.6 depicts the fragment of the 2-D polycluster on which the region of the intercluster boundary and the inner boundary are shown. The existence of boundaries is a result of non-coincidence of regular continuations of two LRC's or two sections of the same cluster, so that the boundaries possess the frustrated structure.

6.4.3 Short-Range Order (SRO) and MRO

Correlations in the relative position of atoms determine the SRO and MRO. The spatial distribution of atoms is described by N-particle ($N = 1, 2, \ldots$) distribution functions. The single-particle distribution functions represent the probability densities of the existence of atoms with different LO: $n_\mu(\Omega_\mu, X)$ is the density of LRC atoms with a μ-type LO and Ω_μ orientation of the polyhedron: n_c is the atom density at coincident sites of boundary layers; n_{uc} is the density of atoms disposed in the complexes of noncoincident sites; n_s is the density of boundary atoms. The density $n_\mu(\Omega_\mu, X)$ is normalized in such a way that the integral

$$\int n_\mu(\Omega_\mu, X) \, d\Omega_\mu = n_\mu(X) \tag{6.5}$$

is equal to the density of atoms with a μ-type LO. It is evident that the sum of the above densities is equal to the density of atoms

$$\sum_{\mu=1}^{s} n_\mu(X) + n_c(X) + n_{uc}(X) = n(X) . \tag{6.6}$$

The sum of the concentrations $C_\mu = n_\mu/n$, $C_c = n_c/n$ and $C_{uc} = n_{uc}/n$ equals unity.

A two-particle distribution function can be represented as a product of a single-particle distribution function $n_\alpha(X_1)$ by the density of the conditional probability $f_{\alpha\beta}(X_1, X_2)$, meaning that at point X_2 there is an atom with a β-type LO if at point X_1 there is an atom with an α-type LO

$$n_{\alpha\beta}(X_1, X_2) = n_\alpha(X_1) f_{\alpha\beta}(X_1, X_2) = n_\beta(X_2) f_{\beta\alpha}(X_2, X_1) . \tag{6.7}$$

For a homogeneous random network, $n_{\alpha\beta}$ depends only on $X_2 - X_1$, and if the network is isotropic, then it depends only on $|X_2 - X_1|$. In this case, we have

$$n_{\alpha\beta}(X_1, X_2) = n_\alpha(X_1) f_{\alpha\beta}(r) = n_\beta(X_2) f_{\beta\alpha}(r); \; r \equiv |X_2 - X_1| . \tag{6.8}$$

The functions

$$\tilde{f}_{\alpha\beta}(r) = f_{\alpha\beta}(r) - n_\beta$$

are the partial radial-distribution functions (RDF).

In the absence of a long-range order, the two-particle distribution functions and RDF possess evident asymptotic properties

$$\lim_{r \to \infty} n_{\alpha\beta}(X_1, X_2) = n_\alpha(X_1)n_\beta(X_2); \lim_{r \to \infty} \tilde{f}_{\alpha\beta}(r) = 0 , \qquad (6.9)$$

meaning that the correlations diminish in relative positions of atoms at large distances. At small distances, however, the correlations do exist and determine the SRO in an amorphous solid.

The partial RDFs are related to the functions of radial density distribution around the atom. Describing the density fluctuations around the atom with an α-type local order by the sum

$$\tilde{f}_\alpha(r) = \sum_{\beta=1}^{s} \tilde{f}_{\alpha\beta}(r) , \qquad (6.10)$$

we obtain the following expression for the RDF

$$\tilde{F}(r) = \sum_{\alpha=1}^{s} C_\alpha f_\alpha . \qquad (6.11)$$

From (6.9), it follows that

$$\lim_{r \to \infty} \tilde{F}(r) = 0. \qquad (6.12)$$

Let us denote the first coordination–sphere radius, given by the position of the first maximum $\tilde{f}_\alpha(r)$, by r_α^1. The radius $r_{\alpha\beta}^c$, at which $|\tilde{f}_{\alpha\beta}(r)|$ becomes (and remains at $r > r_{\alpha\beta}^c$) negligibly small, is a pair correlation radius. The correlation radius $r_{\mu s}^c$ is equal to the mean distance between μ-type polyhedra and boundary atoms. The average

$$\sum_{\mu=1}^{s} C_\mu r_{\mu s} = r_{cl} \qquad (6.13)$$

is equal to the average LRC size.

The orientational ordering of coordination polyhedra is described by the functions

$$\Delta_\mu^{\alpha\beta}(r) = \int f_{\mu\mu}(\Omega_\mu(0), \Omega_\mu(r), r) \cdot$$
$$\times (\Omega_\mu^\alpha(r) - \Omega_\mu^\alpha(0))(\Omega_\mu^\beta(r) - \Omega_\mu^\beta(0)) \, d\Omega_\mu(0) \, d\Omega_\mu(r) \qquad (6.14)$$

($\Omega_\mu^{\alpha,\beta}$ are the components of the angles Ω_μ: $\alpha, \beta = 1, 2, 3$) which show how rapidly the rms difference between the orientation angles grows as the distance between the polyhedra increases. The quantity

$$D_\mu^{\alpha\beta} = \Delta_\mu^{\alpha\beta}(r_\mu^1)/r_\mu^1 \qquad (6.15)$$

is the tensor *diffusion coefficient* in the orientation space.

6.4.4 Structure Defects

Polyclusters as well as crystals may contain point and extended defects. We will consider them here. The boundaries are 2-D defects. To establish the atomic structure of boundaries, let us consider the section of the boundary between two clusters (see Fig. 6.6). Continue regularly the site network of each cluster into the boundary layer and denote through m^1 and m^2 the site sets of these continuations. Two sites, $i \in m^1$ and $j \in m^2$, will be called coincident if they may be brought into coincidence by elastic deformation of regular continuations. As the elastic deformations are limited by $\mathscr{D}_\mu$ regions, then the coincident sites must be closer to each other than on some critical distance ρ_{ij}^c: $|X_i - X_j| \leqslant \rho_{ij}^c$. Denote by ρ_{12} the value ρ_{ij}^c averaged over all atomic configurations. We will use this value to estimate the critical elastic displacements in the boundary layer. Besides the coincident sites, there are noncoincident boundary ones that cannot be brought to coincidence with regular continuation sites of the opposite cluster by elastic displacements. Figure 6.6 separates coincident and noncoincident boundary sites. It is easy to find the density estimate of coincident sites, n_c, and noncoincident ones n_{uc}:

$$n_c \approx n^2 a^2 \sigma_{12}, \quad n_{uc} \approx 2na(1 - na\sigma_{12}); \quad \sigma_{12} = \pi\rho_{12}^2, \tag{6.16}$$

where n is the average bulk density of atoms.

From simple geometry considerations, it follows that the mean volume per coincident site is $v_a \approx a^3$, while it is about $v_a/2$ for each noncoincident site. Thus, the volume of the hole containing a noncoincident site is essentially larger than the interstitial hole, but is approximately half as large as the vacancy in the LRC body. The arrangement of the atom in the noncoincident site generates comparatively large elastic compression fields. This atom is a partial interstitial ($\tilde{i}$). A vacant noncoincident site represents the partial vacancy ($\tilde{v}$). Numerical estimates show [6.27, 29] that the formation energy of partial vacancy $E_{\tilde{v}}$ equals about half of the formation energy of the regular vacancy E_v: $E_{\tilde{v}} \approx E_v/2$, and the partial interstitial energy $E_{\tilde{i}}$ is much less than the energy of interstitials in LRC E_i: $E_{\tilde{i}} \sim 10^{-1} E_i$. Since E_i equals usually about several eV for metals and E_v equals about 1 eV, then $E_{\tilde{i}} \approx 0.3$–$0.7$ eV and $E_{\tilde{v}} \approx 0.5$ eV.

The question of atom location in boundary sites requires special consideration. The noncoincident sites form some connected sets, i.e., the complexes of noncoincident sites lying in boundary layers. Let us denote the complex that comprises N noncoincident sites with m atoms in them by $C_N(m)$. To find the most probable number of atoms arranged in the complex, we introduce the mean free energy of the complex

$$G_N(m) = E_N(m) - S_N(m)T . \tag{6.17}$$

Here $E_N(m)$ is the energy of the complex formation

$$E_N(m) \approx mE_{\tilde{i}} + (N - m)E_{\tilde{v}} - E_{int}(N, m) ,$$

where E_{int} is the interaction energy of partial point defects, which is essentially contributed by the interaction through elastic fields. The entropy $S_N(m)$ is contributed by the oscillation and configurational entropies

$$S_N^{\text{conf}}(m) \approx k_B \ln \left[\frac{N!}{m!(N-m)!} \right] \tag{6.18}$$

It can be demonstrated [6.27] that $E_{\text{int}}(N, m)$ has its maximum at $m \approx N/2$, when the elastic compression fields generated by partial interstitials essentially compensate for the elastic extension fields surrounding partial vacancies. At $m \approx N/2$, the configurational entropy of the complex is also maximum. Therefore, $m \approx N/2$ is the most probable number of atoms disposed in the complex of N noncoincident sites, so that nearly one half of the noncoincident sites is occupied by atoms, and another half is vacant. The existence of two-level systems, high diffusion mobility of atoms along non-coincidence sections, low-energy structural fluctuations in polyclusters is connected with this circumstance (Sect. 6.6).

Point defects. Vacancies, i.e., vacant regular sites and the atoms in interstitial cavities, i.e., interstitials are point defects of LRC. There are no reasons to consider these defects to be unstable. Their stability is provided by the local ordering of surrounding atoms. The nonelastic displacements of surrounding atoms as a result of point defect formation are possible in places of strong, close-to-critical, elastic distortions of the LRC network.

Note that in DRP-structures, the vacancies are unstable as numerical simulation shows [6.35] the property of DRP structures. Because of the difference in LO and elastic deformations, the energies of vacancy formation are different in different sites. Let $E_{v\mu}^0$ be the energy of vacancy formation in the μ-type site without the elastic deformations account, $U_{v\mu}$ the correction to the energy of vacancy formation that is due to elastic deformations present, and $f_\mu(U_{v\mu})$ the distribution function of $U_{v\mu}$ values normalized to unity. Then the vacancies concentration in μ-type sites is determined by the relation

$$C_{v\mu} = \int_{-\infty}^{\infty} \exp \left[\frac{-(E_{v\mu}^0 + U_{v\mu})}{k_B T} \right] f_\mu(U_{v\mu}) \, dU_{v\mu} . \tag{6.19}$$

If the C_μ is the μ-type sites concentration, then the concentration of vacancies in LRC is, obviously, equal to

$$C_v = \sum_{\mu=1}^{s} C_\mu C_{v\mu} . \tag{6.20}$$

Similar expressions are valid also for the concentration of vacancies in the boundary layer. The binding energy of atoms in coinciding boundary sites is somewhat lower on the average than in the LRC volume. Therefore the vacancies concentration in coincident sites is higher than in the LRC body. If one accounts for the fact that on the boundaries the concentration of nonthermal partial vacancies is high, then we may state that the diffusion of atoms by the vacancy mechanism on the boundaries is considerably higher than in LRC.

The differences in the atom coordination around the interatomic cavities manifest themselves in the distribution of the interstitial formation energies. In LRC's as well as in crystals, the formation of dumbbell interstitial configurations is possible. The expressions for equilibrium concentrations of interstitials are similar to those given above for vacancies (6.19 and 20), with the difference, however, that one accounts in them for the difference of interstitial cavities and the possibility of dumbbell configuration formation.

Note that in LRC, the stable Frenkel pairs may be formed (e.g., under irradiation). The energy spectrum of Frenkel pair formation is somewhat spread due to the spread in energies of vacancies and interstitials formation. The width of this spectrum as well as variations in energy of vacancies and interstitials formation may amount to some eV, and the typical values of the threshold energy of Frenkel pair formation in metallic glasses as well as in crystals may amount to about 25–30 eV. To point defects of a cluster one may attribute also the interstitial and substitutional impurities that locally break the topological and compositional order.

Dislocations. The question about the presence of dislocations in solids becomes crucial, first of all, when analyzing the nondiffusional mechanisms of plastic deformation. Dislocations in polyclusters exist, but their structure differs from the one in crystals. Because of the lack of translational symmetry, there exist no Burgers vectors that are commensurable with the lattice period even in the LRC body. On the other hand, in a polycluster, one may form the Volterra dislocation cutting along a plane and inelastically shifting one part of the polycluster relative to the other in that plane (glide plane). The value of the displacement vector (Burgers vector) must exceed ρ_{12} (and must equal, approximately a) in order to have all sites in the glide layer experience inelastic displacements. The lack of translational invariance and of the natural glide planes in LRC leads to the LO breaking in the glide plane, and it appears to be approximately that in the boundary layer. Therefore, the dislocation structure in LRC is such as in Fig. 6.7, where the glide plane is locally disordered and its edge is the dislocation core. The field of elastic deformations around such a dislocation may be calculated in the continual approximation. Note that a somewhat similar dislocation structure is realized also in some crystals when the Burgers vector does not coincide with the lattice period and in the glide plane, the so-called antiphase boundary, is formed with the broken LO.

In LRC, one may form prismatic loops by removing the limited plane section and subsequently *gluing* opposite surfaces. The obtained loop is the edge of the inner boundary layer formed along the glued layers. The field of elastic deformations around such a loop is described in the continual approximation by the same expressions as in crystals [6.36]. The prismatic loops obtained by locating the *additional* material in the cut along the layer bounded by the loop have in general similar structure. In clusters as well as in crystals under irradiation, the dislocational loops of vacant and interstitial types may be formed. It is important that the formation of such loops lead to an increase in the density of cluster boundaries.

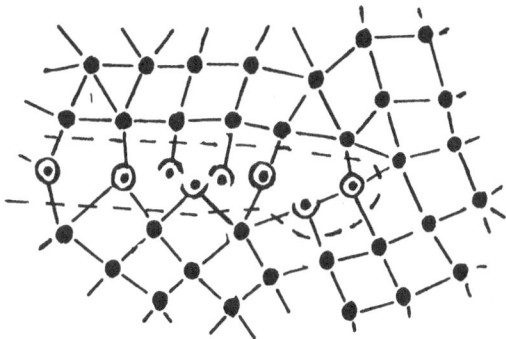

Fig. 6.7. Dislocation structure in LRC: The disordered slip layer and core are shown

Note that the energy of the shift dislocational loop formation in LRC is comparatively large because of the large contribution of the local disordering of atoms in the loop glide layer. At the same time, in the boundary layer where the shift is not accompanied by the increase of local disorder, the energy of loop formation is considerably lower. Due to this, small dislocational loops may be formed on the boundaries fluctuationally and play a considerable role in the *melting* of the boundaries (Sect. 6.8).

6.4.5 Polycluster and Other Models

It is of interest to compare the polycluster model with other models used in the physics of metallic glasses. The DRP-model, not assuming *ab initio* any correlations in the mutual location of atoms and being determined by the formation algorithm, does not admit the following comparison with the polycluster model. Some other models, in which the presence of correlations in the mutual arrangement of atoms is assumed, admit such a comparison. Here we consider 2-D and 3-D structures admitting normal partition on simple structure elements, i.e., polyhedra. We consider a partition to be normal when the polyhedra are adjacent by whole faces, each pair of them having no common inner points and covering all of space. Note that the Voronoi polyhedra compose the complete separation and the CP's do not.

The structures we will now discuss are made up of polyhedra that do not contain atoms and are, therefore, interatomic cavities (H-polyhedra). The H-polyhedra are usually chosen in such a way that the local atom density is as large as possible. From the finite number of H-polyhedra, one may construct some number (may be a very large one) of CP's (P-polyhedra). Among them, one should distinguish those that provide the smallest values of free energy of atoms confined inside. Then, one should consider the possibility of not leaving gaps and not allowing large local distortions when packing polyhedra. The local coordination of atoms in H- and P-polyhedra involves the correlations in the mutual location of polyhedra adjacent by whole faces, i.e., it leads to the

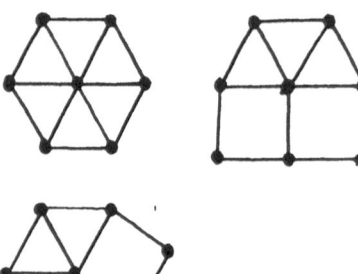

Fig. 6.8. Coordination polyhedra of the triangle-square network

formation of some MRO. Such is the general scheme of model structure construction which we will now consider briefly. In this scheme, one finds the search for ways to form noncrystalline cluster essential in the kinetics of the solid–body formation already discussed in Sect. 6.2.

Triangular-square lattices are model 2-D structures made up of equilateral triangles and squares introduced and studied primarily by *Collins* [6.37]. Considerable progress in studying these structures was achieved then by *Kawamura* [6.38, 39].

As is seen, triangles and squares were chosen as H-polygons. The regular triangle is the most compact configuration of three identical spheres. From regular triangles, a single P-polygon may be formed, i.e., the hexagon. The dense packing of hexagons exists and it represents a translationally invariant lattice. To form the topologically disordered structure, one must include other H-polygons. The square is the second most compact configuration after the triangle. *Collins* showed that the networks may be made up with the arbitrary ratio of triangle–to–square concentrations. From triangles and squares one may make up four types of P-polygons: a hexagon, two pentagons (Fig. 6.8), and a square. The 4-coordinated atom contained in the square has low binding energy and a comparatively high free energy. Therefore, to locally regular atoms one may attribute only 6- and 5-coordinated ones. Thus the random triangular-square lattice contains 3 types of P-polyhedra. The admissible elastic deformations of P-polyhedrons were not studied.

Figure 6.9 pictures the example of the triangular-square network taken from [6.38]. Some 4-coordinated sites are seen comprising inner boundaries of this LRC. It is easy to notice that there are two 5-coordinated atoms in the first coordination sphere of each 5-coordinated atom in LRC. This circumstance follows from the fact that, in P-polyhedra, we have an even number of squares leading to the formation of the MRO, which manifests itself in the formation of chains of 5-coordinated atoms. *Collins* and *Kawamura* studied the thermodynamic properties of triangular-square lattices. *Kawamura* established the existence of the first-order phase transition connected with the transformation of the crystalline structure into a topologically disordered one.

Tetrahedic-octahedric structures that are in some sense 3-D analogs of 2-D triangular-square lattices were introduced and studied analytically by *Ninomia*

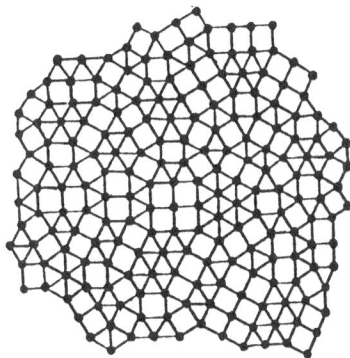

Fig. 6.9. Part of the triangle-square network

[6.40]. The tetrahedron, the analog of the triangle, is the most compact H-polyhedron in 3-D space, and the next after it is the octahedron, the 3-D analog of a square. As is known, the tetrahedron is related to noncrystalline polyhedra. No structure can be normally split into regular tetrahedrons. Including octahedrons and allowing for deformation of H-polyhedrons, one may make up dense structures admitting a normal partition.

From tetrahedrons and octahedrons, ring clusters are constructed among which one choses those with the smallest deformations of octahedrons and tetrahedrons. Related to these are, e.g., ring clusters consisting of 5 tetrahedrons with one common edge. Ring clusters are H-polyhedra. Cells (P-polyhedra) are built from ring clusters and one chooses those with the smallest local distortion of volume angles comprising H-polyhedra. The requirement of connectivity and dense packing of clusters leads to the correlation in their mutual location. Noncrystalline rings (not met in close-packed crystals) appear to be arranged along some lines similar to 5-coordinated sites in triangular-square networks. One calculates the average value of coordination numbers, number of faces, and edges of polyhedra of Voronoi in such a structure and also the average number of face edges. These numbers depend on the relative densities of tetrahedra and octahedra in the structure and for some ratio of them they appear to be close to those for RDP-models obtained by numerical simulation.

Note that the 3-D model is considerably more complicated than its 2-D analog. It seems desirable to add this model with a consideration of cluster energetics (free energy) and to perform the natural generalization including the extended defects, i.e., dislocations and boundaries, as a result of which an example of polycluster structure would be obtained.

6.5 LO in Real Metallic Glasses

The presence of one or several types of LO for the dominating number of atoms is, essentially, a sufficient condition for the existence of a polycluster structure. Therefore, the experimental data on LO in real metallic glasses becomes important.

In [6.41–43] we find reports of the experimental investigations of partial distribution functions of atoms in amorphous alloys. The strong LO is discovered in the alloys of the metal–metalloid and metal–metal types. The compositional order is closely (but not always unambiguously) connected with the topological order, so that the presence of the former testifies to the existence of the definite local topological order.

By studying the *Mössbauer* spectra and the nuclear magnetic resonance of a series of amorphous alloys, it was established [6.44] that their LO differs slightly from that in crystals.

As indirect evidence of the presence of perfect LO in amorphous Fe–B and Pd–Si alloys, there serve the data on the threshold energy of defect formation under electron irradiation at low temperatures. Distinct thresholds of defect formation appear to exist, which are very close to the thresholds of Frenkel pairs formation in corresponding crystals [6.45, 46].

The data on photoemission from some metallic glasses may be satisfactorily explained on the basis of the assumption of a perfect LO order [6.47].

The analysis of the effect of hydrostatic pressure and electron irradiation on the diffusion-controlled crystallization process of amorphous alloys of the metal–metalloid and metal–metal types in [6.48, 49] leads to the following conclusions.

i) Atomic diffusion responsible for crystallization proceeds by the vacancy mechanism and the vacancy volume is approximately equal to the average volume per atom.

ii) As in crystals, the electron irradiation accelerates the diffusion due to the point defect pair generation, and fair quantitative description of the obtained results is provided on assuming a high densities of point defect sinks, $\sim 10^{-2}$ per atom in alloys. These results may be regarded as indirect evidence of the polycluster structure of the alloys studied with perfect LO (providing the stability of vacancies) and cluster sizes $\sim 10^2$ Å; this explains the observed sink density because boundaries are sinks for point defects.

It is very difficult even to give the entire list of papers containing results on the presence of some or the other LO in amorphous solids. Not the very existence of the LO, but its character and reasons for its existence have been subject of numerous studies [6.15, 50].

6.6 Low Energy Excitations and Tunneling States

The peculiarity of glasses (not only metallic ones), is the presence of a wide spectrum of low-energy excitations (with energies from 10^{-5} to 1 eV) exhibiting themselves in low-temperature anomalies of specific heat and thermoconductance, in the temperature dependence of the velocity and absorption coefficient of sound, in internal friction, and in a number of other relaxation phenomena. In describing the phenomena listed, various phenomenological models are involved among which the conception of two-level systems proved most fruitful. It was first formulated in [6.51, 52] and then developed in [6.59, 54]. The connection of two-level systems with the glass structure is not usually discussed if one does not take into account the consideration of this problem in the frame of the free volume model [6.55], where one phenomenological description is connected with another. In polyclusters, the structure elements in which localized energetic excitations take place may be pointed out; one may describe as well the character of atomic-configuration rearrangement connected with these excitations and obtain estimates of the characteristic energies and activation energy of such excitations.

In the body of a cluster, the displacements of an atom into a neighboring vacancy (such displacements lead to the diffusion of atoms) and the Frenkel-pair generation are elementary nonelastic rearrangements of atomic configurations. The first has a comparatively high (~ 1 eV) energy barrier and the energy threshold of the other one is about $20-30$ eV. As is seen, these rearrangements are comparatively high energetic. The first of them reveals itself at comparatively high temperatures close to the metallic-glass–crystallization temperature, and the latter usually occurs at intensive external actions. The energies of some inelastic rearrangement of atomic configurations on cluster boundaries where the LO is not perfect are lower.

(i) *The atom redistribution in complexes of noncoincident sites.* As in connected complexes of noncoincident sites $C_N(m)$, the number of located atoms m is equal approximately to half the number of sites N (6.4.4); every such complex is a potential well with N minima, approximately one half of which is occupied by atoms. Because of the random character of the site networks conjoined along the boundaries, the differences of energy atoms in minima are distributed continuously in the range from zero to some tenths eV (this is the scale of partial interstitial formation energies, see 6.4.4). Approximately in this range lie also the barrier heights separating the wells. The concentration of the noncoincident sites is proportional to the boundary density and for an LRC size $\sim 10^2$ Å is on the order of 10^{-2} per atom. The displacement of atoms in a complex is possible as a result of overcoming potential barriers under the influence of thermal fluctuations as well as due to atoms tunneling through low barriers. It is not difficult to estimate the concentration of tunneling states. If one takes into account that, at low temperatures (~ 1 K), the main contribution in tunneling rearrangements are made by transitions through low barriers with the height $\sim 10^{-4}-10^{-3}$ eV, then one finds easily that tunneling atoms comprise $10^{-4}-10^{-3}$ of the total number of atoms located in the noncoincident atom sites. With LRC sizes $\sim 10^2$ Å, the concentration of tunneling atoms is $10^{-5}-10^{-6}$ per atom, a point which agrees with the estimates of two-level system concentrations in various glasses obtained from experimental data.

(ii) *Splitting of coincident sites, coupling and displacement of noncoincident sites.* The elementary excitations listed are shown schematically in Fig. 6.10. The

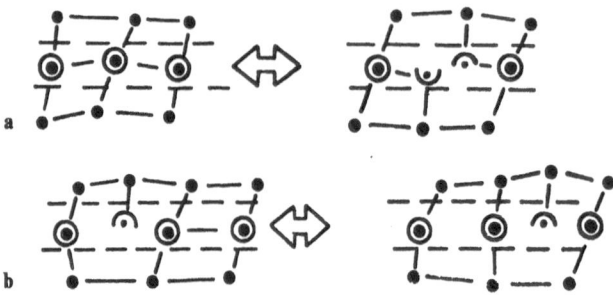

Fig. 6.10. Splitting of coincident and couplings of noncoincident sites (a), displacements of noncoincident sites (b)

distribution function of energy changes of atomic configurations changes smoothly in the broad energy range $[0, E_2]$ with the splitting of coincident sites (Fig. 6.10a). Besides, according to estimates obtained in [6.29] for metallic glasses, E_2 amounts to 1.5–2 eV. The energy threshold values of these rearrangements lie approximately in the same range.

Displacements with the number of coincident sites conserved (Fig. 6.10b) are comparatively low energetic. Energy changes and activation barriers lie in the range from hundredths to some tenths of an eV. The density of structure elements where the rearrangements considered occur amount to 10^{-2} per atom with LRC sizes $\sim 10^2$ Å.

(iii) *Cooperative rearrangements (shifts) without changing the number of coincident and noncoincident sites.* Cooperative rearrangements are depicted schematically in Fig. 6.11. The boundary sections occupied by noncoincident sites are shaded, whereas the sections with coincident sites are left unshaded. At the local shift of the section of one cluster relative to the other, a number of elementary rearrangements may occur as depicted in Fig. 6.10, yet in such a way that the general numbers of coincident and noncoincident sites are equal before and after the shift. Cooperative rearrangement of the type depicted is essentially the shift dislocational loop with its Burgers vector lying in the loop plane and equal to the average value of the site displacements vector of regular continuations on the section rearranged. In the formation energy of such a loop, inelastic rearrangements and the elastic deformation contribute as follows: $E_d = E_d^{ne} + E_d^e$. The contribution value from inelastic rearrangements in the loop plane E_d^{ne}, where the number of coincident and noncoincident sites are conserved, may amount to a tenth of an eV and be either positive or negative. The elastic energy E_d^e may be estimated by using the known expression for the dislocation loop energy.

$$E_d^e = \mu_s b^2 R(2 - v)(1 - v)^{-1}(-\ln tg a/4R - 2)/4 , \qquad (6.21)$$

where R is the radius of the loop, b the length of the Burgers vector, μ_s the shift modulus in the boundary layer, v the Poisson coefficient.

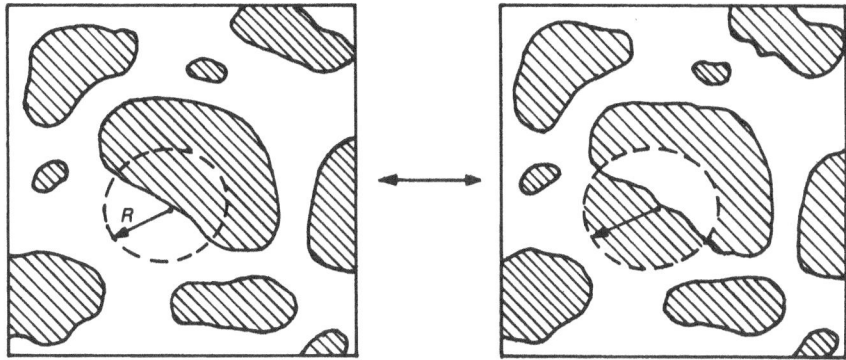

Fig. 6.11. Cooperative rearrangements (shifts) with the number of coincident and noncoincident sites conserved. The broken line surrounds the region R where the shift occurs. The sections of the boundary filled with noncoincident sites are shaded

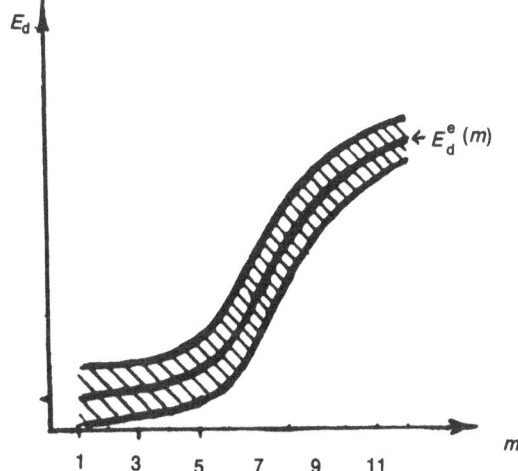

Fig. 6.12. Energies of cooperative structure rearrangements versus the number of non-coincident sites m

The activation energies may approach 1–2 eV. Moreover, the rearrangement itself is connected with a lowering of the configuration entropy inherent to cooperative rearrangements. Figure 6.12 shows the energy band E_d depending on the number of coincident sites m on the section under rearrangement. The average value of E_d is determined by the contribution from elastic deformations and the band width δE_d by the spread of atom binding energies and Burgers vector values. The value δE_g amounts to tenths of an eV according to estimates. Figure 6.13 shows the distribution function of the cooperative rearrangements described over energies $f_d(E_d)$ borrowed from [6.29]. As is seen, at low energies there is the maximum in the density of states. The average frequency of

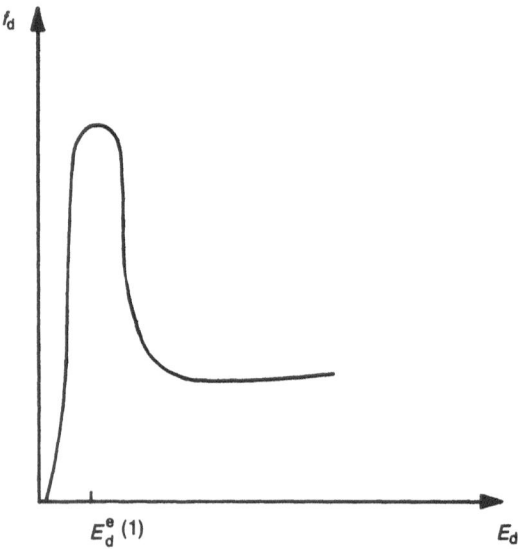

Fig. 6.13. Energy distribution of cooperative rearrangements

cooperative rearrangements $v_d(m)$ under thermal-fluctuation action is described by the following expression [6.29]

$$v_d(m) = (\pi m)^{1/2} 4^{-m} [\delta E_d(m)\beta]^{-1} v_c \exp[-E_d^*(m)\beta] , \qquad (6.22)$$

where v_c is the atom oscillation frequency in coincident sites, $E_d^*(m)$ the height of the energy barrier, and $\beta \equiv k_B T$. The group of factors before v_c describes the contribution to the frequency of cooperative arrangements by entropy changes and, as is seen, diminishes noticeably this frequency.

The low-energy excitations described above contribute essentially to the reversible relaxation processes [6.32], to the internal friction [6.33], and to the specific heat. The expressions for the excitation contributions ii)–iii) to the specific heat are given in [6.29, 30], where it is also shown that cooperative rearrangements iii) contribute greatly to the melting of cluster boundaries which is treated as the glass–liquid transition in polyclusters (Sect. 6.8).

There are a number of experimental data, for example [6.56], with respect to distributions of energies and activation barriers of excitations contributing to reversible relaxation processes. These data agree with the estimates given above.

6.7 Diffusion

In polyclusters, atoms can diffuse on vacant cavities, viz., regular vacancies, interstitial holes in the LRC bulk, and partial vacancies or vacant coincident sites at the boundaries. Since the formation and migration energies of regular vacancies are higher than those of boundary vacancies in coincident and noncoincident sites, the diffusional atom transport occurs primarily along cluster boundaries. Neglecting the contribution of cooperative structural fluctuations to the diffusion, the boundary diffusion coefficient can be written as

$$\bar{D}_s = \tfrac{1}{2}[q^2 + (q + 4D_{uc}D_c)^{1/2}], \quad q = (2w_{uc} - 1)(D_{uc} - D_c) , \qquad (6.23)$$

where D_c, D_{uc} are, respectively, the boundary diffusion coefficients on coincidence and noncoincidence boundary segments, and w_{uc} is the boundary part occupied by noncoincidence segments. At $w_{uc} > 1/2$, among noncoincidence segments there is an infinite segment where a fast diffusion of atoms on partial vacancies occurs within large distances.

At a sufficiently high temperature, $(T \approx T_g)$, the boundary diffusion is essentially contributed by cooperative structural fluctuations (see Sect. 6.6), i.e., thermally activated displacements which may cause, within relatively small segments, the splitting of coincident sites into noncoincident ones, while noncoincident sites couple into coincident ones. With these displacements, the energy varies within tenths of an eV, if the total number of coincident and noncoincident sites remains the same. Because of these structure rearrangements, each boundary atom appears alternately in the coincidence and noncoincidence zones, and the diffusion rate is determined by the frequency of these cooperative rearrangements. Density and frequency of the cooperative rearrangements essentially grow when boundary *melting* takes place and it provides the diffusional viscosity decrease (Sect. 6.9).

In amorphous alloys, the diffusional atom transport takes place over rather large distance, 10^3 Å, whereas the crystallization may proceed with small (~ 1 Å) atomic replacements. This *diffusional paradox* can be explained easily, if the observed diffusion transport occurs mainly along cluster boundaries and, consequently, does not lead to crystallization. The polycluster crystallization develops together with the diffusion enchancement in the LRC bulk. In investigations of the effect of hydrostatic pressure on crystallization kinetics, *Limoge* revealed that the crystallization was controlled by diffusion that occurs through the vacancy mechanism [6.48, 49], which confirms the explanation of the *diffusional paradox* given above.

Interstitial impurities (e.g., hydrogen of helium atoms) can easily move over the interstitial cavities in LRC but comparatively large size partial vacancies on the boundaries are traps for small-size atoms. It was established in a series of experiments that amorphous alloys contain traps for hydrogen and helium atoms and that the concentration of these traps is rather high [6.56]. The concentration dependence of interstitial diffusivity in disordered structures was considered phenomenologically [6.56] and by mean-field approximation [6.57].

Cluster boundary decoration by metalloid atoms may occur due to small-atom locations in noncoincident sites. It is similar to the decorating of the Fe–Ni–B alloy boundaries with boron atoms (Fig. 6.2).

6.8 Boundary Melting and the Glass–Liquid Transition

In this section, we discuss the polycluster thermodynamics and the glass–liquid transition nature. This transition might in any case mean restoring the ergodicity and the liquid phase formation. Generally speaking, the polycluster structure is essentially nonuniform on the microscopic level, and it contains the regions of local disorder, i.e., boundaries. The structure inhomogeneity leads to the inhomogeneity of structural fluctuation distributions. The low-energy structural fluctuations described in Sect. 6.4.6 occur at the boundaries and, in boundary layers, the ergodicity is restored first. This process assumes the features of a phase transition when it has collective character due to interactions between structure fluctuations. The boundary melting process occurs as follows [6.29, 30]. The density of low-energy excitations (ii and iii) (Sect. 6.4.6) increases due to thermal fluctuations with temperature growing. Due to the elastic fields, the excitations interact between themselves and, as a result, the correlations occur between them. The elastic interaction is short-range and the correlations of structure fluctuations are considerable at the sufficiently high density of the latter. Evidently, those correlations which lead to lowering the energy of excitations are amplified.

As an order parameter ξ in the shift loop system, the parameter of orientation correlation of adjacent loops Burgers vectors is chosen. The average energy of loop interaction W depends on ξ and is calculated by averaging the interaction energy of two loops W_{12}. Note that the magnitude W_{12} is comparable with the elastic energy E_{d}° of the loop when the radius of the latter amounts to several atomic radii [6.36], so that collective interactions are not weak for low-energy loops.

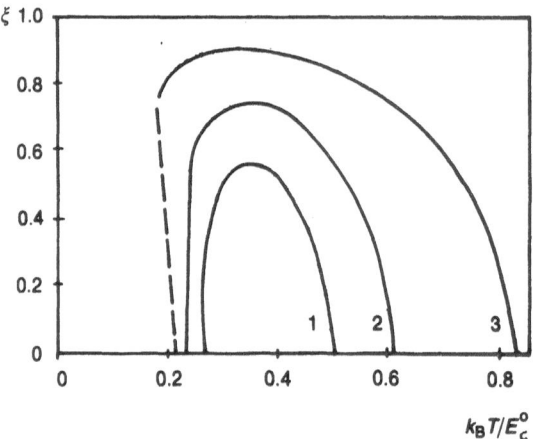

Fig. 6.14. Order parameter vs $k_{\mathrm{B}}T/E_{\mathrm{d}}^{\circ}$ at $L = 1/36$, $W < W_1 (1)$, $W = W_1 (2)$, $W > W_1 (3)$

Let R_0 be the characteristic radius of small loops ($R_0/a \approx 2$–3), and n_0 the density of places in the boundary layer where the loops may be formed ($n_0 \sim 1/a^2$). Then in the loop system there occurs the phase transition meaning the excitation of the correlated loops. The transition takes place if the W value exceeds some critical value W_0 ($W_0 \approx 0.5\, E_d^c$), besides at $W_0 < W < W_1$ ($W_1 \approx 3E_d^c/\ln(2/\alpha)$, $\alpha = (\pi n_0 R_0^2)^{-1}$) this is the second-order phase transition and, at $W > W_1$, it is the first-order one. Figure 6.14 shows the temperature dependence of the order parameter at three values of W. It is seen that correlations in the loop system take place in the finite interval of temperatures. The lower boundary of this interval is around $k_B T_{c1} \approx E_d^c/4$, and the upper one at $k_B T_{c2} \approx W$. At $T > T_{c2}$, the correlations of the loop Burgers vectors disappear. If E_d^c amounts to some tenths of an eV, then $T_{c1} \lesssim 10^3$ K, which corresponds to the interval of characteristic temperatures T_g of the metallic glasses vitrification. The specific-heat temperature dependence at the same W values is presented in Fig. 6.15.

The ergodicity restoring in boundary layers during the transition described occurs due to high atom mobility and fast changes of atomic configurations at high density and frequency of cooperative structure fluctuations. The process described must be accompanied by the increase of the diffusional mobility of atoms in the layers adjacent to the boundary and may lead to a comparatively fast diffusional rearrangement of clusters.

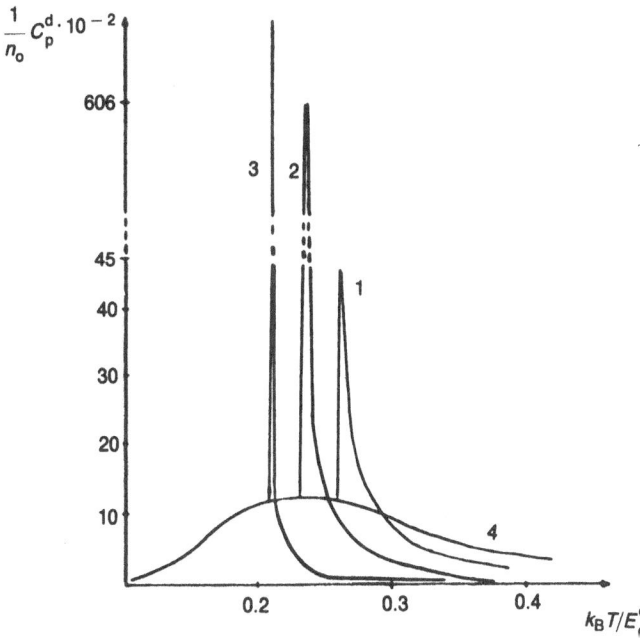

Fig. 6.15. Specific heat of structure fluctuations C_p^d vs temperature at $L = 1/36$ and $W < W_1$ (1), $W = W_1$ (2), $W > W_1$ (3) and $\xi = 0$ (4)

6.9 Mechanical Properties of Polyclusters

Polycluster structure of disordered solids determines the microscopic mechanisms of plastic deformation and fracture. These mechanisms have been studied extensively, for crystals. For every crystalline material an Ashby map may be constructed, i.e., regions on the plane (σ, T) may be pointed out where one or the other mechanism of plastic deformation prevails (σ is the macroscopic stress). For the metallic glasses case, the macroscopic mechanical characteristics are described; it is established that, at low temperatures, the inhomogeneous plastic deformation occurs with shear-bonds formation and, at temperatures close to T_g under a load, the homogeneous viscous flow of glasses takes place. Several model and phenomenological descriptions of metallic glass plastic deformation are suggested based on some model ideas about the microscopic mechanisms (such as in the free-volume model, the dislocational, disclinational models, the adiabatic deformation model). The reviews and original communications on these subjects are contained in [6.10, 21, 22, 59–68]. The mechanism of plastic deformation and mechanical properties of polyclusters are considered in detail in [6.28–31].

6.9.1 Anelastic Deformation

The local inelastic deformation which disappears as external loading is removed is called anelastic. It is contributed by elastic and inelastic deformations of the solid. In the LRC body, the inelastic rearrangements may occur when stresses approach their theoretical limit $\sigma_{th} \sim 10^{-1}\mu$. On boundaries under the action of shift stresses, elementary and cooperative inelastic rearrangements may occur that are essentially boundary slippings localized on small segments. In [6.28, 29] the problem of the boundary slipping and the glide layers is considered (the atomic structure of the glide layer along which one part of LRC shifts relative to another is just the same as for cluster boundaries).

It is easily seen that, in every boundary site under shift stresses, the inelastic rearrangement of type-ii) may occur (Fig. 6.10) if the stress achieves some critical value for the given site. The threshold of the athermal slipping and the slipping kinetics with the account of thermal fluctuations is determined by the distribution function of critical stresses in the boundary layer or in the glide layer $g(\sigma)$. Athermal slipping along boundaries occurs when the shear stress exceeds some critical value σ_s^* depending essentially on the distribution function $g(\sigma)$. Not dwelling upon the details of calculating this magnitude and of the expressions obtained for it [6.29], we give only its rough estimate $\sigma_s^* \approx \bar{\sigma}_s/2$, where $\bar{\sigma}_s = \int_0^\infty \sigma g(\sigma)\,d\sigma$ is the average value of the critical shear stress.

If the applied stresses are much less than σ_s^*, then the inelastic rearrangements occur only for a small part of the sites with the lowest values of critical stress. These rearrangements may be revealed in experiments on the internal friction, the theory of which was suggested in [6.33]. If, however, the external stresses σ^e exceed σ_s^*, then the athermal slip along cluster boundaries occurs. Around the slip sections, the stress fields are redistributed as around the equilibrium cracks and the stress concentration in LRC occurs. If the length of the boundary sections along which the slipping occurs is comparable to the cluster sizes, then for the mean stress value in LRC the following estimate is applicable

$$\sigma_{LRC} \approx \begin{cases} 1.5 \, \sigma^e, & \sigma^e > \sigma_s^* \\ \sigma^e, & \sigma^e < \sigma_s^*. \end{cases} \tag{6.24}$$

On the macroscopic level, this must show itself as a deviation from the stress–strain linear dependence.

6.9.2 Yield Stress of a Polycluster

As a result of stress concentration on the edges of boundary shift sections, the stress may achieve the LRC strength limit

$$\sigma_{edge} \approx \sigma^e [1 + (r_{cl}/2a)^{1/2}]. \tag{6.25}$$

In this case, the glide layer propagates from the boundary into the cluster body, which leads to the nonuniform plastic deformation of a polycluster. Thus the macroscopic yield stress of a polycluster σ^*, as it follows from (6.25), is determined by the relation

$$\sigma^* = \max \{ \sigma_s^*, \sigma_{th} [1 + (r_{cl}/2a)^{1/2}]^{-1} \} . \tag{6.26}$$

6.9.3 Shear Band Formation

At low temperatures, with nearly all thermally activated processes suppressed, the slip is the only possible mechanism of a plastic deformation. Since the polycluster has no natural plane glide layers, these should be formed under the action of stresses. Figure 6.16 shows two types of shear-band formation. The first one (Fig. 6.16a) is realized by breaking the material continuity in the layer having a thickness comparable with the mean cluster size. In the process of shear, the slip occurs along the boundary segments oriented so that the vectors of applied forces are tangential to the boundaries. The normal components of tensile stresses relax in this case due to the crack formation. The formation of this shear band can easily turn into fracture because of the formation of unstable cracks propagating along the cluster boundaries, if

$$\sigma^e > \sigma_f = (2\tilde{\gamma}E/\pi l)^{1/2}, \quad \tilde{\gamma} = \gamma - \gamma_s , \tag{6.27}$$

where E is the Young modulus and γ, γ_s are the densities of surface and boundary energies, respectively. The expression given above is the Griffiths

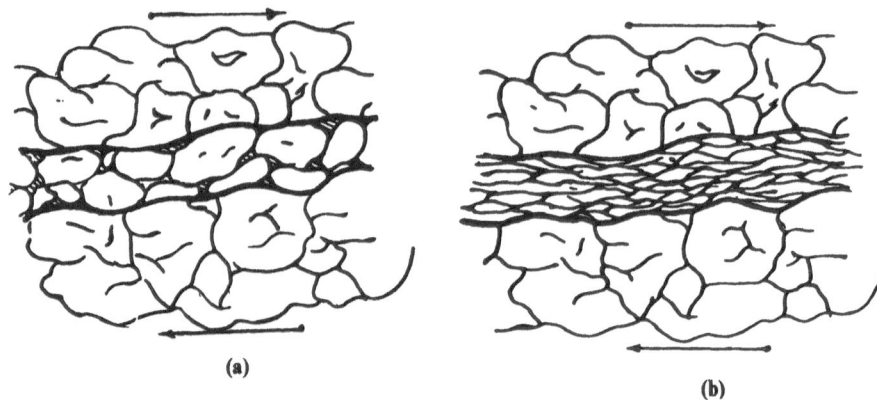

Fig. 6.16. Shear band structure at $\sigma \gtrsim \sigma_f$ (a) and $\rho^* < \rho < \rho_f$ (b) (see text)

criterion of crack growth. Here we must take into account that the size of intercluster cavities l is comparable with r_{cl}. The condition (6.27) determines the stress of polycluster fracture along boundaries, when the cluster contains microcracks with l being the microcrack size.

If $\sigma^* < \sigma^e < \sigma_f$, then the applied stresses lead to the formation of new glide layers and the shear band has the structure depicted in Fig. 6.16b. During this process, the glide-layer branching due to structure microinhomogeneities in the LRC may occur. However, the preferred orientation of the layers depends on the direction of the forces resulting from shear stresses.

6.9.4 Thermally Activated Slip

The formation of a new glide layer due to the motion of the internal boundary edge under stress may occur for $\sigma^e < \sigma^*$ in a thermally activated way. In [6.29] the expression has been derived for the boundary-edge velocity, v_s:

$$v_s \approx a v_c \exp\left\{ - [\sigma_{th} - 2(r_{cl}/a)\sigma^e] v_a \beta \right\} . \tag{6.28}$$

The expressions for the velocity of a homogeneous, thermally activated slip in the layer under applied stress which is less than σ_s^* are somewhat cumbersome [6.29]. However, if the relative area of the glide layer occupied by coincident sites is equal to w_c, and the critical stresses in noncoincident sites are distributed nearly uniformly within the range $[\sigma_{c0}, \sigma_{cm}]$, then the slip velocity is given by

$$v_{sl} \approx a v_c (2\sigma^e \delta_c v_a^2 \beta/w_c)^{1/4} \exp\left\{ [\sigma_{cm} - 2(\sigma^e \delta_c/w_c)^{1/2}] v_a \beta \right\} , \tag{6.29}$$

$$\delta_c \equiv \sigma_{cm} - \sigma_{c0} .$$

It follows from this expression that the activation volume of a homogeneous slip may be as large as 2–$3v_a$. The activation volume of the slip-layer edge motion, as derived from (6.28) is approximately equal to $(r_{cl}/a)^{1/2} v_a$ and makes up about $10 v_a$ for $r_{cl} \approx 10^2 a$.

6.9.5 Diffusion-Viscous Flow

In polyclusters, at a sufficiently high temperature, when the atomic diffusion at the boundaries becomes appreciable, there occurs a diffusion-viscous flow, similar to that of polycrystals. In this case, in a macroscopically homogeneous solid, the rate of the plastic deformation, $\dot{\varepsilon}_{ik}$, and the tensor $\tilde{\sigma}_{ik} = \sigma_{ik} - \mathrm{Sp}\sigma_{ik}/3$ are related as

$$\dot{\varepsilon}_{ik} = \eta^{-1} \tilde{\sigma}_{ik} \; , \tag{6.30}$$

where η is the viscosity,

$$\eta = \alpha_{sl} \eta_{sl} + \alpha_d \eta_d \; ,$$

$\eta_{sl} = \sigma r_{cl}/v_{sl}(\sigma)$ is the slip resistance,

$$\eta_d = (3 \bar{D}_s a \beta)^{-1} (r_{cl}/a)^3 \tag{6.31}$$

is the diffusional viscosity, D_s is the boundary diffusion coefficient, and α_{sl}, α_d are the numerical factors dependent on the cluster shape and boundary orientation. The bulk diffusion is neglected here because its activation leads to the crystallization of the amorphous polycluster. In amorphous alloys, the effective boundary diffusion coefficient entering into (6.31) turns out to be equal to the lowest diffusion coefficients of alloy component atoms. It should be noted that in the process of a flow, alloy segregation occurs. Thus, for most mobile atoms of the species A, the deviation of the concentration, δC_A, from its average value C_A is

$$\delta C_A \approx C_A \sigma v_a \beta, \tag{6.32}$$

which is not too low for reasonable σ values and $T \sim 10^2$–10^3 K.

The diffusion-viscous flow changes to an inhomogeneous slip at a rather high strain rate or with temperature decrease. The boundary between homogeneous and inhomogeneous plastic deformations of a polycluster on the (σ, T) plane is given by [6.29]

$$k_B T = E_s [\ln(v_c/\dot{\varepsilon}) + \ln(\sigma^* v_a/k_B T) - 3\ln(r_{cl}/a)] \; . \tag{6.33}$$

In deriving this expression, we assumed that $\bar{D}_s \sim a^2 v_c \exp(-E_s \beta)$. For $E_s = 2 \, \mathrm{eV}$ [6.73], $v_c \sim 10^{12} \, \mathrm{sec}^{-1}$, $r_{cl} \sim 10^2 a$, the above expression is in fair agreement with the experimentally established boundary of the homogeneous–inhomogeneous flow transition for the amorphous Pd–Si alloy [6.63].

If $\sigma < \sigma^*$, then a mixed flow may occur, at which the diffusion-viscous plastic-deformation is accompanied by a thermally activated formation of new glide layers.

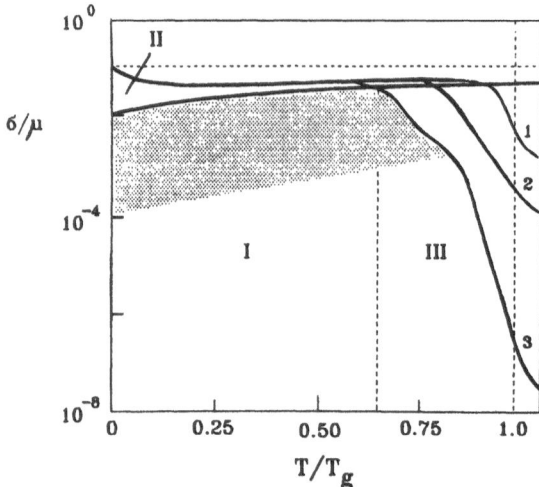

Fig. 6.17. Map of polycluster mechanical states. Region I: elastic and anelastic (shaded area) deformations; Region II: inhomogeneous plastic deformation; Region III: homogeneous diffusional-viscous flow. Curves 1–3 show the temperature dependence of the stress at different constant strain rates

6.9.6 The Map of Mechanical States of Polyclusters

Based on the plastic deformation mechanism descriptions presented above, one may schematically depict the map of mechanical states of polyclusters (Fig. 6.17) similar to those made for crystals [6.74].

On the map of mechanical states in region I, elastic and anelastic (shaded areas) deformations take place. In the region II, the inhomogeneous plastic deformation with the formation of shear bands takes place. The horizontal broken line corresponds to the theoretical yield stress of LRC. In the region III, the homogeneous diffusional-viscous flow takes place and, in the region IV, the mixed viscous flow is realized. Curves 1, 2, 3 show the temperature dependence of the stress at different constant strain rates. The continuations of these curves in regions IV and II correspond to the mixed nonuniform plastic deformation.

A map of similar type (but without some essential details) is contained in [6.75], where the plastic deformation of metallic glasses was considered in the free-volume model.

6.10 Theory of Supercooled Liquid Solidification

This section presents the theory of liquid solidification, in which both crystalline and noncrystalline heterophase fluctuations in supercooled liquids are taken into account [6.76]. A general discussion of this problem is presented in Sect. 6.2

and the main conjecture is the following: competitive nucleation and growth of both the crystalline and noncrystalline embryos in supercooled liquids are important, and glass formation with a polycluster structure occurs when the cluster nucleation rate and (or) the density of noncrystalline embryos in a liquid are predominant. To verify these ideas, the analysis of the thermodynamics and kinetics of clusters and crystallites in liquids must be carried out. The main difficulties arise from a multiplicity of cluster atomic configurations (in kinetics) and from the cluster ergodicity breaking by solidification (in thermodynamics). Similar difficulties are inherent to the spin-glass theory also [6.77] and are typical for considerations of the systems with broken ergodicity.

To describe the thermodynamics and kinetics of clusters with due regard for the multiplicity of the structure states (SS) of these nuclei, it is necessary to introduce the SS sets. The mutual transformations of clusters by atom attachment or detachment, or by atom rearrangements are possible. It means that the SS set may be ordered to form a space. Then, it is possible to consider the nucleation of noncrystalline nuclei (clusters) as a diffusion process in an extended space of cluster structure states (CSS), which is, in fact, the join of hierarchy of ultrametric subspaces. The kinetic criterion of glass transition can be formulated as a condition of the overwhelming transformation of a liquid into a noncrystalline amorphous solid.

The *liquid-cluster* system appears to have equilibrium metastable or stable heterophase states, in which the volume fraction of solid clusters may vary from negligibly small values up to unity. To find these states, the thermodynamics of the *liquid-cluster* systems will be considered.

6.10.1 The Set of Structure States of Solid-State Clusters

The solid-state clusters are distinguished from liquid ones by the rigidity of their structure. As a defining criterion for rigidity we shall choose the stability of point defects, i.e., vacancies and interstitial atoms, in the body of the cluster.

Clusters exhibit the multiplicity of SS. If $U(q_1, \ldots, q_N) \equiv U(q, N)$ is the potential energy of the ensemble of N atoms with the coordinates $(q_1, \ldots, q_N) \equiv (q, N)$, then each stable CSS has the corresponding minimum $U(q, N)$. We denote the set of minima $U(q, N)$ corresponding to the CSS by $\{q, N\}$. The set $U(q, N)$ maps $\{q', N\}$ onto the set of minimal potential energy values $\{u', N\}$. Since the binding energy of each atom in a solid is greater than zero and is limited from below, then $\{u', N\}$ also has the upper limit 0 and a certain lower limit $u^0(N)$.

Based on the assumption of the homogeneity of the random network of cluster nodes (atom sites), we shall assume that at $N \Rightarrow \infty$ there exists the limit

$$\varepsilon_0 = \lim_{N \to \infty} u^0(N)/N \ . \tag{6.34}$$

6.10.2 Structure State Density (SSD)

An important role in thermodynamics and kinetics of cluster formation is played by SSD, $\Psi(U, N)$:

$$\Psi(u, N) = (2\pi r)^{-1} \sum_r w^r(N) \delta[u - u^r(N)] , \qquad (6.35)$$

where $w^r(N)$ is the number of SS of the cluster with the potential energy $u^r(N)$.

We shall assume that the shapes of clusters are nearly spherical, and their surfaces are two-dimensional. The rate of the number of atoms in the surface layer, N_s, to N is approximately equal to

$$p_s(N) = N_s/N \approx 1 - (1 - 1/\rho)^3, \quad \rho = R/a . \qquad (6.36)$$

Here R is the cluster radius, a is the interatomic distance.

We can separate the contributions of the body from those of the surface layer to the potential energy and the SSD [6.78]. This permits us to represent (6.36) as

$$\Psi(u, N) = \int_0^\infty \int_0^\infty \Psi_v(u_v, N_v) \Psi(u_s, N_s) \delta(u - u_v - u_s) du_v du_s , \qquad (6.37)$$

where the subscripts v and s denote the quantities relating to the body and to the surface layer, respectively.

It can be demonstrated that at $N_v, N_s \to \infty$ the SSD takes on the following asymptotic form

$$\Psi_{s,v}(\varepsilon, N_{s,v}) = (N_{s,v}/\pi \delta_{s,v}^2)^{1/2} \exp\{N_{s,v} [\zeta_{s,v} - (\varepsilon - \bar\varepsilon)^2/2\delta_{s,v}^2]\} \qquad (6.38)$$

The quantities $\zeta_{s,v}$, $\varepsilon_{s,v}$, $\delta_{s,v}^2$ are given by the following relations:

$$\exp(\zeta N) \equiv W(N) = \int_{N\varepsilon_0}^0 \Psi(u, N) du ,$$

$$\bar\varepsilon = [NW(N)]^{-1} \int_{N\varepsilon_0}^0 u \Psi(u, N) du , \qquad (6.39)$$

$$\delta^2 = [NW(N)]^{-1} \int_{N\varepsilon_0}^0 (u - \bar\varepsilon)^2 \Psi(u, N) du .$$

In the vicinity of ε_{0v} and ε_{0s} for $N \to \infty$ the SSD can be represented as power series expansions $(\varepsilon - \varepsilon_{0v,s})$. Let l_v^c and l_s^c be the lengths of structure correlations in the body and in the surface layer of the cluster, and $n_{v,s}^c = 4\pi(l_{v,s}^c)^3/3$. Then it can be shown that the expansions start with the terms of order $(N_{v,s}/n_{v,s}^c) - 1$:

$$\Psi(\varepsilon, N_{v,s}) \sim (\varepsilon - \varepsilon_{0v,s})^{(N_{v,s}/n_{v,s}^c) - 1}. \qquad (6.40)$$

6.10.3 Space of Structure States

Since the CSS may change by adding or splitting off atoms, or due to structure rearrangements of the surface layer under thermal fluctuations (rearrangements

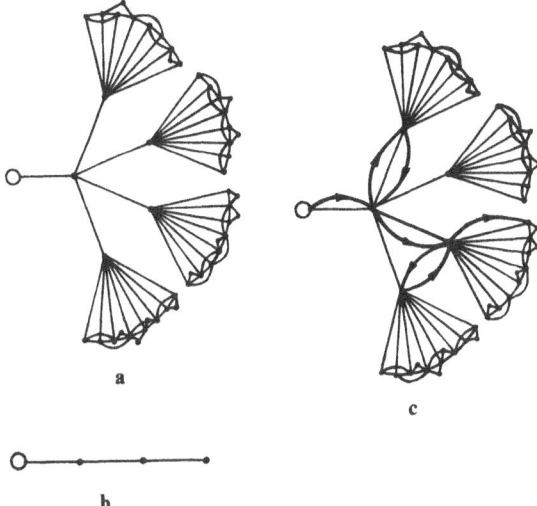

Fig. 6.18. Structure state space of the cluster (a, c) and the crystal (b). The oriented structure trajectory is shown in (c)

in the cluster body are unlikely because of high energy barriers), different CSS appear to be genetically related. These correlations of the clusters with different CSS's allow to order the set of CSS.

Let us consider the cluster that comprises m spherical atomic layers surrounding the "seeded" microcluster. We shall call as adjacent those structure states of two clusters which undergo mutual transformations with the change of the number of atomic layers by unity or (if the transformations are due to the change in the surface layer structure) without this change.

The rigidity of the cluster structure manifests itself in that the structure of subsurface layers experiences insignificant elastic deformations as the structure of the surface layer changes. Note that in the surface layer there may appear some "virtual" atom configurations vanishing on "walling up". Therefore, in the general case, the number of atomic configurations of the subsurface layer does not remain the same as the subsurface layer transforms into the surface one. The role of "virtual" configurations reduces, mainly, to an additional contribution to the configuration entropy of the surface layer.

It is convenient to represent the set of CSS, ordered by the adjacency criterion, as a graph. We assign a vertex to each CSS and to each structure state of the cores containing, respectively, $1, 2, \ldots, m - 1$ atomic layers, and connect the vertices of the adjacent SS by edges. As a result, we obtain the graph, the general view of which is shown in Fig. 6.18a. The graph comprises a hierarchical sequence of levels. The zero level is assigned to the liquid, the first one to the "seeded" microcluster. The vertices of the core with one atomic layer surrounding the microcluster lie at the second level. The last, third, level maps the set of CSS. Here also depicted are some vertices showing the virtual CSS. The vertex adjacency of this level is determined by possible transformation of the surface layer configurations without changing the cluster body configuration. For comparison, Fig. 6.18b shows the graph illustrating the SS of a perfect crystallite.

The line connecting the zero level with any vertex of the subsequent levels is called the line of structure realization or the valley (as adopted in the theory of spin glasses [6.77]). The vertices lying along the valley show the structure states of the microcluster and the cluster core.

We denote the i-th vertex of the k-th level by $\alpha_i(k)$ (index i numbers the vertices of the same level), and the set of k-th level vertices ($k < m + 1$) by M_k. The measure of closeness as well as the measure of distinction between two CSS or two cluster core structures can be determined by the position of the point of junction of their valleys. Any two valleys, joining at a level $m > 1$, have a common section. The measure of closeness of two structures ($\alpha_1(k)$, $\alpha_2(k) \in M_k$) is the number m, and the measure of their distinction is the number

$$d[\alpha_1(k), \alpha_2(k)] = d(k, m) = 2(k - m) . \tag{6.41}$$

This relationship defines the metric on the set M_k, transforming the latter into the metric space. It can be easily seen that the metric (6.41) has the property

$$d(\alpha_1, \alpha_2) \leq \max [d(\alpha_1, \alpha_3), d(\alpha_2, \alpha_3)] , \tag{6.42}$$

which is the sign of ultrametric spaces.

In the description of cluster evolution accompanied by a change in the number of atomic layers, it is convenient to employ the extended set of SS (ESSS), M, which is the join of the sets M_k, $k \in [0, \infty]$.

The ESSS can be provided with the metric

$$d[\alpha_1(k_1), \alpha_2(k_2)] = d(k_1, k_2; m) = k_1 + k_2 - 2m , \tag{6.43}$$

which has the property (6.42), if $k_1 = k_2$. So, the ESSS provided with the metric (6.43) is the join of ultrametric spaces, m.

The succession of structure changes experienced by the cluster during evolution can conveniently be depicted as oriented trajectories (we shall call them structure trajectories (ST) or, simply, trajectories) in m. The ST are the analogues of the phase trajectories of the dynamic system in phase space. Note that any trajectory starting at the zero level and reaching the i-th vertex of the k-th level passes necessarily through all the vertices lying along the valley $\alpha_i(k)$.

The cluster nucleation is the process of random walks in m, while the crystallite nucleation is the diffusion process in the one-dimensional size space. The probabilities of transitions from the vertex to either of the adjacent vertices depend on the free energy of clusters in different structure states.

6.10.4 Free Energy of the Cluster

The free energy of the cluster is calculated by the use of the partition function

$$Z(\beta, N) = \int \exp(- H\beta) \, dq \, dp, \quad p = p_1, \ldots, p_N , \tag{6.44}$$

$\beta = 1/k_B T$, k_B being the Boltzmann constant; p_i are the atom momenta. The Hamiltonian H is the sum of kinetic and potential energies (note that the potential energy takes into account the interaction between cluster atoms and the surrounding liquid).

When calculating $Z(N, \beta)$ we take into consideration that: (i) on integrating over q the main contribution to (6.44) comes from the vicinities of the potential energy minima of the cluster; (ii) the contributions from atoms of the surface layer and the cluster body can be separated asymptotically; (iii) the SSD at $N \gg 1$ is given by the expressions of types (6.38–40). In view of the above, after not complicated calculations, we obtain the approximate expression for $Z(\beta, N)$ valid for great N:

$$Z(\beta, N) = Z_s(\beta, N_s) \int Z_v(\varepsilon, \beta, N_v) \, \Psi(\varepsilon, \beta) \, d\varepsilon \; , \tag{6.45}$$

where

$$Z_v(\varepsilon, \beta, N_v) = \Omega_{pv}(\beta, N_v) \, \Omega_{qv}(\beta, N_v) \exp(-\varepsilon\beta) \; , \tag{6.46}$$

$$Z_s(\beta, N_s) = \Omega_{ps}(\beta, N_s) \, \Omega_{qs}(\beta, N_s) \int \exp(-\varepsilon\beta) \, \Psi_s(\varepsilon, N_s) \, d\varepsilon \tag{6.47}$$

Here Ω_{ps}, Ω_{pv} are the results of integration in (6.44) with respect to atom momenta of the surface layer and the cluster body, respectively; Ω_{qs} and Ω_{qv} are the integration results with respect to atom coordinate deviations from equilibrium values; they contain, in particular, the phonon contribution.

The estimates show that the characteristic time of change of the surface layer configuration is much shorter than that of cluster formation, whereas the characteristic time of cluster body rearrangement involving tens of atoms and more is much longer than other time scales. Therefore, in the process of cluster formation the structure of the surface layer is ergodic, and the structure of the cluster body has a broken ergodicity. From (6.46) and (6.47) we can derive the following expression for the free energy of the cluster

$$\tilde{G}(\tilde{\varepsilon}, \beta, N) = -\beta^{-1} \ln \tilde{Z}_v(\tilde{\varepsilon}, \beta, N_v) + \beta^{-1} \ln \tilde{Z}_s(\beta, N_s)$$
$$\equiv \tilde{G}_v(\tilde{\varepsilon}, \beta, N_v) + \tilde{G}_s(\beta, N_s) = N_v \tilde{\mu}_v(\tilde{\varepsilon}) + N_s \mu_s \; . \tag{6.48}$$

Here and further the tilde denotes the cluster quantities.

As is evident, the free energy of the cluster is the function of the random variable ε with the density

$$g_v(\tilde{\varepsilon}) = \exp(-\tilde{\varepsilon}\beta) \, \tilde{\Psi}_v(\tilde{\varepsilon}, N_v). \tag{6.49}$$

In view of (6.39) and (6.40), it is not difficult to derive from (6.38) the configurational entropy of the surface layer

$$\tilde{S}_{sc}/N_s = \tilde{s}_{sc} = \begin{cases} \zeta_s & \text{at } \tilde{\varepsilon}_{0s} > \bar{\varepsilon}_s - \beta\delta^2 \\ (n_s^c)^{-1} \ln[a_0\beta^{-1} + a_1\beta^{-2}] & \text{at } \tilde{\varepsilon}_{0s} < \bar{\varepsilon}_s - \beta\delta_s^2, \end{cases} \tag{6.50}$$

where a_0 and a_1 are the first two coefficients of the expansion $\tilde{\Psi}_s(\tilde{\varepsilon}, N_s)$ in powers of $(\tilde{\varepsilon} - \bar{\varepsilon}_{0s})$.

A similar quantity $\tilde{s}_{vc}$ can be calculated for the cluster body as well. In accordance with the definition given in [6.11], $\tilde{s}_{vc}$ represents the complexity of the structure displaying the degree of the SS uncertainty for the system with a broken ergodicity. In addition to the free energy of the cluster, we shall also introduce the effective free energy

$$G_{eff} = -\beta^{-1} \ln Z(\beta, N) \equiv (\beta, N) \equiv N\mu_{eff} . \tag{6.51}$$

It will be shown below that, because the clusters are in equilibrium with the liquid, this function determines the rate of cluster ensemble formation. Note that $\mu_{eff} < \mu_{v}$.

It is not difficult to find from (6.38) the melting temperature of a noncrystalline solid:

$$\tilde{T}_m \approx T_m(1 - \delta\tilde{\mu}/H_m), \quad \delta\mu = \tilde{\mu}_v - \mu_v , \tag{6.52}$$

where T_m, H_m and μ_v are, respectively, the temperature, the heat of melting and the chemical potential of the crystal. It can be demonstrated that $\delta\mu$ is practically independent of the temperature.

We also introduce

$$T^* = T_m(1 - \delta\mu_{eff}/H_m), \quad \delta\mu_{eff} = \mu_{eff} - \mu_v . \tag{6.53}$$

The role of this temperature will be discussed below.

6.10.5 The Rate of Cluster Nucleation and the Kinetic Criterion of the Glass Transition

The rate of cluster nucleation can be found when considering the diffusion process in m as similar to the diffusion in the crystallite size space, which is leading to crystallization. The frequency ratio of atom attachment to and detachment from the cluster is generally calculated by using the equations of detailed balance. Then the expression for the rate of cluster nucleation is obtained as [6.12]

$$\tilde{I}(\beta) = \tilde{I}_0(\beta)(N_{eff}^*)^{2/3} \exp(-\Delta G_{eff}^* \beta) , \tag{6.54}$$

where Z_1 is the number of atoms in the unit volume of the liquid, $\tilde{I}_0$ is the factor dependent on the atom attachment frequency, N_{eff}^* is the critical size of the nucleus at which ΔG_{eff} reaches its maximum, ΔG_{eff}^*, $\Delta G_{eff}(\beta, N) = G_{eff}(\beta, N) - N\mu_L$, μ_L being the chemical potential of atoms in the liquid.

The rate of crystallite nucleation $I(\beta)$ is described by a (6.54)-type expression, where N^* and ΔG_{eff}^* are replaced by the critical crystalline nucleus size, N^*, and the free energy of nucleation, ΔG^*.

The comparison between $I(\beta)$ and $\tilde{I}(\beta)$ reveals that the process of solidification of a liquid leads with the overwhelming probability to the formation of

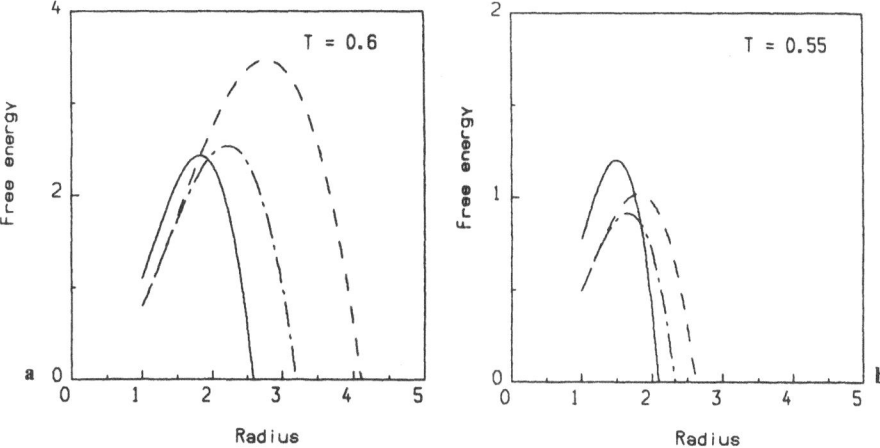

Fig. 6.19. Free energies of: crystal, (solid line), cluster, (dashed line) and cluster in an ensemble, (dash-dotted line) at different temperatures, $T = 0.6$ (a) and $T = 0.55$ (b). Kinetic vitrification temperature is close to 0.6. Temperature, free energies and radius are scaled by T_m, H_m, and a, respectively

a polycluster amorphous body, provided that there is a temperature T_g^k below which

$$\tilde{I}(\beta) > I(\beta). \tag{6.55}$$

Figure 6.19 shows the dependence of ΔG (solid line), $\Delta \tilde{G}$ (dashed line) and ΔG_{eff} (dash-dotted line) on ρ at temperatures $T/T_m = 0.60$ (a) and $T/T_m = 0.55$ (b). The dependence of $\mu_L - \mu_v$ on T was chosen in the form of the empirical relation introduced by *Thompson* and *Spaepen* [6.79]. One should take into account that $\mu_L - \tilde{\mu}_v = \mu_L - \mu_v - \delta\mu$. The calculations were made at $\delta\mu = 0.02$ eV and $s_v = s_s \equiv \zeta = 0.16$. The $\delta\mu$ value was chosen to be close to the measured metal glass crystallization heat values. In the calculations we took into account that the crystallization heat is considerably (nearly 50%) contributed by the cluster boundaries, whose density in metal glasses is high.

The ζ value was chosen close to the value derived in numerical experiments simulating the clusters with Lennard-Jones interaction between the atoms [6.80].

The temperature T_g^k is the kinetic temperature of vitrification. The liquid cooled down to temperatures below T_g^k transforms into a polycluster amorphous solid even at stationary temperature. Since the crystallization and clusterization are the competing processes, it is of importance to know how the volume fraction and the cluster size distribution depend on the cooling rate. The answer to this question has been obtained by integrating the Avrami-Kolmogorov equations [6.81] generalized to the case of simultaneous formation of two new phases. The consideration of these results is beyond the scope of the present communication and will be published elsewhere [6.82].

6.10.6 A Mixed Heterophase State and Structure Relaxation Times

From (6.54) it follows that the ensemble average of the cluster radius grows at $N > N_{eff}^*$, the critical size N_{eff}^* being finite at $T < T^*$. On the other hand, each cluster melts with the overwhelming probability at $T > \tilde{T}_m$. The fact that $T^* > \tilde{T}_m$ suggests at first sight a paradox situation, namely, in the temperature range $[\tilde{T}_m, T^*]$ the liquid is metastable with respect to the ensemble of clusters and should solidify, while a noncrystalline solid (with a broken structural ergodicity) should melt. This paradox situation arose because we considered the thermodynamics of isolated, noninteracting clusters in liquids. Of course, one can ignore these interactions in order to obtain a nucleation rate estimate. In [6.76] this paradox was resolved by taking into account both nucleation and growth of cluster in liquids and liquid nuclei in clusters. This kinetic approach allowed us to estimate the equilibrium volume fraction of clusters in a liquid and the relaxation times, but actually, in order to consider equilibrium states of supercooled liquids, it is necessary to develop the thermodynamic approach taking into account noncrystalline heterophase fluctuations.

To derive the free energy of a supercooled liquid, we shall consider the set of clusters immersed in the gaseous liquid. From the technical point of view, different approaches to the solution of this problem can be applied. For example, one can use the generalized droplet model (it was proposed many years ago by *Frenkel* [6.83] and by *Band* [6.84] and reconsidered and developed by *Fisher* [6.85]), taking into account the volume fraction of clusters in a liquid and the nucleation of liquid clusters in a solid fraction. The analysis in accordance with this approach shows that small-scale clusters are predominant in the size distribution function of clusters. So, if the cluster volume fraction, x, is large enough ($x \simeq 0.1$), the large-scale clusters can be formed as a coagulation of small-scale clusters. This result allows us to consider the mixed state as the one formed by solid and liquid clusters with average scales l_{cl} and l_L, respectively. The l_{cl} is, in fact, the minimal number of atoms in a solid cluster imbedded into a liquid and l_L is the minimal number of atoms of the liquid-like cluster in a solid matrix. Obviously, l_{cl} and l_L are comparable with the number of atoms in coordination polyhedra, thus, l_{cl} and l_L are about 10. For example, an icosahedron is a typical noncrystalline microcluster consisting of 13 atoms.

To formulate the phenomenological free energy of the mixed state, one should take into account that

i) the fraction of cluster surface atoms in contact with the liquid is proportional to $(1 - x)$;

ii) the configurational entropy of clusters decreases with increasing x.

The last statement must be commented. Because the cluster body rearrangements are possible only during remelting or reclusterization, the cluster, when immured in surrounding clusters, loses its ability of altering the structure. Therefore, when the cluster configurational entropy ζ is positive, $\zeta = \zeta_0$ at $x \to 0$ and tends to zero at $x \to 1$. So

$$\zeta = \zeta(x) = \zeta_0 r(x) , \tag{6.56}$$

where $r(x)$ is a smooth, monotonic interpolation function, $r(0) = 1$ and $r(1) = 0$. One can choose $r(x) = 1 - x^\alpha$ with $\alpha \geq 1$.

Note that $r(x)$ is introduced rather from the kinetic considerations than from thermodynamic ones, so it is tempting to relate α with the cooling rate (it is natural to choose $\alpha = \alpha(q)$ in the form of a certain function decreasing with the cooling rate q. Of course, other functions $r(x)$ can be applied. For example, one can use $r(x) = 1 - (x/x_g)^\alpha$ at $x < x_g$, and $r(x) = 0$ at $x \geq x_g$, where x_g is the cluster fraction above which the percolated liquid clusters do not exist.

After that, for $\mu_{mix} - \mu_L \equiv G_{mix}/N - \mu_L$ one can write:

$$\mu_{mix} - \mu_L = [\tilde\mu_v - \mu_L - T\zeta_0 r(x) + P\Delta v]x + p(l_{cl})\Delta\mu_s x(1 - x)$$
$$- T[x \ln x/l_{cl} + (1 - x)\ln(1 - x)/l_L] , \qquad (6.57)$$

where Δv is the difference of volumes per atom in clusters and in the liquid, P is pressure, $p(l_{cl})$ is the fraction of surface atoms in a cluster with size l_{cl},

$$\Delta\mu_s = \tilde\mu_s(l_{cl}) - \mu_{eff} \approx (\varepsilon_L - \tilde\varepsilon_v)/2 - T\zeta_0[1 - r(x)] , \qquad (6.58)$$

and ε_L is the energy per atom in the liquid. The last two terms in (6.57) describe a mixing free energy

It is likely that the difference $\mu_L - \tilde\mu_v = \mu_L - \mu_v - \delta\mu$ can be determined by using the Thomson–Spaepen empirical expression [6.79] for $\mu_L - \mu_v$. It was introduced for simple metals, but we apply it here to the gaseous fraction of liquids because the thermodynamical properties of gaseous liquids must be very similar to those of simple metal melts.

In the considerations of the problems of structure and properties, and vitrification of different liquids, several modifications of the so-called two-species model [6.86–88] based on the free energy expression resembling (6.57) were used.

One can generalize expression (6.57) to take into account the crystallite fraction. The same-type formula can be used for analysis of the liquid-crystallites system. For this purpose, it is necessary to put $\delta\mu = \zeta_0 = 0$ in (6.57).

It follows from (6.57, 58) that $\mu_{mix} - \mu_L$ can have one or two minima at $0 < x < 1$, thus one or two (stable or stable + metastable) states of supercooled liquids can exist. At high temperatures, $T > T^*$, and at low temperatures, $T < \tilde T_m$, only one equilibrium state exists. If two equilibrium states coexist, they differ by the degree of clusterization. If a clusterized fraction is large enough, the state must be treated as solid one. Indeed, in the system at $x > x_c \approx 0.16$ an infinite (percolated) solid cluster is formed and at $(1 - x) > (1 - x)_c \approx 0.16$ a percolated liquid cluster appears. So, at $x > x_g \approx 0.84$ the mixed state is really a solid with heterophase liquid fluctuations. The temperature at which the stable state with $x > x_g$ exists, is the thermodynamic glassing temperature, T_g^{th}.

It is evident that three types of liquids exist.

1) Liquids with only one equilibrium (noncrystalline) state. (Clusters have a comparatively large configurational entropy.) The glass transition always takes place near T_g^{th} when crystallization is prevented.

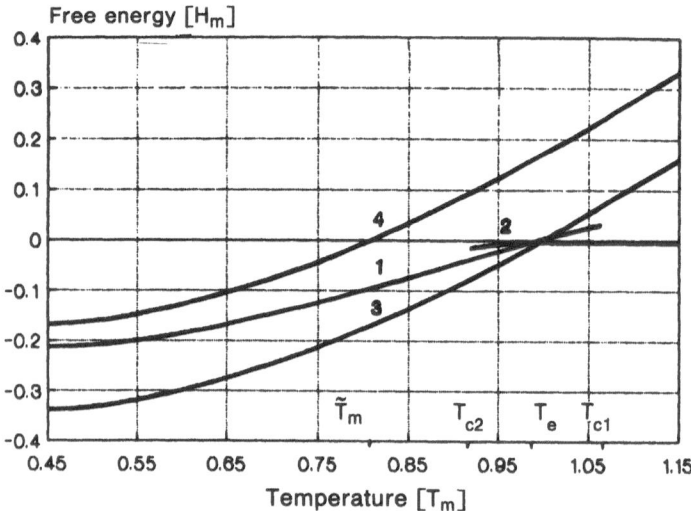

Fig. 6.20. Free energies (in H_m units) vs temperature (in T_m units) for liquid LTS (1), liquid LTS (2), crystal (3) and monocluster (4). Melting temperatures of cluster, $\tilde{T}_m$, as well as T_e and the end-points of LTS, T_{c1} and HTS, T_{c2}, are marked

2) Liquids with two equilibrium states coexisting in a temperature interval $[T_{c1}, T_{c2}]$ with the cluster fractions $x_1(T)$ and $x_2(T)$, moreover, $x_1(T)$, $x_2(T) < x_g$ with $T \in [T_{c1}, T_{c2}]$. For this type of liquids the vitrification takes place near T_g^{th} or near T_g^k and can be preceded by the first-type phase transition "liquid–liquid".

3) Liquids with two equilibrium states, one of which (the low-temperature one) is really the glass state, because $x_1(T) \geq x_g$ for this state at any temperature. (Clusters have a low configurational entropy.) The vitrification of this type of liquids is possible near T_g^k.

The first-type liquids have a good glass-forming ability, while the third-type liquids are poor glass-formers. Apparently, metallic melts belong to second-type liquids. Probably, the so-called "strong liquids" [6.88] are the first-type liquids, and "fragile" liquids are the second-type liquids.

The cluster fractions versus temperature for the second-type liquid is shown in Fig. 6.21, and the free energies of the liquid-clusters system and the liquid-crystals system are depicted in Fig. 6.20. The calculations were performed with the following parameters: $P = 0$, $\Delta S_m = 0.9$, $\delta \mu = 0.14 \, \Delta H_m$, $\zeta_0 = 0.25 \, \Delta S_m$, $p = 1, l_{cl} = 18, l_L = 14$ and $r(x) = 1 - x^3$. It can be seen from Fig. 6.21 that there exist two liquid states one of which, the low-temperature state (LTS), exists in the temperature interval $[T_g^{th}, T_{c1}]$ and the other, the high-temperature state (HTS), exists at $T > T_{c2}$. These states are in equilibrium at $T = T_e$. In LTS, the liquid is highly clusterized while the cluster fraction in HTS does not exceed $x_{c2} \approx 0.10$. At $x \approx x_{c2}$ almost each of the clusters of size $l_{cl} \approx 10$ is in contact with some other small-scale cluster, and therefore, small-scale clusters effectively coagulate and form large-scale clusters. The crystal fraction in the liquid and the

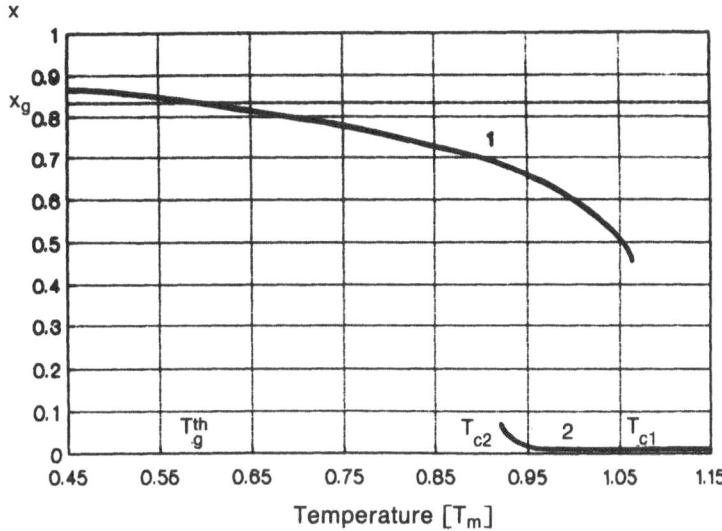

Fig. 6.21. Cluster fraction temperature dependences $x(T)$ for LTS (1) and HTS (2). At thermodynamical vitrification temperature T_g^{th}, we have $x(T) = x_g$ and the percolated liquid cluster disappears at $T < T_g^{th}$

liquid heterophase fluctuations in the crystal at $T_g^{th} < T < T_m$ are small as compared with the cluster fraction in the liquid. The end-point of the high-temperature liquid-crystal state also exists, but it lies at a much lower temperature than T_{c2}. Note that the free energy of LTS is essentially lower than the free energy of a gaseous-type liquid. Thus liquid clusterization suppresses the crystal nucleation and growth.

6.10.7 Discussion

Our main aim was to clarify the role of clusters in the process of liquid solidification and to show that formation of the polycluster glass structure is a typical result of this process. On the basis of the developed approach, such important problems as "entropy crisis" and relaxation phenomena in super-cooled liquids can be considered in detail, but this is far from the topic of this paper and will be considered elsewhere.

It will be recalled that in the dynamic theory of solidification [6.90, 91], the liquid – glass transition is related to the appearance of infinite relaxation time of the liquid. It must be noted that in our case the infinite relaxation times appear near T_e (here slow transformations of the metastable liquid into a stable one take place) and near T_g^{th}, where the relaxation processes are connected with cluster melting and reclusterizations which occur with overcoming large energetic barriers.

Stein and *Palmer* [6.92] have discussed the scenario of supercooled liquid solidification, which is based on the assumption that the line of metastable liquid states (HTS in our model) terminates at a fixed point (similar to T_{c2}) below the freezing temperature (T_g^{th}). In accordance with our model, the fixed end-points of liquid metastable states can or cannot exist, and, if they do exist, then they lie, as a rule, more closely to T_m than to T_g^{th}.

It is of interest to note that the crystal melting temperature can lie in the interval $[T_{c2}, T_{c1}]$, or be below or above this interval. If $T_e \geq T_m$ then the stable LTS exists at $T > T_m$. *Popel* et al. [6.93] have investigated the Pb–Sn melts nearly of eutectic composition and revealed very large relaxation times (about ten hours) and different hysteresis phenomena with heating and subsequent cooling of melts. Probably, the melts under study have $T_e > T_m$ and, therefore, during melting, the crystal transforms into a highly clustered liquid (LTS). Then, when heated, this liquid near T_{c1} transforms into a "normal", low clusterized liquid. During the following cooling, this liquid in HTS crystalizes at $T < T_m$. It is likely, that some of the glass-forming metallic melts possess similar properties. A lot of structural studies witnessed the existence of noncrystalline clusters in supercooled liquids [6.6, 94–96]. However, it is desirable to investigate more carefully the structure, thermodynamic and kinetic properties of the glass-forming melts near T_m.

6.11 Conclusion

A data analysis of the formation kinetics and structure of metallic glasses shows that their structure is a polycluster one. The polycluster model suggested offers the constructive foundation of the description of structural, mechanical, kinetic, and thermodynamic properties of metallic glasses and allows one to obtain the quantitative description of a number of phenomena. Beyond the frame of this chapter falls the consideration of a number of actual problems of metallic–glass physics such as kinetic and structural transformation under irradiation, relaxation processes, electron and phonon processes. The deciphering of all details of the metallic–glass structure, which may be achieved in the nearest future, will serve not only to clarify degree of adequacy of the polycluster model to real objects but also to its development.

Acknowledgements. I thank heartily N.P. Lazarev and I.M. Mikhailovskij for their active cooperation in investigating some problems of metallic–glass physics that found reflection in the present chapter and T. Ninomia for discussing the details of the tetrahedron-octahedron model. Helpful discussions with V. Baryakhtar and F. Hensel are most appreciated.

References

6.1   A.A. Lebedev: Trudy Gos. Opt. Inst. **2**, N.10, 1 (1921)
6.2   A.A. Lebedev, E.A. Porai-Koshits: Izv. Sectora Fiz.-Khim. Analiza **16**, 51 (1948)
6.3   W.H. Zachariasen: J. Am. Chem. Soc. **54**, 3841 (1932)

6.4   A.A. Lebedev: Izv. Akad. Nauk SSSR, Ser. Fiz. **4**, 584 **(1940)**

6.5   J. Stewart, R. Morrow: Phy. Rev. **30**, 232 (1927)

6.6   A. Ubbelohde, *Melting and Crystal Structure*, (Clarendon, oxford 1965)

6.7   J. Bernal: Proc. Roy. Soc. **A280**, 299 (1964)

6.8   P.H. Gaskell: In *Glassy Metals II*, ed by H.-J. Güntherodt, H. Beck. Topics Appl. Phys., Vol. 53 (Springer, Berlin, Heidelberg 1983) Chap. 1

6.9   V.A. Polukhin, N.A. Vatolin, *Modelirovanie amorfnych metallov* (Nauka, Moscow 1985)

6.10 K. Suzuki, K. Fujimori, K. Hashimoto: *Amorfnye metally* (Metallurgiya, Moscow 1987) (In Russian)

6.11 R.G. Palmer: Adv. Phys. **31**, 669 (1982)

6.12 G.W. Christian, *Transformations in Metals and Alloys* (Pergamon, Oxford 1965)

6.13 K. Samwer: *Amorphization in Solid Metallic Systems*, Phys. Repts **161**, N1, (1988)

6.14 D. Turnbull: J. Non-cryst. Solids **75**, 197 (1985)
      P.H. Gaskel: J. Non-cryst. Solids **75**, 329 (1985)

6.15 S.R. Elliot: *Physics of Amorphous Materials* (Longman, London 1984)
      M.I. Klinger: Phys. Rep. **165**, 5 (1988)

6.16 N.I. Noskova, N.F. Vil'danova, A.A. Glazer, A.P. Potapov: In *Fizika metallicheskikh tverdykh tel* (Udmurt State University, Izhevsk 1989) p. 83 (In Russian)

6.17 Y. Hirotsu, M. Uehara, M. Ueno: J. Appl. Phys. **59**, 3081 (1986)

6.18 A.S. Bakai, I.M. Mikhailovsky, P. Ya. Poltinin, L.I. Fedorova: In *Voprosy atomnoi nauki i tekhniki* Ser. Radiation Damage Physics and Radiation Technology (KFTI, Akad. Nauk Ukr.SSR) 3(45), 44 (1988), in: *IMAM-3 Rapid Quenching/Powder Preparation* (MRS, Pittsburg 1989)

6.19 J. Piller, P. Haasen: Acta Met. **30**, 1 (1982)

6.20 A. Melmed, R. Klein: Phys. Rev. Lett. **56**, 1478 (1986)

6.21 J. Gilman: J. Appl. Phys. **46**, 1625 (1975)

6.22 J.C.M. Li, *Frontiers in Materials Science* (Decker, New York 1976)

6.23 M. Cohen, D. Turnbull: J. Chem. Phys. **31**, 1164 (1959)

6.24 D. Turnbull, M. Cohen: J. Chem. Phys. **52**, 3038 (1970)

6.25 M. Cohen, G. Grest: Phys. Rev. **B20**, 1077 (1979)

6.26 A.S. Bakai: Pis'ma Zh. Ehksp. Teor. Fiz. **9**, 1477 (1983)

6.27 A.S. Bakai: Polycluster amorphous structures and their properties. Preprint KFTI, 84-33 (TSNII Atominform, Moscow 1984) (In Russian)

6.28 A.S. Bakai: Polycluster amorphous structures and their properties II. Preprint KFTI, 85-28 (TSNII Atominform, Moscow 1985) (In Russian)

6.29 A.S. Bakai: *Polycluster amorphous solids* (Energoatomizdat, Moscow 1987) (In Russian)

6.30 A.S.Bakai: Z. Phys. Chem. Neue Folge **158**, 201 (1988)

6.31 A.S. Bakai: In *IMAM-3 Rapid Quenching/Powder Preparation* (MRS, Pittsburg 1989)
      A.S. Bakai: Mater. Sci. Forum **123–125**, 145 (1993)

6.32 A.S. Bakai: J. Non-Cryst. Solids **117 & 118**, 252 (1990)

6.33 A.S. Bakai, N.P. Lazarev: In *Phonon '89*, ed. by H. Hunklinges (World Scientific, Singapore 1990)

6.34 G.F. Voronoi: J. Reine Angew. Math. **134**, 198 (1908)

6.35 P. Chaudhari, S. Spaepen, P.J. Steinhardt: In *Glassy Metals II*, ed by H.-J. Güntherodt, H. Beck, Topics Appl. Phys., Vol. 53 (Springer, Berlin, Heidelberg 1983) Chap. 5.

6.36 J.P. Hirth, J. Lothe, *Theory of Dislocations* (McGraw-Hill, New York 1968)

6.37 R. Collins: Proc. Phys. Soc. **83**, 553 (1964)

6.38 H. Kawamura: *Topological Disorder in Condensed Matter*. ed. by F. Yonesawa, T. Ninimia, Solid-State Sciences, 46 (Springer, Berlin, Heidelberg 1983) pp. 181–190

6.39 H. Kawamura: Progr. Theor. Phys. **70**, 352 (1983)
      H. Kawamura: Physica **A177**, 73 (1991)

6.40 T. Ninomia: In *Topological Disorder in Condensed Matter* ed. by F. Jonesawa, T. Ninomiya, Springer Sci Solid-State Sci. Vol. 46 (Springer, Berlin, Heidelberg 1983) pp. 40–50

6.41 J. Wong: In: *Glassy metals I*, ed. by H.-J. Güntherodt, H. Beck, Topics Appl. Phys. Vol. 46 (Springer, Berlin, Heidelberg 1981) Chap. 4

6.42 E. Svab. N. Kroo, S.N. Ishmaev, I.P. Sadikov, A.A. Chernyshov, Solid State Commun. **46**, 351 (1983)

6.43 P. Lamparter, S. Steeb: In *Rapidly Quenched Metals V*, ed. by S. Steeb, H. Warlimont (North-Holland, Amsterdam 1985)

6.44 P. Pannisod, I. Bakonyi, R. Hasegawa: Phys. Rev. **B28**, 2374 (1983)

6.45 A. Audouard, J. Balogh, J. Dural, J.C. Jonset: Radiat. Eff. **62**, 161 (1982)

6.46 S. Klaumünzer, G. Schumacher: Acta Metall. **30**, 1493 (1982)

6.47 S. Nagel: Adv. Chem. Phys. **51**, 227 (1982)

6.48 Y. Limoge: Z. Phys. Chem., Neue Folge **156**, 391 (1988)

6.49 Y. Limoge: J. Non-Cryst. Solids **117 & 118**, 708 (1990) 1989, to be published in J. Noncryst. Solids

6.50 J. Hafner: In *Amorphous Solids and Nonequilibrium Processing*, ed. by M. von Allmen (Les Editions de Physique, 1984) pp. 219–229

6.51 P. Anderson, B. Halperin, C. Varma: Phil. Mag. **25**, 1 (1972)

6.52 W. Phillips: J. Low Temp. Phys. **7**, 351 (1972)

6.53 H. Löhneysen: Phys. Repts. **79**, 161 (1981)

6.54 J.L. Black: In *Glassy Metals I*, ed. by H.-J. Güntherodt, H. Beck, Topics Appl. Phys. Vol. 46 (Springer, Berlin, Heidelberg 1981) Chap. 8

6.55 M. Cohen, G. Grest: Phys. Rev. Lett. **45**, 1271 (1980)

6.56 R. Kirchheim, F. Sommer, G. Schluckebier: Acta Met. **30**, 1059 (1982)
      R. Kirchheim: Acta Met. **30**, 1069 (1982)
      R. Kirchheim, U. Stolz: J. Non-Cryst. Solids **790**, 323 (1985)

6.57 N.P. Lazarev, M.P. Fateev: Fizika Tverdogo Tela **33**, 521 (1991)

6.58 R. Maddin, T. Masumoto: Mat. Sci. Eng. **9**, 153 (1972)

6.59 C. Pampillo: J. Mat. Sci. **11**, 2209 (1975)

6.60 F. Spaepen: Acta Met. **25**, 407 (1976)

6.61 A.S. Argon: Acta Met. **27**, 47 (1979)

6.62 A.M. Glezer, B.V. Molotilov, O.L. Utevskaya: Metallofizika (Ukrainian SSR) **5**, N1, 29 (1983) (In Russian)

6.63 A.S. Argon: J. Phys. Chem. Solids **43**, 945 (1982)

6.64 A.S. Argon, H.J. Kuo: J. Non-Cryst. Solids **37**, 241 (1980)

6.65 A.S. Argon, L.T. Shi: Acta Met. **31**, 499 (1983)

6.66 F. Spaepen, O. Turnbull: Scripta Met. **8**, 120 (1974)

6.67 P. Donovan, P. Stobbs: Acta Met. **29**, 1419 (1981)

6.68 A.S. Argon, J. Megusar, N.J. Grant: Scripta Met. **19**, 591 (1985)

6.69 H.-U. Künzi: In *Glassy Metals II*, ed. by H.-J. Güntherodt, H. Beck, Topics Appl. Phys., Vol. 53 (Springer, Berlin, Heidelberg 1983) Chap 6

6.70 C. Herring: J. Appl. Phys. **21**, 437 (1950)

6.71 R.L. Coble: J. Appl. Phys. **34**, 1679 (1963)

6.72 I.M. Lifshits: Zh. Ehksp. Teor. Fiz. **44**, 1349 (1963)

6.73 A. Taub, F. Spaepen: Acta Met. **28**, 1781 (1980)

6.74 M.F. Ashby: Acta Met. **20**, 887 (1972)

6.75 F. Spaepen: In *Physics of Defects*, Les Houches Lectures XXXV. ed. by R. Balin et al. (North-Holland, Amsterdam 1981) p. 133

6.76 A.S. Bakai: Proc. of VIII Intl Conf. Liquid and Amorphous Metals, Vienna, 1992, to be published in Fizika Nizkikh Temperatur

6.77 K. Binder, A.P. Young: Rev. Mod. Phys. **58**, 801 (1986)

6.78 A.S. Bakai: to be published in Fizika Nizkikh Temperatur

6.79 S.V. Thompson, F. Spaepen: Acta Met. **27**, 1855 (1979)

6.80 H. Stillinger, T. Weber: Phys. Rev. **A28**, 2408 (1983)

6.81 A.N. Kolmogorov: Izv. Acad. Nauk SSSR, Ser Mat. **3**, 355 (1937)

6.82 N.V. Alekseechkin, A.S. Bakai, N.P. Lazarev: to be published in Sov. Sol. State Phys

6.83 Ya. I. Frenkel: J. Chem. Phys. **7**, 200 (1939)

6.84 W. Band: J. Chem. Phys. **7**, 324, 927 (1939)

6.85   M.E. Fisher: Physics **3**, 225 (1967)

6.86   E. Rapoport: J. Chem. Phys. **48**, 1433 (1968)

6.87   O.V. Mazurin: *Vitrification* (Nauka, Leningrad 1986) (in Russian)

6.88   P.-J. de Gennes: *Scaling Concepts of Polymer Physics* (Cornel Univ. Press, London 1978)

6.89   C.A. Angel et al. J. Chem. Phys. **82**, 773 (1958)

6.90   E. Leuthauser: Phys. Rev. **B29**, 2765 (1984)

6.91   U. Bengtzelius, W. Götze, A. Sjolander: J. Phys. C **31**, 595 (1984)

6.92   D.L. Stein, R.G. Palmer: Adv. Phys. **34**, 669 (1982)

6.93   P.S. Popel et al.: Izv. Acad. Nauk SSSR, Metally **2**, 53 (1985); **4**, 198 (1985); **3**, 52 (1987)

6.94   N. Mattern, A.G. Ilinskii, H. Hermann, A.V. Romanova: phys. stat. sol. (a) **97**, 397 (1986)

6.95   W. Weiss, H. Alexander: J. Phys. F: Metal Phys. **17**, 1983 (1987)

6.96   L.E. Mikhailova, Chen Si-Shen, A.V. Romanova, A.G. Ilinskii: Metallofizika **13**, 116 (1991)

Subject Index

Topics in Applied Physics   Founded by Helmut K. V. Lotsch

Springer-Verlag
and the Environment

We at Springer-Verlag firmly believe that an international science publisher has a special obligation to the environment, and our corporate policies consistently reflect this conviction.

We also expect our business partners – paper mills, printers, packaging manufacturers, etc. – to commit themselves to using environmentally friendly materials and production processes.

The paper in this book is made from low- or no-chlorine pulp and is acid free, in conformance with international standards for paper permanency.